Coal Geology

Coal Geology

Larry Thomas
Dargo Associates Ltd

JOHN WILEY & SONS, LTD

Copyright © 2002 by   Larry Thomas, Annedd Bach, Michaelchurch Escley, Herefordshire HR2 0JW

                              Telephone (+44) 1981 510373

Published 2002 by   John Wiley & Sons Ltd, The Atrium, Southern Gate, Chichester,
                              West Sussex PO19 8SQ, England

                              Telephone (+44) 1243 779777

Email (for orders and customer service enquiries): cs-books@wiley.co.uk
Visit our Home Page on www.wileyeurope.com or www.wiley.com

Other Wiley Editorial Offices

John Wiley & Sons Inc., 111 River Street, Hoboken, NJ 07030, USA

Jossey-Bass, 989 Market Street, San Francisco, CA 94103-1741, USA

Wiley-VCH Verlag GmbH, Boschstr. 12, D-69469 Weinheim, Germany

John Wiley & Sons Australia Ltd, 33 Park Road, Milton, Queensland 4064, Australia

John Wiley & Sons (Asia) Pte Ltd, 2 Clementi Loop #02-01, Jin Xing Distripark, Singapore 129809

John Wiley & Sons Canada Ltd, 22 Worcester Road, Etobicoke, Ontario, Canada M9W 1L1

British Library Cataloguing in Publication Data

A catalogue record for this book is available from the British Library

ISBN 0-471-48531-4

Typeset in 9/11pt Times by Laserwords Private Limited, Chennai, India
Printed and bound in Great Britain by Antony Rowe Ltd, Chippenham, Wiltshire
This book is printed on acid-free paper responsibly manufactured from sustainable forestry
in which at least two trees are planted for each one used for paper production.

Contents

Preface

The *Handbook of Practical Coal Geology* (Thomas 1992) was intended as a basic guide for coal geologists to use in their everyday duties, whether on site, in the office or instructing others. It was not intended as a definitive work on all or any particular aspect of coal geology, rather as a handbook to use as a precursor to, or in conjunction with, more specific and detailed works.

This new volume is designed to give both the coal geologist and others associated with the coal industry background information regarding the chemical and physical properties of coal, its likely origins, its classification and current terminology. In addition I have highlighted the currently known geographical distribution of coal deposits together with recent estimates of world resources and production. I have also outlined the exploration techniques employed in the search for, and development of, these coal deposits and the geophysical and hydrogeological characteristics of coal-bearing sequences, together with the calculation and categorisation of resources/reserves.

Chapters are devoted to the mining of coal, to the means of extracting energy from coal other than by conventional mining techniques, and to the environmental concerns associated with the mining and utilisation of coal.

Also covered is the development of computer technology in the geological and mining fields, and the final chapter is a condensed account of the marketing of coal, its uses, transportation and price.

Many sources of information have been consulted, the majority of which are listed in the reference section. A set of appendices contains information of use to the reader.

I would like to thank all those colleagues and friends who have helped and encouraged me with the book from conception to completion. In particular special thanks are due to Steve and Ghislaine Frankland of Dargo Associates Ltd, Alan Oakes, Rob Evans, Dr Keith Ball, Professor Brian Williams, Mike Coultas, Reeves Oilfield Services, IMC Geophysics Ltd, Datamine International and Palladian Publications, as well as the staff at John Wiley & Sons Ltd.

I should also like to thank those authors and organisations whose permission to reproduce their work is gratefully acknowledged.

Finally I would like to thank my wife Sue and my family for their support, encouragement and assistance with the manuscript.

Larry Thomas
Dargo Associates Ltd

1

Preview

1.1 SCOPE

The object of this book is to provide both geologists and those associated with the coal industry, as well as teachers of courses on coal, its geology and uses, with a background of the nature of coal and its varying properties, together with the practice and techniques required in order to compile geological data that will enable a coal sequence under investigation to be ultimately evaluated in terms of mineability and saleability. In addition, the alternative uses of coal as a source of energy, together with the environmental implications of coal usage, are also addressed. Modern computer techniques and the marketability of coal are briefly covered.

Each of these subjects is a major topic in itself, and the book only covers a brief review of each, highlighting the relationship between geology and the development and commercial exploitation of coal.

1.2 COAL GEOLOGY

Coal is a unique rock type in the geological column; it has a wide range of chemical and physical properties, and has been studied over a long period of time. This volume is intended to be a basic guide to understanding the variation in coals, their modes of origin, and to the techniques required to evaluate coal occurrences.

The episodes of coal development in the geological column are given, together with the principal coal occurrences worldwide. It is accepted that this is not totally exhaustive and that coal does occur in small areas not indicated in the figures or tables.

Current estimates of global resources and reserves of coal, together with coal production figures, are listed, and whilst these obviously become dated, they do serve to indicate where the major deposits and mining activity are currently concentrated.

In relation to the extraction of coal, the understanding of the geophysical and hydrogeological properties of coals is an integral part of any coal mine development, and these are reviewed, together with the principal

methods of mining coal. The increasing use of computer technology has had a profound impact on geological and mining studies. Some of the applications of computers to these are discussed.

An important development in recent years has been the attempts to use coal as an alternative energy source by either removing methane gas from the coal or by liquefying the coal as a direct fuel source, or by gasification of coal *in situ* underground. These technologies together are particularly significant in areas where conventional coal mining has ceased or where coal deposits are situated either at depths uneconomic to mine, or in areas where mining is considered environmentally undesirable.

1.3 COAL USE

The principal uses of traded coals worldwide is for electricity generation and steel manufacture, with other industrial users and domestic consumption making up the remainder.

Lack of environmental controls in the use of coal in the past has led to both land and air pollution as well as destruction of habitat. Modern environmental guidelines and legislation are both repairing the damage of the past and preventing a reoccurrence of such phenomena. An outline is given of the types of environmental concerns that exist where coal is utilised, together with the current position on the improvements in technology in mining techniques, industrial processes and electricity generation emissions.

The marketing of coal is outlined, together with the contractual and pricing mechanisms commonly employed in the coal producer/coal user situation.

1.4 BACKGROUND

In most industrial countries, coal has historically been a key source of energy and a major contributor to economic growth. In today's choice of alternative sources of energy, industrialised economies have seen a change in the role for coal.

Originally coal was used as a source of heat and power in homes and industry. During the 1950s and 1960s cheap oil curtailed the growth of coal use, but the uncertainties of oil supply in the 1970s led to a resumption in coal consumption and a rapid growth in international coal trade. This in turn was followed by an increasingly unfavourable image for coal as a contributor to greenhouse gas (GHG) emissions, closely identified with global warming. The coal industry has responded positively to this accusation and modern industrial plants have much lower emissions levels than in previous years. However, there is still a question mark over coal use in the public's conception of energy sources, even though coal accounts for less than 13% of all GHG emissions, and electricity generation from coal contributes to around 10% of man-made sources of CO_2. As an extreme example, the elimination of coal from electricity generation in the European Union would only reduce existing CO_2 emissions by less than 2% (Knapp 1997).

The world consumption of fossil fuels, and thus emissions of CO_2, will continue to increase, and by the year 2010 fossil fuels will still meet around 90% of primary energy requirements. The objectives of the 'United Nations Framework Convention on Climate Change' (UNFCCC), signed at the 1992 Earth Summit in Rio de Janeiro, is to 'stabilise GHG concentrations in the atmosphere at a level that would prevent dangerous anthropogenic interference with the climate system'. No set levels were identified but emissions in developed countries were expected to be reduced to 1990 levels. A series of annual meetings by the international body under UNFCCC, the Conference of the Parties (COP)

has taken place, notably COP-3 in Kyoto, Japan in 1997, at which the Kyoto Protocol was drawn up, setting emissions targets for all the countries attending. However, Government Ministers at COP-6 in The Hague in November 2000 have failed to agree on the way forward to meet the Kyoto Protocol targets. This has placed the whole of the Kyoto Protocol's ambitious and optimistic plan for a global agreement on GHG emissions reduction in an uncertain position (Knapp 2001). This could be an indication of over-ambitious goals rather than any failure in the negotiations, and it is up to the parties concerned to establish a realistic set of targets for emissions reductions in the future.

It remains a fact that many economies still depend on coal for a significant portion of their energy needs. Coal currently accounts for 23% of the world's consumption of primary energy, and, importantly, coal provides fuel for the generation of around 38% of the total of the world's electricity. Traded black coal amounts to 573 Mtpa, of which 381 Mt is steam coal and 192 Mt is coking coal.

Coal reserves are currently estimated to be around 985 billion tonnes, and the world coal reserves-to-production ratio is nearly six times that for oil, and four times that for natural gas. This, together with the globally democratic distribution and secure nature of coal deposits, will ensure that coal will continue to be a major energy resource for some considerable time to come.

With this scenario in mind, this volume is intended to assist those associated with the coal industry, as well as educationalists and those required to make economic and legislative decisions about coal.

The philosophy and views expressed in this book are those of the author and not the publisher.

2

Origin of Coal

2.1 INTRODUCTION

Sedimentary sequences containing coal or peat beds are found throughout the world and range in age from Upper Paleozoic to Recent.

Coals are the result of the accumulation of vegetable debris in a specialised environment of deposition. Such accumulations have been affected by synsedimentary and post-sedimentary influences to produce coals of differing rank and differing degrees of structural complexity, the two being closely interlinked.

Remarkable similarities exist in coal-bearing sequences, due for the greater part to the particular sedimentary associations required to generate and preserve coals. Sequences of vastly different ages from areas geographically separate have a similar lithological framework, and can react in similar fashions structurally.

It is a fact, however, that the origin of coal has been studied for over a century and that no one model has been identified that can predict the occurrence, development and type of coal. A variety of models exist which attempt to identify the environment of deposition, but no single one can adequately give a satisfactory explanation for the cyclic nature of coal sequences, the lateral continuity of coal beds, and the physical and chemical characteristics of coals.

2.2 SEDIMENTATION OF COAL AND COAL-BEARING SEQUENCES

During the last 35 years, interest has grown rapidly in the study of sedimentological processes, particularly those characteristic of fluviatile and deltaic environments. It is these in particular that have been closely identified with coal-bearing sequences.

It is important to give consideration both to the recognition of the principal environments of deposition, and to the recent changes in emphasis regarding those physical processes required in order to produce coals of economic value. In addition, the understanding of the shape, morphology and quality of coal seams is of fundamental significance for the future planning and mining of coals. Although the genesis of coal has been the subject of numerous studies, models that are used to determine the occurrence, distribution and quality of coal are often still too imprecise to allow such accurate predictions.

2.2.1 Depositional models

The recognition of depositional models to explain the origin of coal-bearing sequences and their relationship to surrounding sediments has been achieved by a comparison of the environments under which modern peats are formed and ancient sequences containing coals.

Cecil *et al* (1993) suggest that the current models often concentrate on the physical description of the sediments associated with coal rather than on the geological factors that control the genesis of coal beds. They also suggest that models that combine sedimentation and tectonics with eustasy and chemical change have not yet been fully developed. Such integrated models would give an improved explanation of physical and chemical processes of sedimentation. It should be noted that the use of sequence stratigraphy in facies modelling is based on physical processes and does not take into account chemical stratigraphy. This will prove a deficiency when predicting the occurrence and character of coal beds.

The traditional depositional model used by numerous workers was based on the 'cyclothem', a series of lithotypes occurring in repeated 'cycles'. This concept has been modified to a model that relates lateral and vertical sequential changes to depositional settings that have been recognised in modern fluvial, deltaic and coastal barrier systems. The traditional model is based on the work carried out in the USA by Horne *et al* (1978, 1979), Ferm *et al* (1979), Ferm and Staub (1984), Staub and Cohen (1979) in a series of studies in the 1970s. The sequences or lithofacies are characterised by the sedimentary features listed in Table 2.1. Other workers include Thornton (1979) and Jones and Hutton (1984)

Table 2.1   Sedimentary features used to identify depositional environments. From Horne *et al* (1979)

Recognition characteristics	Fluvial & upper delta plain*	Transitional lower delta plain*	Lower delta plain*	Back-barrier*	Barrier*
I Coarsening upwards					
A Shale & siltstone sequences	2–3	2	1	2–1	3–2
(i) >50 ft	4	3–4	2–1	2–1	3–2
(ii) 5–25 ft	2–3	2–1	2–1	2–1	3–2
B Sandstone sequences	3–4	3–2	2–1	2	2–1
(i) >50 ft	4	4	2–1	3	2–1
(ii) 5–25 ft	3	3–2	2–1	2	2
II Channel deposits					
A Fine grained abandoned fill	3	2–3	1–2	2	3–2
(i) Clay & silt	3	2–3	1–2	2	3–2
(ii) Organic debris	3	2–3	1–2	2–3	3
B Active sandstone fill	1	2	2–3	2–3	2
(i) Fine grained	2	2	2–3	2–3	2
(ii) Medium & coarse grained	1	2–3	3	3	2–3
(iii) Pebble lags	1	1	2	2–3	3–2
(iv) Coal spars	1	1	2	2–3	3–2
III Contacts					
A Abrupt (scour)	1	1	2	2	2–1
B Gradational	2–3	2	2–1	2	2
IV Bedding					
A Cross beds	1	1	1	1–2	1–2
(i) Ripples	2	2–1	1	1	1
(ii) Ripple drift	2–1	2	2–3	3–2	3–2
(iii) Trough cross beds	1	1–2	2–1	2	2–1
(iv) Graded beds	3	3	2–1	3–2	3–2
(v) Point bar accretion	1	2	3–4	3–4	3–4
(vi) Irregular bedding	1	2	3–2	3–2	3–2
V Levee deposits					
A Irregularly interbedded sandstones & shales, rooted	1	1–2	3–2	3	4
VI Mineralogy of sandstones					
A Lithic greywacke	1	1	1–2	3	3
B Orthoquartzite	4	4	4–3	1–2	1
VII Fossils					
A Marine	4	3–2	2–1	1–2	1–2
B Brackish	3	2	2	2–3	2–3
C Fresh	2–3	3–2	3–4	4	4
D Burrow	3	2	1	1	1

*1-Abundant, 2-Common, 3-Rare, 4-Not present.

on coal sequences in Australia, and Guion *et al* (1995) in the UK.

More recent studies have compared such established depositional models with modern coastal plain sedimentation, e.g. in equatorial Southeast Asia, and have concentrated in particular on modern tropical peat deposits; see Cecil *et al* (1993), Clymo (1987), Gastaldo *et al* (1993), McCabe and Parrish (1992). Studies by Hobday (1987), Diessel (1992), Lawrence (1992), Jerzykiewicz (1992), Dreesen *et al* (1995), Cohen and Spackman (1972, 1980), Flint *et al* (1995), McCabe (1984, 1987, 1991) have all further developed the model for coal deposits of differing ages, using the traditional model but relating it to modern sedimentary processes.

In parallel with this work, detailed studies of peat mires have both raised and answered questions on the development of coal geometry, i.e. thickness and lateral extent, together with the resultant coal chemistry.

The traditional model is still a basis for modern coal studies, but linked to a better understanding of peat development and preservation.

2.2.2 The traditional model

2.2.2.1 *Coastal barrier and back barrier facies*

The coastal end of the depositional model is characterised by clean barrier sandstones, which, in a seaward direction become finer grained and intercalate with red and green calcareous shales and carbonate rocks, the latter containing marine faunas. Landwards they grade into dark grey lagoonal shales with brackish water faunas, and into marginal swamp areas on which vegetation was established. The barrier sandstones have been constantly reworked and are therefore more quartzose than those sandstones in surrounding environments with the same source area.

They exhibit a variety of bedding styles: firstly, extensive sheets of plane-bedded sandstones with rippled and burrowed upper surfaces, interpreted as storm washover sands; secondly, wedge-shaped bodies that extend landward, can attain thicknesses of up to 6 metres, and contain landward dipping planar and trough cross beds, interpreted as floodtide delta deposits; and thirdly, channel-fill sandstones which may scour to depths of over 10 metres into the underlying sediments, interpreted as tidal channel deposits.

A depositional reconstruction is shown in Figure 2.1(a) based on studies by Horne *et al* (1979).

The lagoonal back barrier environment is characterised by upwards coarsening, organic-rich grey shales and siltstones overlain by thin and discontinuous coals. This sequence exhibits extensive bioturbation zones, together with bands and concretions of chemically precipitated iron carbonate (sideritic ironstone). The extent of such sequences is considered to be in the order of 20–30 metres in thickness and 5–25 kilometres in width. A typical vertical sequence of back barrier deposition is shown in Figure 2.1(b).

2.2.2.2 *Lower delta plain facies*

Lower delta plain deposits are dominated by coarsening upwards sequences of mudstone and siltstone, ranging from 15 to 55 m in thickness, and 8 to 110 km in lateral extent. The lower part of these sequences is characterised by dark grey to black mudstones with irregularly distributed limestones and siderite (Figure 2.2(a)).

In the upper part, sandstones are common, reflecting the increasing energy of the shallow water as the bay fills with sediment. Where the bays have filled sufficiently to allow plant growth, coals have formed. Where the bays did not fill completely, bioturbated, siderite-cemented sandstones and siltstones have formed.

This upwards coarsening pattern is interrupted in many areas by crevasse-splays (Figure 2.2(b)). In the American Carboniferous, crevasse-splay deposits can be 10+ m in thickness and 30 m to 8 km wide.

In many cases, a transitional lower delta plain sequence is characteristic, featuring alternations of channel, interdistributary bay and crevasse-splay deposits; a depositional reconstruction is shown in Figure 2.3(a), and a generalised vertical sequence in Figure 2.3(b).

Overlying and laterally equivalent to the bay-fill sequences are thick lithic sandstones up to 25 m in thickness and up to 5 km in width. These are interpreted as distributary mouth bar deposits; they are widest at the base and have gradational contacts. They coarsen upwards and towards the middle of the sand body. In some places, fining-upwards sequences are developed on the top of the distributary mouth bar and bay-fill deposits.

These distributary channel-fill deposits have an irregular sharp basal contact, produced by scouring of the underlying sediments. At the base, pebble and coal-fragment lag deposits are common.

Because of the rapid abandonment of distributaries, fine-grained mudstone fills are common in lower delta plain deposits. They represent silt and organic debris which has settled from suspension in the abandoned distributary. In some areas, thick organic accumulations filled these channels, resulting in the formation of lenticular coals. Apart from those formed in the abandoned channels, coals are generally relatively thin and widespread. Such coals are oriented parallel to the distributary patterns.

2.2.2.3 *Upper delta and alluvial plain facies*

In contrast to the thick fine-grained sequences of the lower delta plain facies, upper delta plain deposits are dominated by linear, lenticular sandstone bodies up to 25 m thick and up to 11 km wide. These sandstones have scoured bases and pass laterally in the upper part into grey shales, siltstones and coals. The sandstones fine upwards with abundant pebble conglomerates, and coal clasts in the lower part. The sandstones are characterised by massive bedding and are overlain by siltstones.

These sandstone bodies widen upwards in cross-section and are considered to have been deposited in the channels and on the flanks of streams that migrated across the upper delta plain; see Figure 2.4(a).

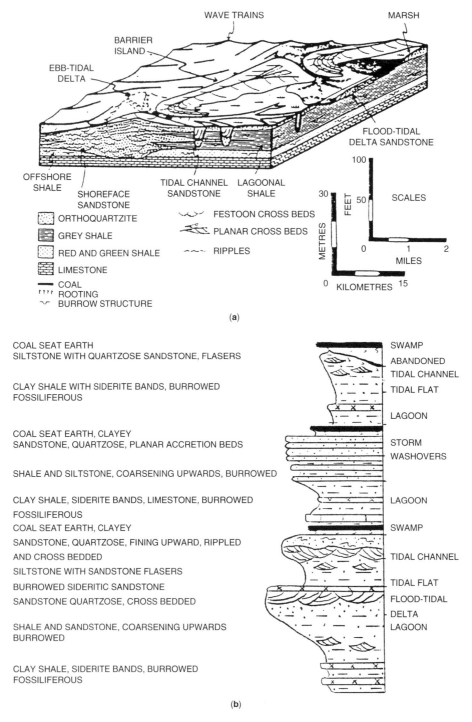

Figure 2.1 (**a**) Barrier and back barrier environments including tidal channels and flood-tidal deltas, based on exposures in Kentucky, USA. From Horne *et al* (1979); (**b**) Generalised vertical section through back barrier deposits in the Carboniferous of eastern Kentucky, USA. From Horne *et al* (1979)

COAL
SEAT EARTH, CLAYEY
SANDSTONE, FINE TO MEDIUM GRAINED
MULTI-DIRECTIONAL PLANAR AND
FESTOON CROSS BEDS
SANDSTONE, FINE GRAINED, RIPPLED
SANDSTONE, FINE GRAINED, GRADED BEDS
SANDSTONE, FLOW ROLLS
SANDSTONE, FINE GRAINED, FLASER BEDDED
AND SILTSTONE

SILTY SHALE AND SILTSTONE WITH CALCAREOUS
CONCRETIONS, THIN-BEDDED, BURROWED
OCCASIONAL FOSSIL

CLAY SHALE WITH SIDERITE BANDS, BURROWED
FOSSILIFEROUS

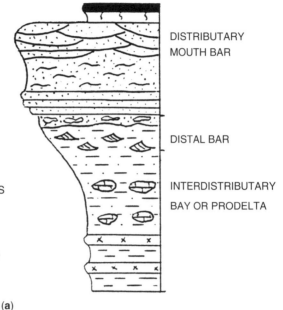

DISTRIBUTARY
MOUTH BAR

DISTAL BAR

INTERDISTRIBUTARY
BAY OR PRODELTA

(a)

COAL
ROOTED SANDSTONE
SANDSTONE, FINE GRAINED, CLIMBING RIPPLES
SANDSTONE FINE TO MEDIUM GRAINED
SANDSTONE, MEDIUM GRAINED, FESTOON CROSS BEDS
CONGLOMERATE LAG, SIDERITE AND COAL PEBBLES
SANDSTONE, SILTSTONE, GRADED BEDS

SANDSTONE, FLOW ROLLS
SANDSTONE, SILTSTONE, FLASER BEDDED
SILTSTONE/SILTY SHALE, THIN BEDDED, BURROWED

BURROWED SIDERITIC SANDSTONE
SANDSTONE, FINE GRAINED
SANDSTONE, FINE GRAINED, RIPPLED

SILTY SHALE, SILTSTONE WITH CALCAREOUS
CONCRETIONS, THIN BEDDED, BURROWED

CLAY SHALE WITH SIDERITE BANDS, BURROWED,
FOSSILIFEROUS

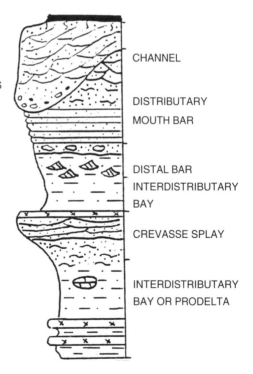

CHANNEL

DISTRIBUTARY
MOUTH BAR

DISTAL BAR
INTERDISTRIBUTARY
BAY

CREVASSE SPLAY

INTERDISTRIBUTARY
BAY OR PRODELTA

(b)

Figure 2.2 Generalised vertical sequences through lower delta plain deposits in eastern Kentucky, USA. (**a**) Typical coarsening-upward sequence; (**b**) Same sequence interrupted by crevasse-splay deposits. From Horne *et al* (1979)

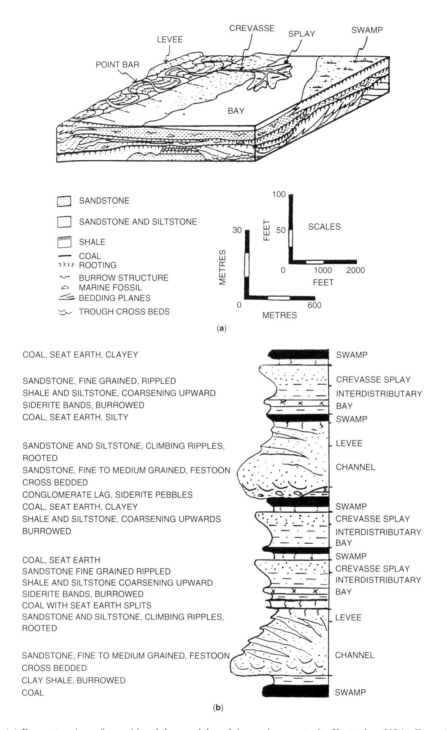

Figure 2.3 (a) Reconstruction of transitional lower delta plain environments in Kentucky, USA. From Horne *et al* (1979); (b) Generalised vertical sequence through transitional lower delta plain deposits of eastern Kentucky and southern West Virginia, USA. From Horne *et al* (1979)

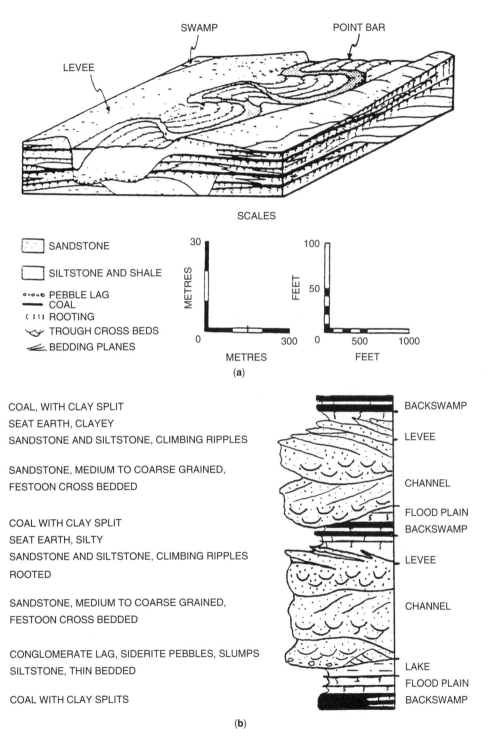

Figure 2.4 (**a**) Reconstruction of upper delta plain-fluvial environments in Kentucky, USA. From Horne *et al* (1979); (**b**) Generalised vertical sequence through upper delta plain-fluvial deposits of eastern Kentucky and southern West Virginia, USA. From Horne *et al* (1979)

Coal seams in the upper delta plain facies may be 10+ m in thickness, but are of limited lateral extent. Figure 2.4(b) illustrates a vertical sequence of upper delta plain facies from eastern Kentucky and southern West Virginia, USA.

Between the upper and lower delta plains, a transition zone exhibits characteristics of both sequences. This zone consisted of a widespread platform on which peat mires were formed. This platform was cut by numerous channels and the sequence disrupted by crevasse-splay deposits. The coals formed on the platform are thicker and more widespread than the coals of the lower delta plain; such a sequence is shown in Figure 2.3(b).

2.2.3 Modern peat analogues

The principal characteristics of a coal are its thickness, lateral continuity, rank, maceral content and quality. Apart from rank, which is governed by burial and subsequent tectonic history, the remaining properties are determined by factors controlling the mire where the peat originally formed. These factors include type of mire, type(s) of vegetation, growth rate, degree of humification, base-level changes, and rate of clastic sediment input (McCabe and Parrish 1992).

About 3% of the earth's surface is covered by peat, totalling 310 million hectares (World Energy Council 1998). This includes the tropical peats (>1 m thick) of Southeast Asia which cover almost 200 000 km².

During the last 15 years, numerous studies have attempted to understand more fully how peat-producing wetlands or mires are developed and maintained, and in particular how post-depositional factors influence the formation of coals.

Diessel (1992) divides peat-producing wetlands into ombrogenous peatlands or mires (owing their origin to rainfall) and topogenous peatlands (owing their origin to a place and its surface/groundwater regime). A great variety of topogenous peats form when waterlogging of vegetation is caused by groundwater, but ombrogenous peats are of greater extent but less varied in character.

Based on this distinction, Diessel (1992) gives a classification of peatlands or mires as shown in Table 2.2. This is illustrated in Figure 2.5, which shows the relationship between ombrotrophic and rheotrophic mires in terms of the influence of rainwater and groundwater in their hydrological input. The inorganic content of mires is seen to increase in the topogenous rheotrophic mires.

The classification of the two hydrological categories of mire lists a number of widely used terms. Moore (1987) has defined a number of these.

- *Mire* is now accepted as a general term for peat-forming ecosystems of all types.
- *Bog* is generally confined to ombrotrophic peat-forming ecosystems.
- *Bog forest* consists of ombrotrophic forested vegetation, usually an upper storey of coniferous trees and a ground layer of *sphagnum* moss.
- *Marsh* is an imprecise term used to denote wetlands characterised by floating vegetation of different kinds including reeds and sedges, but controlled by rheotrophic hydrology.
- *Fen* is a rheotrophic ecosystem in which the dry season water table may be below the surface of the peat.

Table 2.2 Classification of mires. Adapted from Diessel, *Coal-bearing depositional systems*, Springer-Verlag, Berlin, 1992. Reproduced with permission

Peatlands (Mires)		
Ombrogenous		Topogenous
(*ombrotrophic = rain fed*)		(*mineralotrophic = mineral fed*)
(*oligotrophic = poorly fed*)		(*rheotrophic = flow fed*)
		(*eutrophic = well fed*)
Raised bog	Tree cover	Marsh
(*Sphagnum* bog)	increases	Fen
(Bog forest)		Swamps
		(Floating swamps)
		(Swamp forest)
Transitional or mixed mires (*mesotrophic*)		

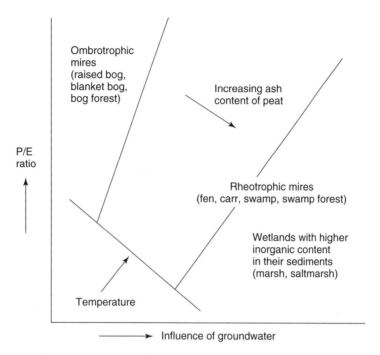

Figure 2.5 Proposed relationship between mires in terms of the relative influence of rainwater and groundwater in their hydrological input (Moore 1987). Reproduced with permission of The Geological Society

- *Swamps* are a rheotrophic ecosystem in which the dry season water table is almost always above the surface of the sediment. It is an aquatic ecosystem dominated by emergent vegetation.
- *Floating swamps* develop around the fringes of lakes and estuaries and extend out over open water. These platforms can be thick and extensive particularly in tropical areas.
- *Swamp forest* is a specific type of swamp in which trees are an important constituent, e.g. mangrove swamps.

The resultant characteristics of coals are primarily influenced by the following factors during peat formation: type of deposition, the peat-forming plant communities, the nutrient supply, acidity, bacterial activity, temperature and redox potential.

In order for a mire to build up and for peat to accumulate, the following equation must balance.

Inflow + Precipitation

= Outflow + Evapotranspiration + Retention

The conditions necessary for peat accumulation are therefore a balance between plant production and organic decay. Both are a function of climate, plant production and organic decay; such decay of plant material within the peat profile is known as humification. The upper part of the peat profile is subject to fluctuations in the water table and is where humification is most active. The preservation of organic matter requires rapid burial or anoxic conditions (McCabe and Parrish 1992), the latter being present in the waterlogged section of the peat profile. In addition, an organic-rich system will become anoxic faster than an organic-poor one as the decay process consumes oxygen. This process is influenced by higher temperatures, decay rates being fastest in hot climates. Rates of humification are also affected by the acidity of the groundwater, as high acidity suppresses microbial activity in the peat.

Peat formation can be initiated by:

(a) terrestrialisation, which is the replacement, due to the setting up of a body of water (pond, lake, lagoon, interdistributary bay) by a mire;
(b) paludification, which is the replacement of dry land by a mire, e.g. due to a rising groundwater table.

As peat is relatively impermeable, its growth may progressively impede drainage over wide areas, so that low-lying mires may become very extensive. In those areas where annual precipitation exceeds evaporation, and where there are no long dry periods, a raised mire may develop. Such mires are able to build upwards because they maintain their own water table. The progression of a peat-forming environment from the infilling of a watercourse or lake, to a low-lying mire and finally to a raised mire should produce zonation in the peat accumulated, as shown in Figure 2.6.

Depositional models may show peat formation adjacent to and intercalated with areas of active clastic deposition. Such peats accumulating on interchannel areas on the delta plain may be disrupted by clastic contamination from crevasse-splays or by subsidence of the interchannel area resulting in submergence of the peat, cessation of peat development and clastic influx. Sediment may also be introduced into low-lying mires by floods, storm surges or exceptionally high tides. The overall result of clastic contamination is an increase in the ash content of the peat. Also, inundation of mires by aerated waters helps to degrade the peat and enrich it with inorganics.

Basin subsidence combined with ombrogenous peat accumulation such that the rise in the peat surface continues to outstrip the rate of subsidence will lead to the formation of thick and clean (low mineral matter content) coals (McCabe 1984). Low ash coals therefore must have formed in areas removed or cut off from active clastic deposition for long periods of time, e.g. centuries. Partings in coals, such as mudstone, indicate the interruption of peat formation and may represent intervals of thousands of years.

For a thick peat layer to form in a topogenous setting it is essential that the rise in the water table and the rate of peat accumulation are balanced. In the case of a slower rise in water level, peat accumulation could be terminated by oxidation, but in a very wet climate peat formation might continue under high moor conditions. Actual rates of peat accumulation or accretion vary in different climates and with the type of vegetation. Assuming a compaction ratio of 10:1 (Ryer and Langer 1980) to operate in the transition from peat to bituminous coal, and considering that some of the coal seams are tens of metres thick, optimum peat-forming

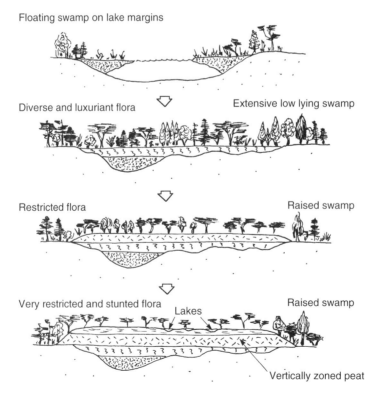

Figure 2.6 Evolutionary sequence of swamp types showing the development of a raised swamp with distinct peat zonations. From McCabe (1984), by permission of the International Association of Sedimentologists

conditions must therefore require the maintenance of a high groundwater table over very long periods of time, i.e. 5–10 Ka for every metre of clean bituminous coal.

As peat accumulation is regulated by temperature and precipitation, tropical and subtropical regions are well suited for large peat development, where rates of decay are higher. Most modern peats are situated in low terrains not far above sea level.

In contrast to the traditional depositional model, studies of modern environments suggest that significant areas of low ash peats are not present on delta plains, and that most mires on coastal or floodplain areas are not sites of true peat accumulation. The exception appears to be those areas where raised mires have developed. Floating mires may also produce low ash peats, but these are thought generally to be of limited extent. Examination of modern delta plain peats shows that they have an ash content of over 50% on a dry basis, and that peats with less than 25% ash on a dry basis rarely exceed one metre in thickness. These peats if preserved in the geological record would form carbonaceous mudstones with coaly stringers.

Studies of raised mires indicate that ash levels can be less than 5%, and over large areas may be as low as 1–2%. Rates of organic accumulation in raised mires outstrip rates of sedimentation from overbank or tidal flooding. However, although some low ash coals have doubtless originated as products of raised mires, many coals are thought to have formed under paleoclimates unsuitable for raised mire development. One suggestion is that low ash coals originated as high ash peats and were depleted in ash during the coalification process. Acidic waters may hasten the dissolution of many minerals, but not all mires are acidic and some may even contain calcareous material. Another concept is that peat accumulation was not contemporaneous with local clastic deposition, suggesting that resulting coals are distinct from the sediment above and below the coal. Those areas of the mire that have been penetrated by marine waters may be identified in the resultant coal by high sulphur, hydrogen and nitrogen contents.

As a corollary to the mechanism of clastic contamination of peats, those raised mires that are able to keep pace with channel aggradation could confine the fluvial sediments to defined narrow courses. If this is so, the presence of thick peats could influence the depositional geometry of adjacent clastic accumulations (Figure 2.7).

The majority of coals are developed from plants which have formed peat close to where they grew. Such coals are underlain by seat earths or rootlet beds, and are known as autochthonous coals. However, coals which have formed from plant remains which have been transported considerable distances from their original growth site are known as allochthonous coals, e.g. large rafts of peat or trees drifting on lakes or estuaries. Allochthonous coals do not have an underlying rootlet bed, but rest directly on the bed below. In the Cooper Basin, South Australia, thick Gondwana (Permian) coals show evidence of both autochthonous and allochthonous deposition. The allochthonous coals are closely associated with lacustrine sediments, and are thick and widespread (Williams B.P.J. pers. com.).

2.2.3.1 Case studies

A number of studies of peat accumulation in tropical Southeast Asia (Neuzil *et al* 1993, Cecil *et al* 1993, Ruppert *et al* 1993, Gastaldo *et al* 1993) show possible similarities between the extensive coastal plain peat deposits in Indonesia and Malaysia and the North American/European Carboniferous coal-bearing sequences, in terms of sediment accumulation, mineralogy, geochemistry and maceral content.

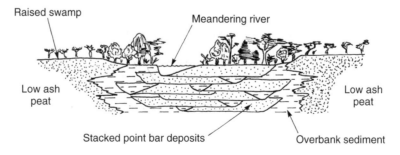

Figure 2.7 Theoretical model of fluvial architecture in areas of raised swamps. The elevated swamp restricts overbank flooding and prevents avulsion, leading to the development of stacked channel sandstones. From McCabe (1984), by permission of the International Association of Sedimentologists

Cecil *et al* (1993) produced an Indonesian analogue for Carboniferous coal-bearing sequences, in which it is suggested that the domed (convex upper surface) ombrogenous peat deposits of the ever-wet tropical area may represent the modern equivalent of Lower to mid Middle Pennsylvanian (Carboniferous) coal deposits of the eastern USA. In their study on the island of Sumatra, they conclude that peat formation has been controlled primarily by the allogenic processes of sea-level change and the modern ever-wet climate. Autogenic processes such as delta switching, channel cutting and barrier-bar migration are considered to be of secondary importance as a control on the formation of peat. This is in contrast to the traditional model described earlier. The tropical climate is seen to favour formation of laterally extensive, thick, low sulphur, low ash peat deposits rather than active clastic sedimentation. Erosion and sediment transportation are restricted in the tropical rainforest environment. Gastaldo *et al* (1993), in their study of the Mahakam delta in East Kalimantan, observed no autochthonous peat formation, but rather accumulations of allochthonous (derived and transported) peat on the lower delta plain tidal flats, the peats having derived from swamps and tropical forest higher on the delta plain.

Neuzil *et al* (1993) studied the inorganic geochemistry of a domed peat in Indonesia. The inorganic constituents in peat are the primary source for mineral matter in coals, and the study showed that large amounts of low ash peat can develop close to marine conditions and above a marine substrate without high sulphur or pyrite contents. In domed ombromorphic peats, the geochemical controls on mineral matter are dominantly autogenic, independent of surrounding depositional environments. Neuzil *et al* (1993) also considered that quality predictions for coal derived from domed peat deposits cannot be based on facies relations with enclosing sedimentary rocks. Rather, prediction of coal quality should be based on autogenic geochemical processes and controls of peat formation, recognised by the composition and distribution of mineral matter in coal.

Grady *et al* (1993) have carried out petrographic analysis on Indonesian peat samples, and found that the optical characteristics of peat constituents are comparable to the maceral content of brown coals. The distribution of maceral types in the modern peat was also found to be analogous with maceral profiles from Carboniferous coals in the USA, and could be used to interpret the changing conditions in the original peat mire. Styan and Bustin (1983) studied the sedimentology of the Frazer River delta peat and used it as a modern analogue for some ancient deltaic coals.

Studies of modern peat-forming environments both in tropical and nontropical areas will continue to improve the understanding of coal-forming environments and more importantly the mechanisms for the retained accumulation of peat, its lateral development and chemical constituents. This will have economic significance when applied to higher rank coals.

2.2.4 Facies correlation

The recognition of the variety of facies types described in the facies model is essential in order that their lateral and vertical relationships can be determined and correlated to produce the geometry of lithotypes within a study area.

In order to achieve this, examination of surface exposures, both natural and man-made, and borehole data is required to establish the particular lithological sequence present at each data point. It is the correlation between data points that is critical to the understanding of the patterns of coal development and preservation in any given area of interest.

It is an unfortunate fact that for a great number of coal-bearing sequences, good recognisable and widespread marker horizons are rare. In part this stems from the very localised patterns of deposition within many coal-forming environments. However, some distinctive deposits may be present; for example marine mudstones, usually overlying a coal seam, may contain marine or brackish-marine fauna and may also have a particular geochemical/geophysical profile. In areas where contemporary volcanic activity took place during coal deposition, deposits of fine-grained volcanic ash intercalated with coal-bearing sediments produce widespread 'tonstein' horizons, which also have a distinctive geochemical/geophysical signature.

Other less reliable lithotypes can be used, certainly on a local scale (e.g. within a mine lease area), such as sandstone complexes, freshwater limestones with their associated fauna, and the coal seams themselves.

Lithotype correlation from boreholes and surface exposures is dependent on the use of identifiable lithological horizons. Figure 2.8 shows correlation of irregular sand bodies within a coal-bearing sequence (Nemec 1992); such complexity makes individual coal seam correlation difficult and in some cases impossible. A widespread coal horizon is commonly used; however, due to the differential rates of sedimentation both above and below the chosen coal horizon, depiction of the sequence can result in distortions of the succeeding coal bed, as shown in Figure 2.9, where borehole sections

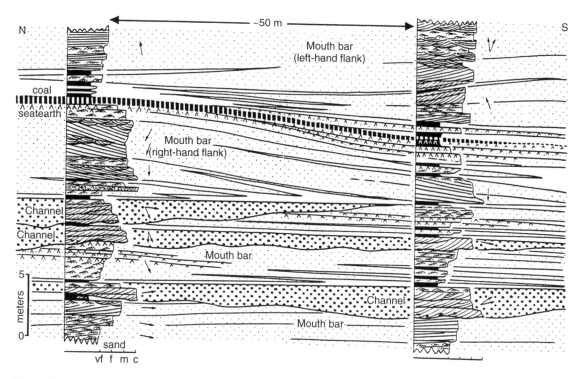

Figure 2.8 Cross-section showing portions of superimposed mouth-bar lobes with associated Seatearths and coal (Drønbreen locality, central Spitsbergen). The lobes prograde to the east (away from the viewer). From Nemec (1992), by permission of the Geological Survey of America

have been used to determine the facies character of coal-bearing strata in the Arkoma basin, USA (Rieke and Kirr 1984). Similar lithotype correlations are shown in Figure 2.10 from the USA and Figure 2.11 from India, the latter illustrated by the much used 'fence' diagram presentation.

In modern coal sequence correlation, increasing use is being made of downhole geophysical logs of boreholes. The individual profile of each borehole can be compared with its neighbouring boreholes. An example of Canadian coal-bearing sequences showing the correlation of lithotypes with their geophysical profiles is given in Figure 2.12. Details of the variety of geophysical logs used in coal sequence correlation are described in Chapter 7.

Once the distribution pattern of the various lithotypes present in an area has been established, it may be possible to predict the likely sequence in adjacent areas. This is particularly important for neighbouring areas with proven coal reserves, which may be concealed beneath younger deposits, or which may lack quantitative geological data. If it is likely that coal is developed at economic thickness and depth, then a facies study of

the known area may guide predictions for drill sites in adjacent areas. In the early stages of exploration this can be an important tool to deploy.

In the example shown in Figure 2.13, Area 1 has a known distribution pattern of coal and noncoal deposits, determined from the correlation of the boreholes present. Area 2 is as yet unexplored, but from the data available in Area 1, together with an appraisal of the topography in Area 2, it is likely that Coal A will be present at similar depth and thickness, at least in that part closest to the nearest known data points in Area 1, i.e. at points 2a, 2b, 2c and 2d. If Area 2 is considered for development, exploratory drill sites would be located at sites 2a to 2u before any close-spaced drilling would be sanctioned. Similarly the areas of split coal and channel sandstone in Area 1 would need to be identified in Area 2 to determine how much coal loss is likely to occur here.

2.2.5 Facies maps

In close association with the correlation of facies, the most significant sedimentological features are those

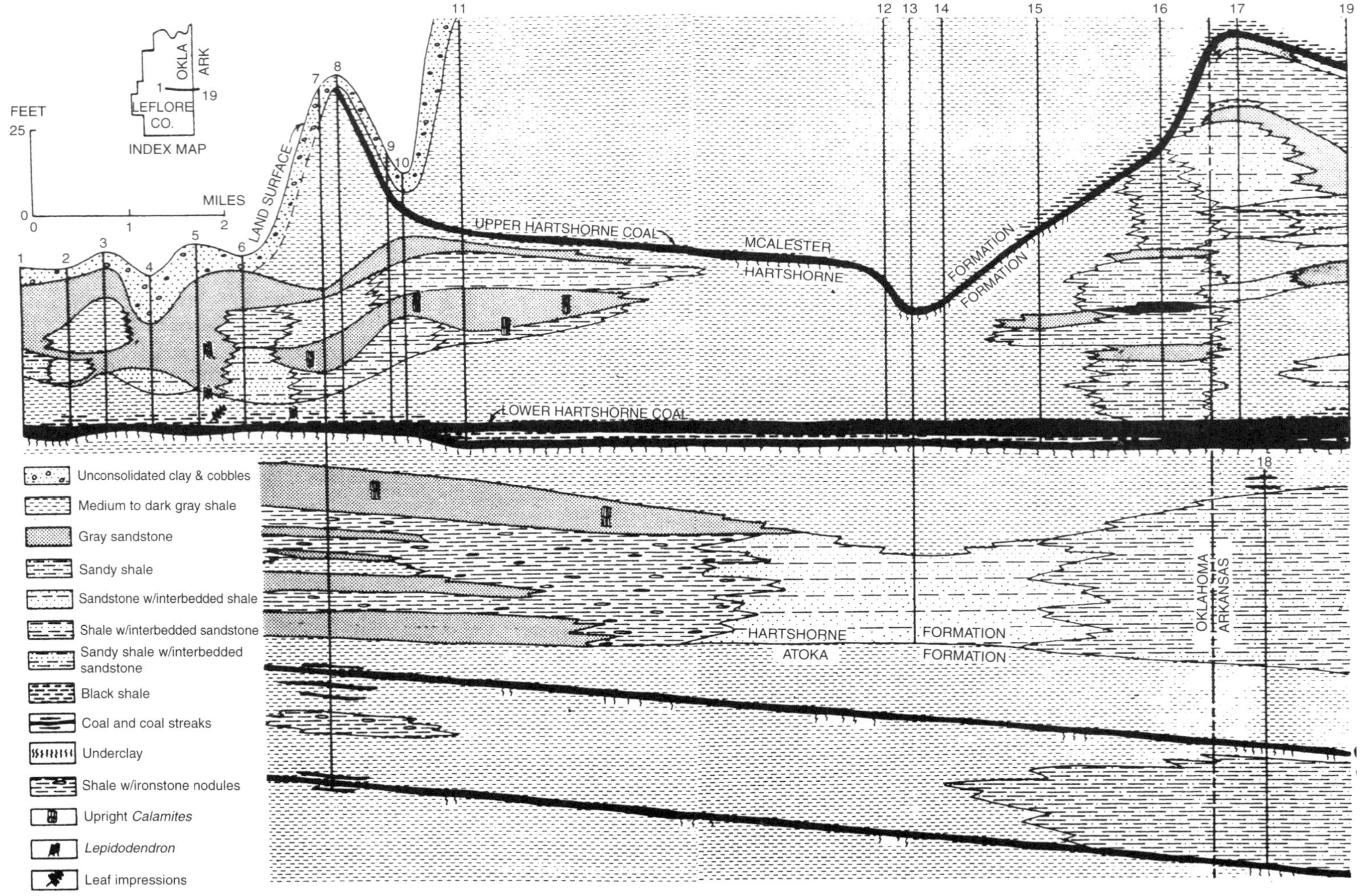

Figure 2.9   Cross-section showing stratigraphic relations of coals and sandstones in the Hartshorne and Atoka Formations, Le Flore County, Oklahoma, and Sebastian County, Arkansas, USA. From Rieke and Kirr (1984). Reproduced by permission of D.R. Donica

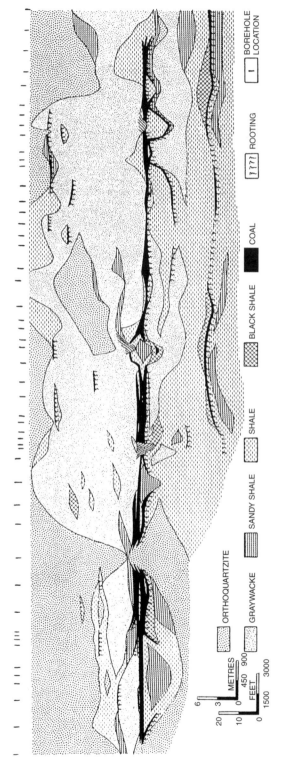

Figure 2.10   Cross-section showing correlation of lithofacies and associated coals above and below the Beckley Seam, West Virginia, USA, based on borehole data. From Ferm *et al* (1979)

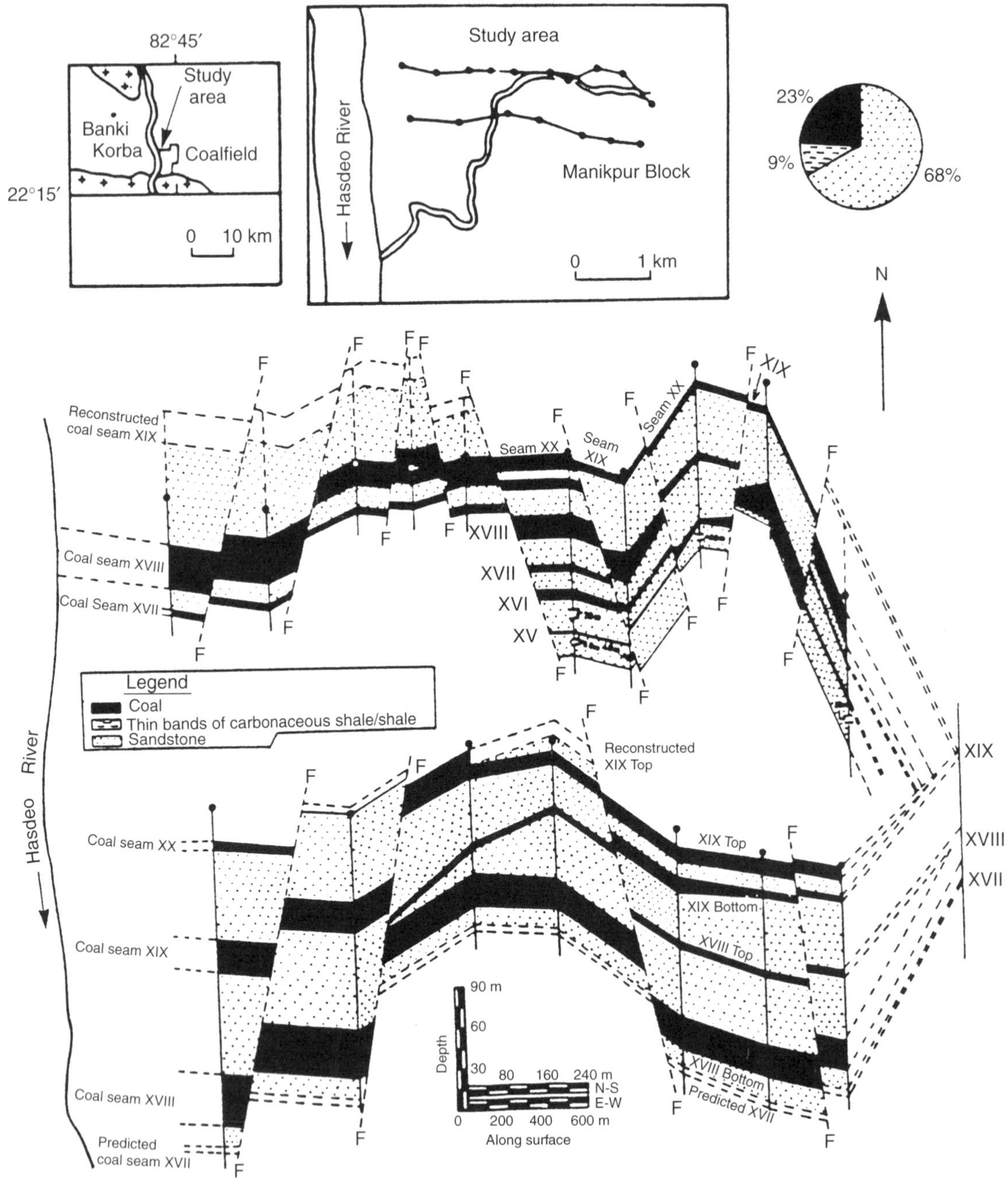

Figure 2.11   Fence correlation diagram showing the geometry of a coal and sandstone sequence in the Lower Permian Barakar Formation, Korba Coalfield, India. From Casshyap and Tewari (1984), by permission of the authors and Blackwell Scientific Publications

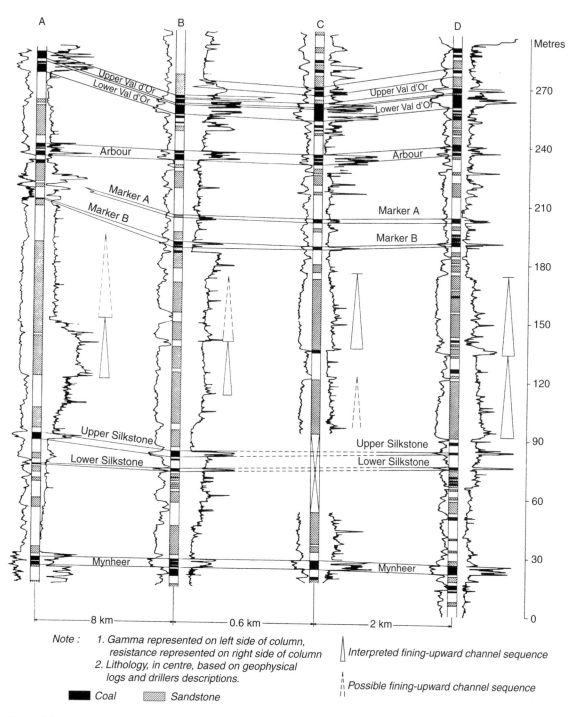

Figure 2.12 Correlation based on geophysical logs, Coalspur Beds, Upper Cretaceous/Lower Tertiary, Alberta, Canada. From Jerzykiewicz and McLean (1980), Geological Survey of Canada, Department of Energy, Mines and Resources paper 79-12. Reproduced with the permission of the Minister of Public Works and Government Services Canada, 2001 and courtesy of the Geological Survey of Canada

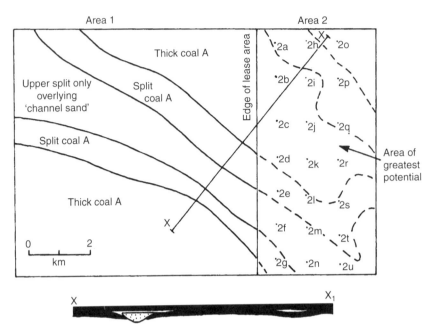

Figure 2.13 Lithofacies map illustrating how such mapping can be extended to an adjoining area to locate an additional area of thick coal and an area of coal split. Reproduced by permission of Dargo Associates Ltd

of seam splitting, washouts and floor rolls, as well as the more obvious variations in seam thickness, seam quality, interburden and overburden nature and thickness, together with the identification of igneous intrusions in the coal-bearing sequence. From borehole and surface data, all of these features can be quantified and portrayed in plan or map form.

Facies maps are usually compiled for the area of immediate interest, i.e. the mine lease area, but plans covering larger areas can be produced which give a useful regional picture of coal development. Such a large-scale study is shown in Figure 2.14, which illustrates a paleogeographic reconstruction of the depositional setting of the Beckley Seam of south West Virginia. The reconstruction is based on 1000 cored boreholes in an area of 1000 km². In this example, coal thickness variations are closely related to the pre-existing topography, produced by depositional environments that existed prior to coal formation. The shape of the coal body also has been modified by contemporaneous and post-depositional environments, such as channels. Consideration of these features during mine planning can maximise the recovery of the thicker areas of coal while avoiding the areas of 'want', i.e. those areas depleted in coal.

Figure 2.15 shows a lithofacies map of part of the Patchawarra formation (Permian) in Australia, on which clastics-to-coal ratio contours are plotted, indicating areas of coal and no coal (or noncoal deposition). Figure 2.16 then shows the paleogeographic reconstruction of the same interval, and the area of high clastics-to-coal ratio represents an area dominated by fluvial channel deposits (Thornton 1979).

2.2.5.1 Seam splitting

This common phenomenon occurs when a coal seam, traced laterally, is seen to 'split' into a minimum of two individual coals or 'leafs' separated by a significant thickness of noncoal strata. Such noncoal materials within a seam are referred to as 'partings' or 'bands', and may be composed of a variety of lithotypes. Such partings and bands are the result of clastic deposition replacing organic accumulation. They may represent crevasse-splay overbank deposits, or, if the partings are well developed laterally, represent either widespread flooding of the mire from adjacent river courses or periodic marine flooding into those mires close to the coast.

Seam splitting can be simple or form a complex series of layered organic and clastic materials. Simple splits occur when organic accumulation is interrupted and replaced for a short period by clastic deposition. Once the influx of detrital material ceases, vegetation is re-established and organic accumulation thus

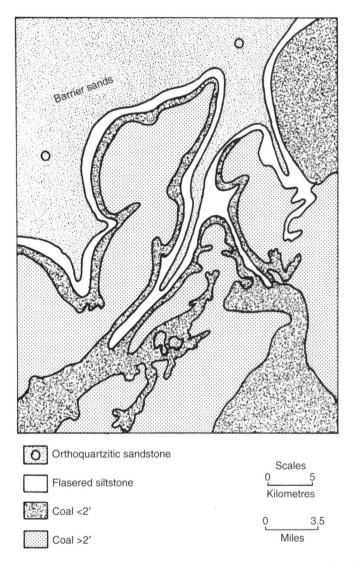

Orthoquartzitic sandstone

Flasered siltstone

Coal <2'

Coal >2'

Scales

0 _____ 5
Kilometres

0 _____ 3.5
Miles

Figure 2.14 Mapped lithotypes compiled from 1000 boreholes over an area of 1000 km^2, illustrating the regional depositional setting of the Beckley Seam, West Virginia, USA. From Horne *et al* (1979)

continues. This may occur once or many times during the deposition of a coal seam. When traced laterally, splits may coalesce or further divide. This has the detrimental effect of reducing good sections of coal that can be mined, particularly if the partings are quartz-rich, thus creating mining difficulties, particularly in underground workings. Figure 2.17 illustrates differential splitting of a coal seam across a mine working; such variations are significant to the economics of coal mining, more particularly in underground operations.

Other types of seam splitting are known as 'S' or 'Z' splits (Figure 2.18). This feature is characterised by two seams usually separated by 20–30 metres of sediment. The upper seam splits and the bottom leaf apparently descends through the clastic interval to unite with the lower seam. Such features, which are well documented in the UK (Fielding 1984) and Australia, are considered to be produced by accelerated subsidence induced by differential compaction of peat, clay and sand-rich lithotypes which have been deposited in two adjacent 'basin' areas on the delta plain, and require continuous

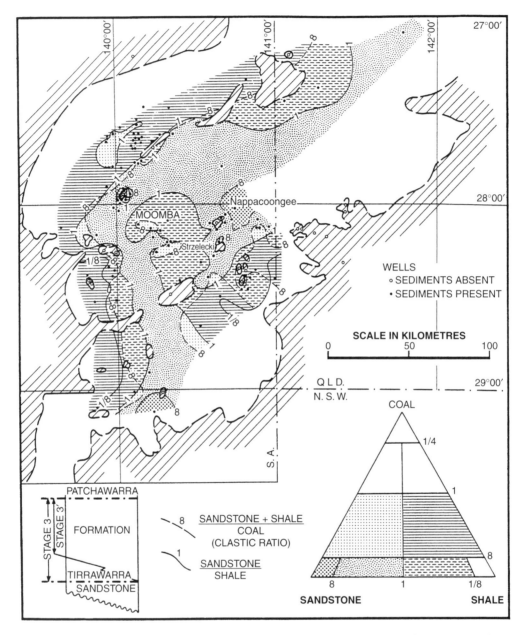

Figure 2.15   Lithofacies map of part of the Patchawarra Formation (Permian) South Australia, showing sandstone/shale and sandstone + shale/coal ratios. From Thornton (1979). Reproduced by permission of Primary Industries and Resources, South Australia

peat formation and accumulation on the abandoned 'basin' surfaces and in interbasin areas. Coal seam splits are also formed by the influence of growth faulting as described in 2.3.1.2 and illustrated in Figure 2.25.

Splits are of considerable significance: coals that have been identified as being of workable thickness

may in one or more areas split into two or more thinner seams that are uneconomic to exploit. Such splitting effectively limits those areas of economically recoverable coal reserves.

High angle splitting can produce instability, particularly in opencast workings, where mudstones or

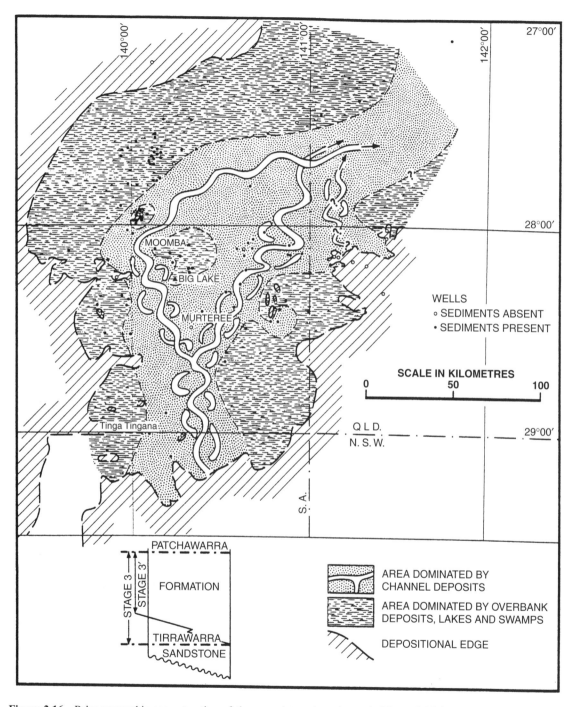

Figure 2.16 Paleogeographic reconstruction of the same interval as shown in Figure 2.15, indicating that the area of high clastics-to-coal ratio represents an area dominated by fluvial channel deposits. From Thornton (1979). Reproduced by permission of Primary Industries and Resources, South Australia

Figure 2.17   Development of a coal seam splitting in the Beckley Seam across a mine working. From Ferm *et al* (1979)

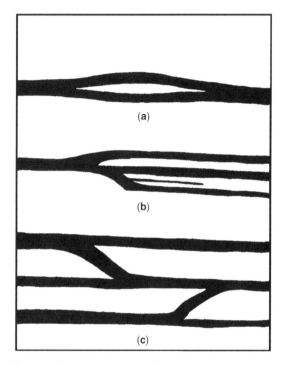

Figure 2.18   Common types of coal seam split. (**a**) Simple splitting; (**b**) multiple splitting; and (**c**) 'Z' or 'S' shaped splitting. Reproduced by permission of Dargo Associates Ltd

fractured sandstones overlying such an inclined split may readily allow passage of groundwater and/or produce slope failure.

2.2.5.2 *Washouts*

Washouts occur where a coal seam has been eroded away by wave or river current action, and the resultant channel is filled with sediment. The coal may be wholly or partly removed by this process. Washouts are usually elongate in plan, and infilled with clastic material either as mudstone, siltstone or sandstone, depending on whether the erosive phase was followed by a reduction in current energy, so reducing the grain size of the sediment transported to infill the channel (Figure 2.19).

Initially the edges of the washout tend to be sharp, but then may have become diffused by differential compaction of the coal and noncoal materials.

Washouts are a major problem in mining operations, particularly in underground workings. Washouts can seriously reduce the area of workable coal; therefore the delineation of such features is an essential prerequisite to mine planning. Detailed interpretation of the sedimentary sequences exposed in outcrops, boreholes and in underground workings allows a facies model to be constructed; this in turn may help to predict the orientation of washouts. Contemporaneous fault and fold influences during sedimentation can result in clastic wedges pinching out against these positive elements, with coal seams tending to merge over the structural highs. Another feature is the 'stacking' or localisation of channelling along the flanks of such flexures, producing elongate sandstone bodies which can influence mine planning operations.

2.2.5.3 *Floor rolls*

These are the opposite phenomenon to washouts, and are characterised by ridges of rock material protruding upwards into the coal seam. Like washouts they reduce the mineable thickness of the coal seam. If they have to be mined with the seam, as is commonly the case, the dilution of the coal quality will result in an increase in the ash content. Floor rolls are often the result of differential compaction of peat around clastic deposits in the lower part of the seam as the upper part of the seam accumulates.

2.2.5.4 *Coal seam thickness variations*

The production of isopach maps of coal seams, sandstone thickness or percentages of lithotypes present in any area of interest is an important guide to the eventual exploitation of coals. Figure 2.20 shows two contour

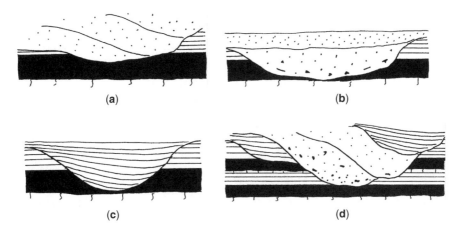

Figure 2.19 Channelling in coal seams. (**a**) Sand filled channel producing a sandstone roof to the coal seam; (**b**) Sand and coal detritus filled channel with coal seam eroded; (**c**) Mudstone filled channel with coal seam eroded; (**d**) Multiple channel sequence with sandstone and mudstone fills; the channel has removed the upper leaf of the coal seam. Reproduced by permission of Dargo Associates Ltd

maps of the northeastern Fuxin Basin, PRC (Wu *et al* 1992) in which sand percentage decreases to the west, which coincides with thicker coal development. Similar maps showing splitting of coal seams, their change in thickness and distribution, and the thickness and trend of the noncoal interburden are essential to the mine planning process (see also Figure 2.15).

The importance of these is self evident since, depending on the economics of the mine site, coal less than a predetermined thickness will not be mined. This means that an area containing significant reserves may not be exploitable due to the thinness of a good mining section, particularly if the coal becomes inferior above and below the good coal section, i.e. makes a poor floor and roof to the seam. Conversely there can be problems with an excessively thick coal in underground conditions producing poor roof or floor conditions. In opencast mines, thick coals are desirable, and the geotechnical nature of the overlying and underlying strata is important, particularly with regard to water movement and collection, as well as ground and slope instability.

Figure 2.21 shows an example of the variations in thickness of a coal in which the areas of thick/thin coal are clearly defined. The areas of coal thinning may indicate the attenuation of the seam, or that the seam is splitting, producing a thinner upper leaf. Such occurrences are influential on the siting of mining panels in underground workings, and also affect the coal/overburden ratios in opencast operations.

2.2.5.5 Interburden/overburden thickness

The amount and nature of the lithotypes present between coal seams and between the uppermost coal seam and the present land surface all have particular relevance to opencast mining operations. If the ratio of the thickness of such sediments to the thickness of workable coal is excessive, then the deposit will be deemed uneconomic. Such ratios are variable and may be dependent on other costs such as labour and transport. Most desirable coal/interburden/overburden ratios are in the order of 2:1–5:1, although they may be higher in certain circumstances, i.e. 10:1–15:1. In addition, if the lithotypes include hard indurated sandstone which will require blasting, then this is an added cost which has to be allowed for in the economic appraisal of the coal deposit.

2.2.5.6 Coal seam quality variations

Variations in the environments of deposition strongly influence the resultant quality of coals. As described in Section 2.2.3, peat mires can intermittently receive influxes of detritus by marine invasion, overbank flooding or from airborne sources such as contemporaneous vulcanism. Such occurrences will cause all or part of the coal seam to contain a higher ash level; this may be local or widespread. If the peat mire has been invaded by marine waters for a long period of time, precipitation of minerals into the uppermost part of the peat is likely; in particular, the sulphur content in those parts of a coal seam so affected may be greatly increased.

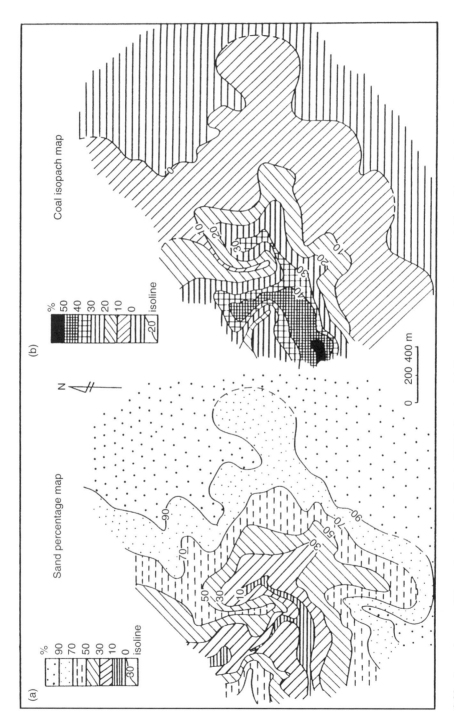

Figure 2.20 Isopach maps of the northeastern Fuxin Basin, PRC, showing: (**a**) Percentage of sandstone, and (**b**) coal isopachs for the middle part of the Haizhou Formation (Lower Cretaceous) showing the relationship of increased sandstone thickness with decreasing coal thickness. From *Wu et al* (1992). Reproduced by permission of the Geological Society of America

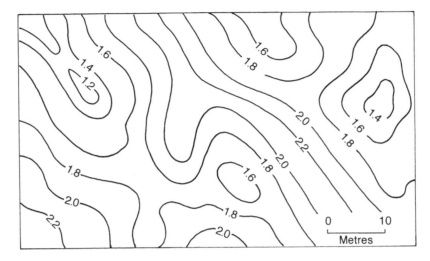

Figure 2.21 Coal seam thickness isopach map, hypothetical example (thickness values are in metres). Reproduced by permission of Dargo Associates Ltd

The plotting of coal seam quality parameters will not only give an indication of their distribution, but also indicate the paleoenvironmental influences that existed during the depositional and post-depositional phases of coal formation. Conversely, the interpretation of the paleoenvironment will help to predict coal quality in selected areas. That is, those areas considered distant from marine influence should have lower sulphur content, and coals deposited away from the main distributary channels and only subjected to low energy currents can be expected to have lower ash contents.

The above relationships have been summarised in the literature as follows: rapid subsidence during sedimentation generally results in abrupt variations in coal seams, but is accompanied by low sulphur and trace element content, whereas slower subsidence favours greater lateral continuity but a higher content of chemically precipitated material.

The examination of the coal quality analyses, particularly from cored boreholes, allows the coal geologist to plot coal quality variations across the study area. The parameters that are particularly relevant are volatile matter, ash and sulphur content. Deficiencies in the volatile matter content (notably in proximity to igneous intrusions) and too high amounts of ash and sulphur can lead to the coal under consideration being discarded as uneconomic due to increased costs of preparation or by simply just not having those properties required for the market that the coal is targeted for.

2.3 STRUCTURAL EFFECTS ON COAL

Any significant lateral or vertical structural change in a coal seam has a direct bearing on its thickness, quality and mineability.

Such changes can be on a small or large scale, affect the internal character of the coal, or simply displace the coal spatially, replacing it with noncoal sediment, or, in certain circumstances, with igneous intrusives. Disruption to coal seam thickness and continuity can lead to the interruption or cessation of mining, which will have economic repercussions, particularly in underground mines where mining flexibility is reduced. Therefore an understanding of the structural character of a coal deposit is essential in order to perform stratigraphic correlation, to calculate coal resource/reserves, and to determine the distribution of coal quality prior to mine planning.

2.3.1 Syndepositional effects

The majority of coal-bearing sediments are deposited in or on the margins of tectonic basins. Such a structural environment has a profound influence on the accumulating sediments in terms of both the nature and the amount of supply of detrital material required to form such sequences, and on the distribution and character of the environments of sedimentation.

In addition, diagenetic effects within the accumulating sediments produce structural deformation; this may be due to downward pressure from the overlying strata, and may be combined with water loss from the sediments while still in a nonindurated or plastic state.

2.3.1.1 Microstructural effects

The combination of thick sediment accumulation and rapid basin subsidence can produce instability particularly along the basin margins.

The effects on coal-bearing sediments are frequently seen in the form of slumping and loading structures, and liquifaction effects, the latter characterised by the disruption of bedding laminae and the injection of sediment into the layer above and below. Under such loading effects, coal may be squeezed into overlying strata and the original seam structure may be completely disrupted. In addition, coals may be injected by surrounding sediment in the form of sedimentary dykes. Interbedded sequences of mudstone, sandstone and coal that have undergone loading deformation exhibit a variety of structures such as accentuated loading on the bases of erosive sandstones, flame structures, distorted and dislocated ripples, and folded and contorted bedding (Figure 2.22).

Instability within environments of deposition, whether induced by fault activity or simply by overloading of accumulated sediment, can produce movement of sediments in the form of gravity flows; Figure 2.23 illustrates such a phenomenon. If a coal is transported in this fashion, the result can be an admixture of coal material and other sediment with no obvious bedding characteristics. Figure 2.24 shows a coal which has become intermixed with the surrounding sediment; it is now in an unworkable state as the ash content is too high and the geometry of the coal seam is irregular.

2.3.1.2 Macrostructural effects

Within sedimentary basins, existing faults in the underlying basement may continue to be active and influence the location, thickness and character of the sedimentary sequence. Many coal-bearing basinal sediments display evidence of growth faulting. In West Virginia and Pennsylvania, USA broad scale tectonic features have caused local thickening of the sequence in response to an increased rate of subsidence, as distinct from more stable platform areas (i.e. less rapidly subsiding), where sedimentation prograded rapidly over the shelf. In South Wales, UK, growth faults have again influenced sedimentation; here, in addition to active basement elements, faults are developed that owe their origin to gravity sliding within the sedimentary pile (Elliott and Lapido 1981). Overpressured, noncompacted argillaceous sediments initiate faults on gentle gradients. Such faults tend to have a curved cross-sectional profile, steep at the top and flattening progressively into bedding plane faults, often along the roof of a coal. In many cases such faults are partially eroded before the succeeding sediments are laid down.

Seam splitting can also, in certain circumstances, be attributed to growth faulting. Reactivation of faults with changes in the sense of movement can result in the downwarping of sections of peat beds; this is then followed by nonpeat deposition on the downwarped section, and then peat deposition resumed at the original level of the first peat. Figure 2.25 shows the possible mechanism for the formation of such a coal seam split.

Figure 2.22   Deformed bedding in Tertiary coal-bearing sediments, East Kalimantan, Indonesia. Photograph by LPT. Reproduced by permission of Dargo Associates Ltd

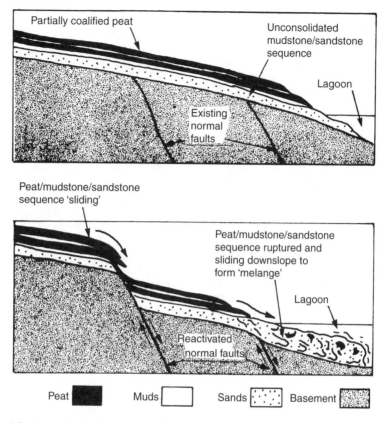

Peat ▉ Muds ☐ Sands ⠿ Basement ▨

Figure 2.23 Normal fault reactivation causing instability in a partially coalified peat sequence, with downslope slumping to produce a 'melange' of coal and intermixed sediment. Reproduced by permission of Dargo Associates Ltd

Periodic changes in base level in deltaic areas through fault activation will result in changes in the development and character of coals. With emergence, coals may become more extensive and, where the influx of detritus is curtailed, have a lower ash content. If submergence occurs, coals may be restricted areally, or receive increased amounts of detritus which may increase the ash content, or even cease to develop at all. Furthermore, submerged coals may be contaminated with marine waters, which could result in a higher sulphur content in the uppermost parts of the seam.

Growth folds also influence the deposition patterns in coal basins; local upwarping can accelerate the rates of erosion and deposition in some parts of a basin, but can also have the effect of cutting off sediment supply by uplift or by producing a barrier to the influx of detritus.

In very thick sedimentary sequences, the continued growth of such folds can result in the production of oversteepened fold axes. Where this occurs, overpressured mudstone at depth may be forced upwards and actually breach the anticlinal axial areas; this can be seen by the breaking up of the surface strata and the intrusion of material from below. Such diapiric intrusion breccia can be found in East Kalimantan, Indonesia, and these are often accompanied by the development of mud volcanoes along the axial region of the anticlines. Development of diapiric structures can disrupt as well as distort coal beds: in the Belchatow opencast coal mine in Poland, a large diapir has intruded into the coal-bearing sequence, dividing the coal reserve into two distinct areas; see Figure 2.26.

In the Kutei Basin in East Kalimantan, Indonesia, the established structural pattern continually evolved throughout the Tertiary Period. In this area, the anticlines are tight, with steep or overturned dips accompanied by steep reverse and normal faults in the complex axial regions. The synclines are broad and wide with very low dips; the transition between the two structures can be abrupt, now represented by steep reverse faults. These growth folds are thought to have been further

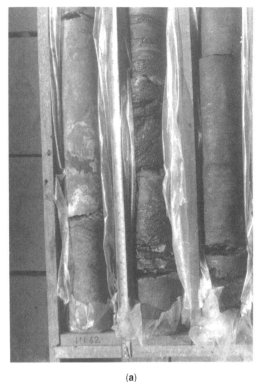

(a)   (b)

Figure 2.24   Cores exhibiting 'melange' or mixing of lithotypes due to gravity sliding in Tertiary coal-bearing sediments, East Kalimantan, Indonesia. Photographs by LPT. (**a**) Left core: mixing of sandstone and siltstone with subordinate coal. Centre core: coal mixed with mudstone and siltstone. Right core: coal and mudstone mixing; (**b**) All cores: sandstone, siltstone and coal mixing. Reproduced by permission of Dargo Associates Ltd

accentuated by gravity sliding associated with very thick accumulation of sediment (up to 9000 m) in the Kutei Basin, and rifting in the Makassar Strait to the east. The structural grain and the paleostrike were roughly parallel in this region, the resultant sequence characterised in its upper part by upper delta plain and alluvial plain sedimentation with numerous coals. This structural pattern is shown in the map (b) in Figure 2.38.

Penecontemporaneous vulcanism can also have a profound effect on the character of coals. Large amounts of airborne ash and dust together with waterborne volcanic detritus may result in the deposition of characteristic dark lithic sandstones, possible increases in the ash content in the peat mires, and the formation of tonstein horizons.

2.3.2   Post-depositional effects

All coal-bearing sequences have undergone some structural change since diagenesis. This can range from gentle warping and jointing up to complex thrusted and folded coalfields usually containing high rank coals.

These post-depositional structural elements can be simply summmarised as faults, joints (cleat), folds, and igneous associations. Mineral precipitation may also produce some changes in the original form and bedding of coal-bearing sequences.

2.3.2.1   Jointing/cleat in coal

Coal, and in particular all ranks of black coal, is noted for the development of its jointing, more commonly referred to as cleat. This regular pattern of cracking in the coal may have originated during coalification; the burial, compaction and continued diagenesis of the organic constituents results in the progressive reduction of porosity and permeability. At this stage, microfracturing of the coal is thought to be generated. The surfaces and spaces thus created may be coated and filled with mineral precipitates, chiefly carbonates and sulphides.

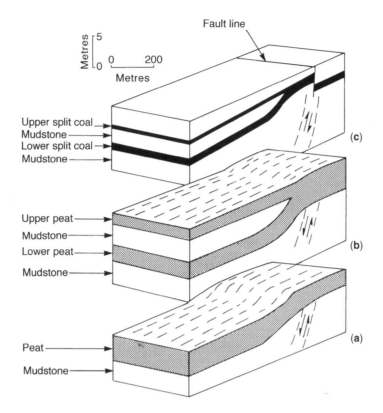

Figure 2.25 Seam splitting caused by differential movement of faults during peat deposition. (**a**) Fault downthrow results in downwarping of the peat; (**b**) Downwarp filled in with mudstone, peat development resumed at original level; (**c**) Fault throw sense reversed, uplifting split coal and downthrowing unsplit section of the coal seam. From Broadhurst and Simpson (1983). Reproduced by permission of University of Chicago Press

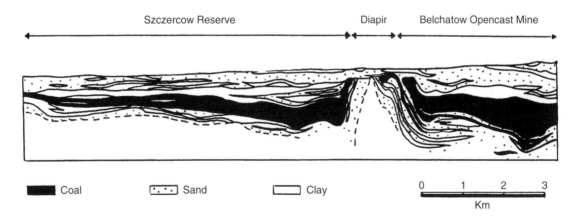

Figure 2.26 Section across Belchatow opencast mine showing effect of a diapiric intrusion into the thick coal seam. The diapir in this instance is a salt dome and effectively divides the coal reserves into two distinct areas. Reproduced by permission of Dargo Associates Ltd

Such microfracture patterns are usually developed in two directions approximately at right angles to each other and to the bedding. Often one cleat direction is better developed than the other; the dominant direction is known as cleat or bord, and the lesser direction as the end or end cleat. Figure 2.27(a) shows a well developed orthogonal cleat pattern in a Carboniferous bituminous coal from the Midlands, UK, and Figure 2.27(b) shows a strong cleat pattern in an anthracite of Triassic age from the Republic of Vietnam. Figure 2.27(c) shows cleat development together with conchoidal fracture in a Tertiary brown coal from Republic of Serbia. Figure 2.27(d) shows a well developed cleat and joint pattern in a Gondwana bituminous coal from Central India. It is noticeable that cleat can be seen in all thicknesses of coal, even in the thinnest films of coal included within other lithotypes.

2.3.2.2 Faulting

The development of strong joint and fault patterns in coal-bearing sequences is the commonest post-depositional structural expression. Briefly, the principal fault types are as follows.

Normal faults are produced by dominantly vertical stress resulting in the reduction of horizontal compression, leaving gravity as the active compression; this results in the horizontal extension of the rock sequence.

This form of faulting is common; movements can be in the order of a few metres to hundreds of metres. Figure 2.28 shows a normal fault with a throw of about two metres; such faulting is not too problematical in opencast workings, but larger throws can result in the cessation of opencast mining either locally or totally. Figure 2.29 shows a local highwall termination due to faulting in an opencast mine in Bosnia-Herzegovina, and Figure 2.30 illustrates a normal fault downthrowing overburden (light colour) against a coal seam in an opencast mine in India. In underground workings even small scale faulting can result in cessation of the mining of fully automated faces, resulting in loss of available reserves.

The dip of normal faults ranges widely: in coalfields most are thought to be in the region of 60–70 degrees. Some normal faults die out along their length by a decrease of throw towards either one or both ends of the fault. Again a fault may pass into a monoclinal flexure, particularly in overlying softer strata. Such faulting also produces drag along the fault plane, the country rock

being pulled along in the direction of movement. Where large faults have moved on more than one occasion, and this applies to all kinds of faulting, a zone of crushed coal and rock may extend along the fault plane and have a width of several metres, such a crush zone can be seen in the highwall of an opencast working in the UK, as shown in Figure 2.31.

Large scale normal faults are produced by tensional forces pulling apart or spreading the crustal layer: where these faults run parallel with downfaulted areas in between, they are known as graben structures. Many coalfields are preserved in such structures: the brown coalfields of northern Germany and eastern Europe, and the Gondwana coalfields of India and Bangladesh are examples.

Low angle faults with normal fault displacements are known as lag faults. They originate from retardation of the hanging wall during regional movement, as shown in Figure 2.32. Lag faults are common in the coalfields of South Wales, UK.

Reverse faults are produced by horizontal stress with little vertical compression; this results in the shortening of the rock section in the direction of maximum compression.

Very high angle reverse faults are usually large structures, associated with regional uplift and accompanying igneous activity. In coal geology, those reverse faults with low angles (<45 degrees) are more significant. A typical reverse fault structure is seen in Figure 2.33, where the fault has dislocated a coal seam by several metres. When the angle is very low, and the lateral displacement is very pronounced, such faults are termed thrust faults. The shape of such low angle reverse faults is controlled by the nature of the faulted rocks, especially when a thrust plane may prefer to follow the bedding plane rather than to cut across it.

In typical sequences of coal, seatearth and mudstone with subordinate sandstone, such low angle faults often follow the roof and/or the floor of coal seams as these allow ease of movement, the seatearths often acting as a lubricant. One detrimental effect is the contamination of the coal seam with surrounding country rock, so reducing its quality and, in some cases, its mineability.

In highly tectonised coal deposits, a great number of coal seam contacts have undergone some movement and shearing; in some cases the whole seam will have been compressed and moved – this may be displayed in coals as arcuate shear planes throughout; Figure 2.34 portrays this feature.

(a)

(b)

Figure 2.27   (**a**) Orthogonal cleat pattern in Meltonfleet Coal, Upper Carboniferous, Yorkshire, UK. Reproduced by permission: IPR/25-6C British Geological Survey. © NERC. All rights reserved; (**b**) Well developed joint and cleat pattern in Triassic anthracite, Republic of Vietnam. Photograph by LPT. Reproduced by permission of Dargo Associates Ltd; (**c**) Cleat development and conchoidal fracture in Tertiary brown coal, Republic of Serbia. Photograph by courtesy of Dargo Associates Ltd; (**d**) Well developed joint pattern in Permian (Gondwana) coal, central India. Photograph by courtesy of Dargo Associates Ltd

(c)

(d)

Figure 2.27 *(continued)*

The development of thrust zones in coal sequences is illustrated in Figure 2.35. Here lateral compression has produced thrusting along preferred lithological horizons, and continued compression has resulted in the upper part of the sequence being more tectonically disturbed than the lower part; this deformation is now termed progressive easy-slip thrusting. Such events are particularly common in coalfields which have suffered crustal shortening, as is the case in South Wales, UK (Gayer

et al 1991), and in the Appalachians, USA. Thrusting is also accentuated where coal and mudstone sequences are sandwiched between thick sequences of coarse clastics, the upper and lower portions of the sequence reacting to compressive forces quite differently to the incompetent coals and mudstones.

Strike-slip faults have maximum and minimum stress in the two horizontal planes normal to one another. This has the effect of producing a horizontal movement

Figure 2.28   Normal fault with downthrow of 2 m to the right. Tertiary coal-bearing sediments, Sumatra, Indonesia. Photograph by LPT. Reproduced by permission of Dargo Associates Ltd

Figure 2.29   Highwall termination due to faulting in Tertiary brown coal opencast mine in Bosnia-Herzegovina. Photograph by courtesy of Dargo Associates Ltd

Figure 2.30   Normal fault downthrowing overburden (light colour) against a coal seam in an opencast mine in central India. Photograph by courtesy of Dargo Associates Ltd

Figure 2.31   Large fault zone exposed in highwall in an opencast mine, South Wales, UK. Photograph by LPT. Reproduced by permission of Dargo Associates Ltd

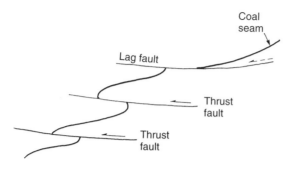

Figure 2.32 Lag fault produced by retardation of the upper part of the sequence during the forward movement of the lower sequence by thrust faulting. From Sherborn Hills (1975) *Elements of Structural Geology*, 2nd edn, Chapman and Hall, reproduced with kind permission of Kluwer Academic Publishers

either in a clockwise (dextral) or anticlockwise (sinistral) sense. Strike-slip faulting is usually found on a regional scale, and although important, has a lesser influence on the analysis of small coal deposits and mine lease areas.

Evidence of faulting on the rock surface can be seen in the form of slickensides; these are striations on the fault plane parallel to the sense of movement. Some fault planes have a polished appearance, particularly where high rank coal has been compressed along the fault plane. Conical shear surfaces are characteristically developed in coal; these are known as cone-in-cone

structures, and are the result of compression between the top and bottom of the coal.

Coal responds in a highly brittle manner to increasing deformation by undergoing failure and subsequent displacement along ever-increasing numbers of fracture surfaces (Frodsham and Gayer 1999). In tectonically deformed coalfields, and in particular in mine workings, it is important that a rapid assessment of the physical state of the coal can be made. Visual assessment of the appearance of coal can be made in hand specimen samples, core samples, or by outcrop observations (see Section 6.4.10). The Average Structural Index (ASI) can be used to assess the relative strength of deformed coal samples on the basis of appearance and the frequency of fractures in the specimen. Table 2.3 shows the coal types and the ASI rating for coals of differing levels of structural intensity. On a larger scale, the Coal Bedding Code is based on the principle that with increasing frequency of tectonic fracturing, the bedding planes in coal seams become progressively obscured. Frodsham and Gayer (1999) describe five categories of bedding plane obscurity and large scale fracture intensity within coal seams from South Wales, UK (Figure 2.36). These categories range from Bedding Code 4 – Excellent, where coal exhibits very clear bedding with fractures moderately spaced at less than one per metre, to Bedding Code 0 – Absent, where the bedding is completely destroyed by tectonic fracturing. The use of Coal Bedding Codes has proved successful in predicting the location of

Figure 2.33 Coal seam dislocated by reverse fault, throw 1.5 m, UK opencast mine. Photograph by M.C. Coultas

Figure 2.34   Highly sheared anthracite coal seam (seam thickness 1.2 m) in opencast mine, South Wales, UK. Photograph by LPT. Reproduced by permission of Dargo Associates Ltd

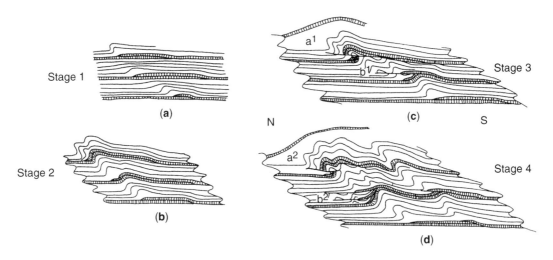

Figure 2.35   Model for four stages in progressive easy-slip thrusting. (**a**) Thrusts develop simultaneously as flats along the floors of overpressured coal seams, cutting up to the roofs of the seams along short ramps; propagation folds grow at the thrust tips; (**b**) Thrusts continue to propagate with amplification of the tip folds, until a lower propagation fold locks up a higher thrust, producing downward facing cut-offs; (**c**) Continued out of sequence movement on higher thrusts results in break back thrusting in the hanging wall and footwall areas; thrusts locally cut down stratigraphy in the transport direction; (**d**) Progressive out of sequence hanging wall break back produces distinctive geometry with the structure in a lower thrust being apparently unrelated to that in a higher thrust slice. Progressive footwall break back produces folded thrusts. From Gayer *et al* (1991). Reproduced by permission of the authors and the Ussher Society

Table 2.3 The Average Structural Index (ASI) for coals. From Frodsham and Gayer (1999)

Basic coal type	ASI value	Type and frequency of fracturing	Structural state of the coal
Normal	1	entirely nontectonic	undisturbed bright hard coal
Normal	2	mainly nontectonic; 1–2 striated fractures	strong, bright and hard but easily split along fractures
Normal	3	mainly nontectonic; 1–2 polished fractures	strong, bright and hard but easily split along fractures
Normal	4	mainly nontectonic, several tectonic of either kind	bright but coal becoming noticeably weakened by tectonic fractures
Abnormal	5	mainly tectonic fractures of either kind but also several nontectonic	coal exhibiting a change in overall structure, has largely dull or shiny lustre and lacks strength
Abnormal	6	striated fracture planes dominant with only a few nontectonic	disturbed and very dull
Abnormal	7	polished fracture planes dominant with only a few nontectonic	disturbed and excessively shiny
Abnormal	8	wholly tectonic fractures, no nontectonic left	disturbed, either dull or excessively shiny
Outburst	9	pervasive microfractures	highly friable, soft sooty texture no *in-situ* strength

coal outbursts as more deformed parts of the mine are approached (Frodsham and Gayer 1999).

2.3.2.3 Folding

Coals in coal-bearing sequences may be folded into any number of fold styles, for example as shown in Figure 2.37. In coalfield evaluation, the axial planes of the folds need to be located and the dips on the limbs of the folds calculated. In poorly exposed country the problem of both true and apparent dips being seen has to be carefully examined. Also in dissected terrain, dips taken at exposures on valley sides may not give a true reflection of the structural attitude of the beds at this locality; many valley sides are unstable areas and mass movement of strata is common, resulting in the recording of oversteepened dips. This is characteristic of areas of thick vegetation cover where a view of the valley side is obscured and any evidences of movement may be concealed. If the field data suggest steeper dipping strata, this will give less favourable stripping ratio calculations for an opencast prospect and may contribute to the cancellation of further investigations. Similarly, in underground operations, if the dip of the coal seams steepens, it can make the working of the coal

difficult, and in the case of longwall mining, prevent further extraction. Therefore it is important to be sure that all readings taken reflect the true nature of the structure in the area of investigation.

Compression of coal seams during folding can produce tight anticlinal folds with thrusting along the nose of the fold, these have been termed queue anticlines. Coal seams can be pinched out along the fold limbs and appear to have flowed into the axial areas of the anticlines. Where this occurs from two directions approximately normal to one another, coals can be concentrated in 'pepperpot' type structures. Such features are usually found only in highly tectonised coalfields, and examples of such intense deformation are illustrated in Figure 2.38(a), and a coal seam squeezed in this way is shown in Figure 2.38(b); here the coal has been compressed in and around the overlying sandstone. In many instances such structural complication will render a coal seam unmineable except by the most primitive of methods, but coal concentrations in the axial regions of folds have been mined in the same fashion as mineral 'saddle reefs'.

Detailed mapping of folded coal deposits is an essential part of the exploration process; examples of folded coal deposits are illustrated in Figure 2.39(a), where a pattern of zig-zag folds characterises the Wurm

Figure 2.36   The Coal Bedding Code, showing five categories of bedding plane obscurity and large scale fracture intensity within coal seams in South Wales, UK. From Frodsham and Gayer (1999). Reproduced by permission

Figure 2.37 Intensely folded Carboniferous coal-bearing sediments, Little Haven, Pembrokeshire, UK. Photograph by courtesy of Dargo Associates Ltd

Coal Basin, Germany, and Figure 2.39(b), which shows a series of asymmetrical folds with associated thrusting from the Kutei Basin, Indonesia. Such examples serve to show the necessity of acquiring a good understanding of the structural elements and style of the coal deposit in order to identify those areas where coal is preserved in such quantities, attitude and depth as to allow mining to develop.

2.3.2.4 Igneous associations

In many coalfields associated igneous activity has resulted in dykes and sills being intruded into the coal-bearing sequence.

The intrusion of hot molten rock into the coals produces a cindering of the coal and a marked loss in volatile matter content which has been driven off by heat. This can have the effect of locally raising the rank of lower rank coals, and can therefore in certain circumstances make the coal attractive for exploitation. Such 'amelioration' of coal seams is a common feature in areas of igneous activity; good examples are found in Indonesia and the Philippines where Tertiary subbituminous coals have been ameliorated up to low volatile bituminous and some even to anthracite rank.

The majority of dykes and sills are doleritic in composition, as in the case of South African and Indian coalfields, but occasionally other types are found. In the Republic of Korea acidic dykes and sills are intruded into the coals; Figure 2.40 shows acidic igneous material intruding a coal seam in an underground working.

In areas where igneous intrusions are prevalent in mine workings, plans showing the distribution and size of igneous bodies are required in order to determine areas of volatile loss where the coal has been baked, and because of the hardness of the igneous material, tunnelling has to be planned with the position of intrusions in mind. Igneous sills have a tendency to jump from one coal seam to another, so that close spaced drilling is often required to identify precisely the nature

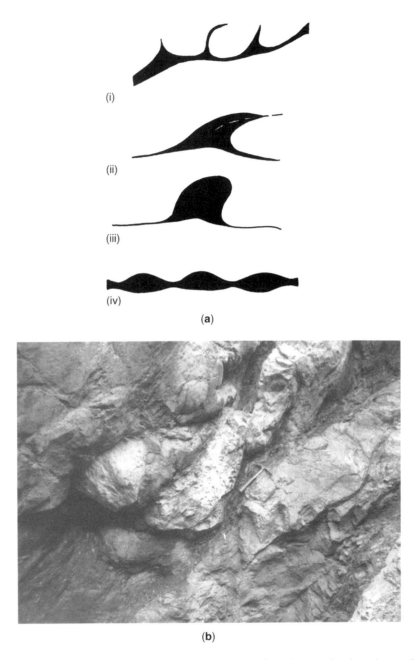

Figure 2.38   (a) Tectonic deformation of coal seams due to compression: (i) squeezing into the overlying formation; (ii) queue anticline; (iii) 'pepper-pot' structure; (iv) 'rosary' structure; (b) Carboniferous anthracite squeezed in and around overlying sandstone, Samcheog Coalfield, Republic of Korea. Photograph by LPT. Reproduced by permission of Dargo Associates Ltd

and position of such intrusions. Igneous intrusions are found in coal sequences worldwide but in particular are a common feature of South African coal workings. Where such igneous bodies exist, the coal geologist must identify the areas occupied by igneous material within the mine area, and also those seams affected by igneous activity.

In addition, methane gas driven off during intrusion may have collected in intervening or overlying porous sandstones. Mine operatives need to investigate this possibility when entering an intruded area of coal.

2.3.2.5 Mineral precipitates

A common feature of coal-bearing sequences is the formation of ironstone, either as bands or as nodules. They usually consist of siderite ($FeCO_3$) and can be extremely hard. Where ironstone nucleation and development takes place either in, or in close proximity to, a coal seam, this can deform the coal, cause mining difficulties, and, because of the difficulty in separating coal and ironstone while mining, will have an effect on the quality of the run-of-mine product.

Iron sulphide (FeS) in the form of iron pyrite may be precipitated as disseminated particles, as thin bands, or, as is more common, as coatings on cleat and bedding surfaces (the coal specimen in Figure 6.7(a) displays pyrite in this form). Inorganic sulphur held in this form in coal can be removed by crushing and passing the coal through a heavy liquid medium. Organic sulphur held elsewhere in the coal cannot be readily removed, and remains an inherent constituent of the coal.

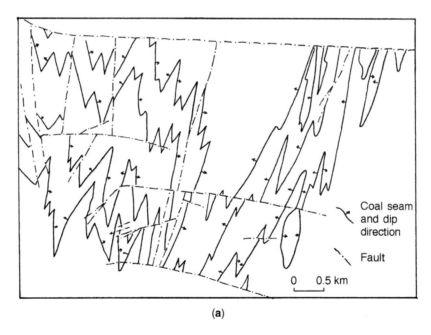

(a)

Figure 2.39 Outcrop patterns in folded coalfields. (**a**) Zigzag folding of coal seams and associated faulting, Wurm Coal Basin, Germany. From Stutzer and Noe (1940), by permission of the University of Chicago Press; (**b**) Asymmetrical folding, broad synclines and sharp anticlines associated with thrusting, East Kalimantan, Indonesia. From Land and Jones (1987), reproduced by permission of the authors and The Geological Society

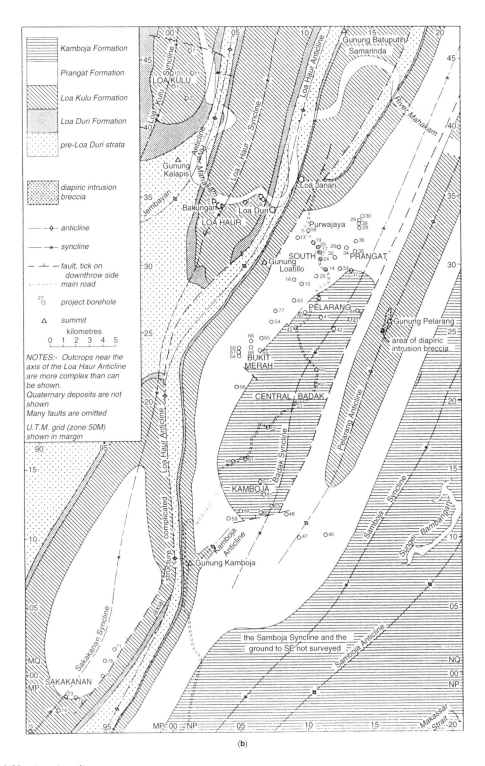

Figure 2.39   (*continued*)

Figure 2.40 Jurassic anthracite (dark colour) intruded by granitic dykes and sills (light colour). Chungnam Coalfield, Republic of Korea. Photograph by LPT. Reproduced by permission of Dargo Associates Ltd

Other mineral precipitates are usually in the form of carbonates, coating cleat surfaces, or occasionally as mineral veins. Where quartz veining occurs, this has the detrimental characteristics of being hard, liable to produce sparks in an underground environment where gas is a hazard, and also, when crushed, is an industrial respiratory health hazard.

3

Age and Occurrence of Coal

3.1 INTRODUCTION

Although land plants first developed in the Lower Paleozoic Era, and coal deposits of Devonian age are the earliest known, it was not until the Upper Paleozoic Era, particularly the Carboniferous and Permian Periods, that sufficient plant cover was established and preserved to produce significant coal accumulations.

Throughout the geological column, i.e. from the Carboniferous Period to the Quaternary Period, coal deposits have been formed. Within this time range, there have been three major episodes of coal accumulation.

The first took place during the Late Carboniferous–Early Permian Periods. Coals formed at this time now form the bulk of the black coal reserves of the world, and are represented on all of the continents. The coals are usually of high rank and may have undergone significant structural change. Carboniferous–Permian coal deposits stretch across the northern hemisphere from Canada and the United States of America, through Europe and the Commonwealth of Independent States (CIS) to the Far East. In the southern hemisphere, the Carboniferous–Permian coals of Gondwanaland are preserved in South America, Africa, the Indian Subcontinent, Southeast Asia, Australasia and Antarctica.

The second episode occurred during the Jurassic–Cretaceous Period, and coals of this age are present in Canada, the USA, China and the CIS.

The third major episode occurred during the Tertiary Period. Coals formed during this period range from lignite to anthracite. Tertiary coals form the bulk of the world's brown coal reserves, but also make up a significant percentage of black coals currently mined. They are characterised by thick seams and have often undergone minimal structural change. Tertiary coals are also found worldwide, and are the focus of current exploration and production as the traditional Carboniferous coalfields become depleted or geologically too difficult to mine.

Figure 3.1 shows the generalised distribution of world coal deposits in terms of geological age and area (modified from Walker 2000).

3.2 PLATE TECTONICS

Evidence of ocean floor spreading and the identification of modern plate margins has enabled the mechanism of plate tectonics to be understood, i.e. the earth's crust and upper part of the mantle consist of a number of mobile plates that have responded to convection currents in the mantle. This has resulted in the amalgamation and fragmentation of plates throughout geological time. du Toit (1937) proposed that the supercontinent *Pangaea* consisted of *Laurasia* in the northern hemisphere and *Gondwanaland* in the southern hemisphere. These two land areas split apart in early Triassic time, followed by further rifting which has produced the various smaller continents that exist today.

During the Carboniferous Period, in the northern part of *Pangaea* (*Laurasia*), the coal basins of western and central Europe, the eastern USA, and the CIS were equatorial in nature and tropical peat mires containing a flora of *Lepidodendron, Sigillaria and Chordaites* were characteristic of coal deposition (Figure 3.2(a)). The climate changed during the Permian Period and coal deposition ceased in the northern area. In the southern part of *Pangaea* (*Gondwanaland*), covering what is now South America, southern Africa, India, Australia and Antarctica, peat mires formed under cooler, more temperate conditions, characterised by the *Glossopteris* flora (Figure 3.2(b)). After the break-up of *Pangaea*, coal deposition continued through the Triassic, Jurassic, Cretaceous Periods (Figure 3.3(a)) and the Tertiary Period (Figure 3.3(b)), where another change in the floral types took place, heralding the onset of *Angiosperm* floras. These changes in vegetation type are reflected in the type and proportion of maceral types present in the coals. The Laurasian coals are rich in the vitrinite group of macerals whereas the Gondwana coals have a much

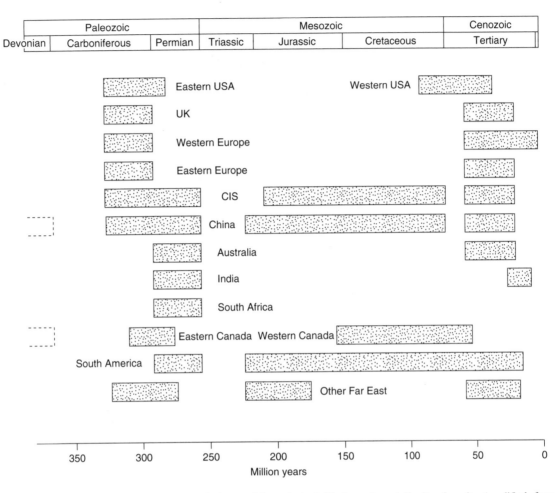

Figure 3.1 Geological age distribution of the world's principal black coal and lignite deposits (modified from Walker 2000) with permission of IEA Coal Research

higher percentage of the inertinite group of macerals with varying amounts of vitrinite. Gondwana coals have a higher content of mineral matter but lower sulphur contents than the Laurasian coals (see Section 4.1.2).

Later formed coals are more complex in their chemical make-up, a likely reflection of the variety of plant types and environments of deposition.

The Tertiary coals are for the most part lignites, although in some areas they have undergone severe temperature changes, which have produced higher rank coal in some areas, e.g. Indonesia, Colombia and Venezuela.

3.3 STRATIGRAPHY

The age of all the major coal deposits is well documented, and the stratigraphy of each deposit has been studied in detail. This is particularly true for those deposits which have an economic potential.

The origin of coal is characterised by deposition in foredeep and cratonic basins. The essentially nonmarine nature of these coal-bearing sequences has meant that detailed chronostratigraphy has often been difficult to apply due to the lack of biostratigraphic evidence, notwithstanding studies of floras and nonmarine faunas.

In the Carboniferous of western Europe, a number of marine transgressions has enabled the coal-bearing sequences to be divided into a number of stratigraphic sections, and where individual coal seams have either an overlying marine mudstone or nonmarine bivalve band, then correlations are possible over large distances. In the Carboniferous of the USA and also of China,

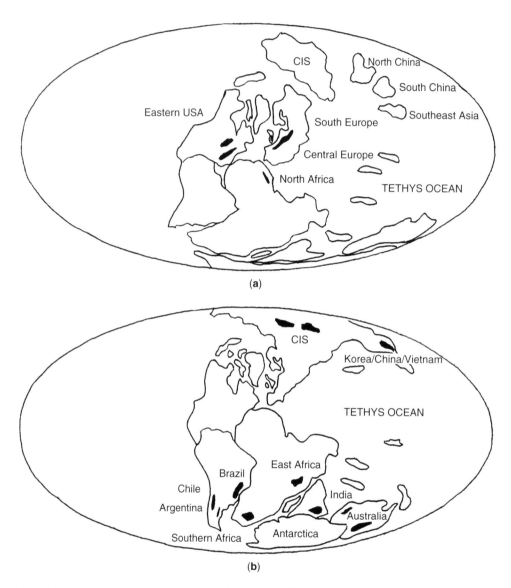

Figure 3.2 (a) Paleogeographical reconstruction of Late Carboniferous showing principal areas of coal deposition; (b) Paleogeographical reconstruction of Permian–Triassic showing principal areas of coal deposition. (Modified from Walker 2000) with permission of IEA Coal Research

discrete limestone beds within coal-bearing sequences have enabled good broad stratigraphic control over large areas. The later Permian, Mesozoic and Tertiary coal deposits all have similar constraints on detailed correlation. In the UK, the long history of studying the Carboniferous (Westphalian) has enabled the chronostratigraphy and biostratigraphy to be classified as shown in Table 3.1.

The establishment of a stratigraphic framework for a coal-bearing sequence can be approached in two ways: an examination of the sedimentary sequence in which the coals occur; and a detailed study of the coals themselves. It is usual to apply a combination of chronostratigraphy (where possible) and lithostratigraphy for individual coal deposits. This may be supported by geophysics and detailed sedimentological studies. In addition, it can be augmented by petrographical analysis of the coals and palynological studies which will aid the identification of individual coals or series

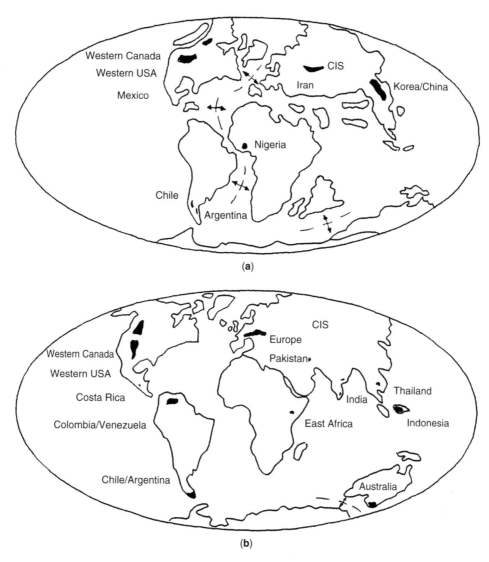

Figure 3.3 (**a**) Paleogeographic reconstruction of Jurassic–Cretaceous showing principal areas of coal deposition; (**b**) Paleogeographic reconstruction of Tertiary showing principal areas of coal deposition. (Modified from Walker 2000) with permission of IEA Coal Research

of coals. The combination of all these studies is the basis on which to build the geological model and to develop a three-dimensional picture of the coal deposit.

In the UK, detailed studies over the last 150 years of the various coalfield areas has resulted in a wealth of geological data being collected and interpreted. For example, in the Yorkshire coalfield, the identification of coals and marine and nonmarine fossiliferous horizons has enabled a detailed stratigraphy to be built up.

Figure 3.4 shows the stratigraphical column of 1120 m of the Westphalian succession for the Wakefield district of Yorkshire (Lake 1999). The compilation of detailed stratigraphic logs from sections, boreholes and mines then allows correlations to be made. Figure 3.5 depicts a series of logs within the Leicestershire coalfield and shows correlations based on faunal horizons and coals (Worssam and Old 1988).

In the USA, a large amount of geological investigation took place in the Pennsylvanian coal-bearing sequence

Table 3.1 Detailed chronostratigraphy and biostratigraphy of the Carboniferous (Westphalian A–C) coal-bearing sequence of the UK (Lake 1999). IPR/23-10C British Geological Survey. © NERC. All rights reserved

CHRONOSTRATIGRAPHY				BIOSTRATIGRAPHY		
SERIES	STAGE	CHRONOZONE	MARINE BAND	NONMARINE BIVALVES	MIOSPORES	MEGAFLORA
WESTPHALIAN	BOLSOVIAN	Phillipsi		*Anthraconauta phillipsi* Zone	*Torispora securis* (X)	*Paripteris linguaefolia*
		Upper Similis-Pulchra	Cambriense			
			Shafton	*Anthraconaia adamsi-hindi*	*Vestispora magna* (IX)	
			Edmondia			
			Carway Fawr			
			Aegiranum			
	DUCKMANTIAN	Lower Similis-Pulchra	Sutton	*Anthracosia atra*		
			Haughton			
			Clown			
			Maltby			*Lonchopteris rugosa*
		Modiolaris	Lowton Estheria	*Anthracosia caledonica*	*Dictyotriletes bireticulatus* (VIII)	
				Anthracosia phrygiana		
				Anthracosia ovum		
			Vanderbeckei			
	LANGSETTIAN	Communis		*Anthracosia regularis*	*Schulzospora rara* (VII)	
				Carbonicola crista-galli		*Lyginopteris hoening-hausii*
			Low Estheria			
				Carbonicola pseudorobusta		
				Carbonicola bipennis	*Radiizonates aligerens* (VI)	
			Kilburn	*Carbonicola torus*		
			Burton Joyce			
			Langley			
		Lenisulcata		*Carbonicola proxima*	*Triquitrites sinani Cirratriradites saturni* (SS)	
			Amaliae	*Carbonicola extenuata*		
			Meadow Farm			
			Parkhouse			
			Listeri			
			Honley			
			Springwood	*Carbonicola fallax-protea*		
			Holbrook			

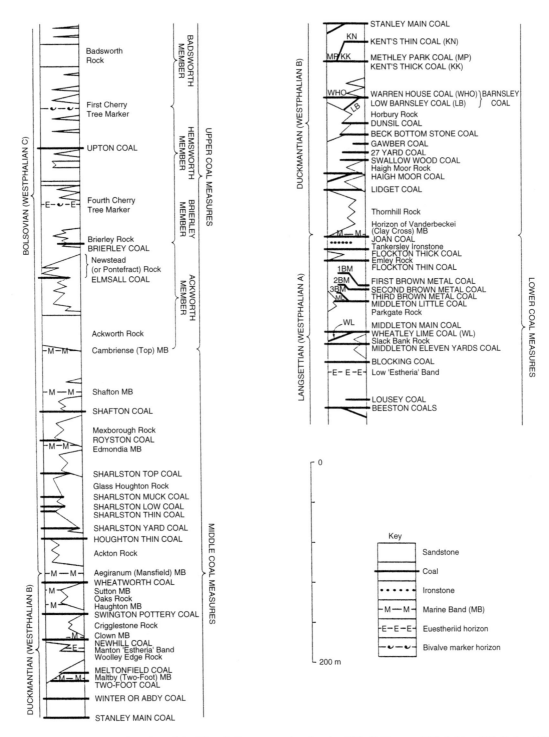

Figure 3.4 Detailed stratigraphy of the Westphalian succession for the Wakefield area of Yorkshire, UK (Lake 1999). IPR/23-10C. British Geological Survey, © NERC. All rights reserved

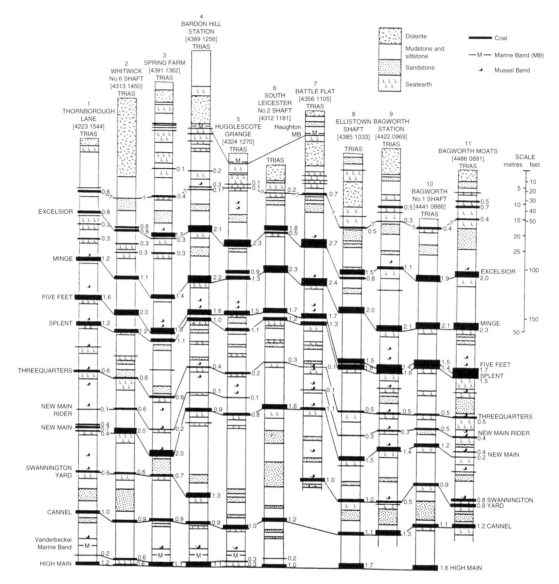

Figure 3.5 Stratigraphical correlation based on faunal horizons and coals (Worssam & Old 1988). IPR/23-10C. British Geological Survey. © NERC. All rights reserved

in the Illinois Basin. Correlation studies have been based on spores (Kosanke 1950) and on sedimentological analysis of the shape and distribution patterns of sand bodies and their effect on the principal economic coal seams in the basin (Potter 1962, 1963, Potter and Glass 1958 and Potter and Simon 1961). Figure 3.6 illustrates the use of lithological and electric logs to show the stratigraphic relationships due to the development of the Anvil Rock Sandstone and its effect upon the Herrin No. 6 seam (Potter and Simon

1961). Similar studies have been carried out in all the major coal deposits worldwide with a view to identifying the effects on coal-mining economics by geological processes.

3.4 AGE AND GEOGRAPHICAL DISTRIBUTION OF COAL

A brief summary is given of the geographic distribution of the known coal deposits of the world. It is designed

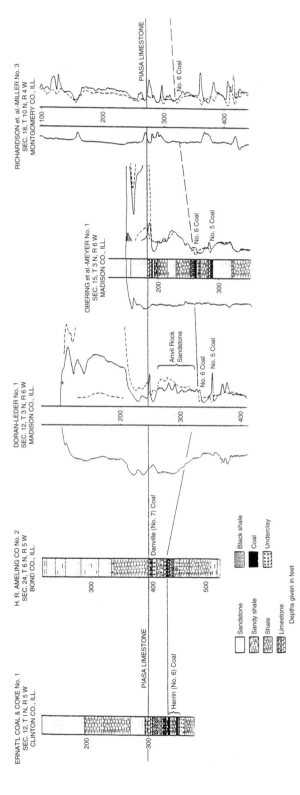

Figure 3.6 Use of lithological and electric logs to show the stratigraphical relationships between the development of the Anvil Rock Sandstone, and its effect on the Herrin No. 6 coal (Potter and Simon 1961) reproduced by permission of the Illinois State Geological Survey

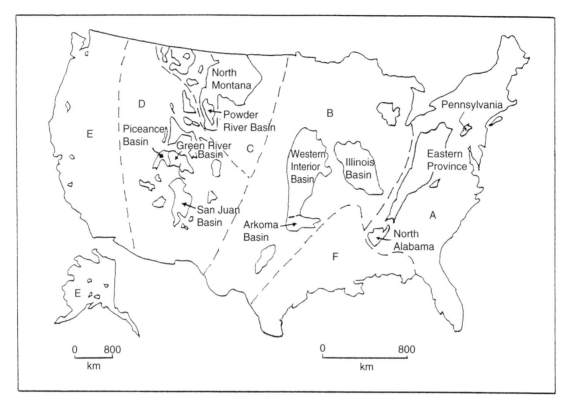

Figure 3.7   Coal deposits of the USA. (**a**) Eastern Province; (**b**) Interior Province; (**c**) Northern Great Plains Province; (**d**) Rocky Mountains Province; (**e**) Pacific Coast Province; (**f**) Gulf Coastal Plain Province. Reproduced by permission of Dargo Associates Ltd

as guide to the location of the principal coalfields throughout the world. The detailed stratigraphical ages of the deposits are not given, usually only the period in which they were formed. The distribution of coal deposits throughout the world is shown in Figures 3.7 to 3.15, and are dealt with in nine geographical regions. Details are based on Walker (2000) in his review of the major coal-producing countries, and from Saus and Schiffer (1999) in their review of lignite in Europe.

3.4.1 United States of America

The coal deposits of the USA have been divided into six separate areas or provinces, based on the findings of the US Geological Survey (Figure 3.7), and Nelson (1987).

The Eastern Province is the oldest and most extensively developed coal province in the USA. The coal is of Carboniferous age (Pennsylvanian), and the province contains two-fifths of the nation's bituminous coal plus almost all the anthracite. Coal rank increases from west

to east so that high volatile bituminous coal gives way to low volatile bituminous and anthracite. The sulphur content of the coals is higher in the west and decreases to the east, the older coals containing the highest sulphur levels.

The Appalachian Basin extends for over 1500 km from Pennsylvania in the north to Alabama in the southwest, with a width of 400 km in the north tapering to 25 km in the south. The Pennsylvanian sequence is around 1000 m thick and the basin is subdivided into a northern region, comprising southwest Pennsylvania, eastern Ohio, northern Virginia, and northeastern Kentucky; a central area which reaches across southern West Virginia, eastern Kentucky, western Virginia and northern Tennessee; the southern region covers southern Tennessee and northern Alabama. The northern region contains over 60 coal seams of varying economic significance, the central area contains around 50 coals, and the south-central area has up to 26 economic coal seams. Seams are usually between 0.5 m and 3.6 m in thickness, with varying degrees of structural intensity. Deep mines characterise the older workings; more recently, opencast mining in the form of contour mining and mountain-top removal has been established.

Most reserves are in high and medium volatile bituminous coal, used chiefly as steam coal. The USA's largest reserves of coking coal are low volatile, low sulphur coals situated in central Pennsylvania and West Virginia.

The Interior Province comprises a number of separate basins containing Carboniferous bituminous coal. The eastern area is the Illinois Basin, which extends across central and southern Illinois, southwest Indiana and western Kentucky. Pennsylvanian sediments reach 1200 m in thickness with around 25 coals of economic interest. In Illinois, coal rank increases from high volatile bituminous C coal in the central and northern areas, to high volatile bituminous B and A in west Kentucky. Coals generally have a high sulphur content of 3–7%, although low sulphur coals are present.

Coal seams range from 0.5 m to 2.5 m in thickness, and two seams, the Springfield and Herrin Coals, are >1.5 m, and cover thousands of square kilometres. These seams have accounted for more than 90% of production from this region. The western Interior Province includes deposits in Iowa, Missouri, Kansas, Oklahoma, Arkansas and Texas, and is characterised by thinner (<1.5 m) but laterally extensive seams, mined exclusively by opencast stripping.

The southernmost part, the Arkoma Basin, has coals of higher rank, including semi-anthracites with low sulphur contents. Some of the low sulphur coal has coking properties.

The Northern Great Plains Province contains coals of Cretaceous to Tertiary age. The chief coal-producing area is the Powder River Basin, which lies in northern Wyoming and southeastern Montana. Coals of Cretaceous age are present, but the major coal development occurred in the Lower Tertiary. Between 8 and 12 important coals averaging 6–30 m in thickness are present in 600–900 m of sediments. These are overlain by a further 300–600 m, containing up to 8 major coal seams. Some bituminous coals are present, but the main reserves are of subbituminous coal and lignite. These coals have assumed a greater significance in recent years because of their low sulphur content. This makes them attractive to users constrained by environmental legislation.

Cretaceous coals also occur in northern Montana; these are bituminous in rank.

The Rocky Mountain Province contains coal preserved in a series of intermontane basins. In northwest Colorado and southwest Wyoming occur the two principal structural basins, the Green River Basin in the west and the Great Divide Basin in the east. The coal-bearing sediments are Upper Cretaceous to Lower Tertiary in age and are more than 900 m in thickness, containing multiple bedded coal seams. In west central Colorado and central Utah, the Piceance Basin contains over 3000 m of sediments of Late Cretaceous to Lower Tertiary age. In southwestern Colorado and northeastern New Mexico, Late Cretaceous coals are present in the San Juan Basin.

Most Cretaceous coal is bituminous and subbituminous, and the Tertiary coals are subbituminous and lignite. The seams are thick, in the range of 3.0 m to 10.0 m. Some are structurally affected and some are ameliorated by igneous intrusives. In general the coals have low sulphur contents (<1%), with some small areas of coal suitable for coking purposes. The usual mining practice is by opencast and contour stripping mines, with some drift mining.

The Pacific Coast Province covers those coal deposits found west of the Rocky Mountains, and those in Alaska. The coal is Tertiary in age, and is found in small widely scattered basins from California in the south to Washington in the north. The coal has been tectonised and metamorphosed.

In Alaska, numerous deposits of subbituminous and high volatile bituminous coal have been identified. As yet the area is not fully explored; however, indications are that the coal resources could be up to 10% of the economically recoverable coal in the world.

Finally the Gulf Coastal Plain Province contains lignites of Tertiary age. The lignite is produced from large opencast pits, where seams range from 1.0 m to 7.5 m in thickness. Production has greatly increased in the last 20 years and is primarily for electricity generation.

3.4.2 Canada

The largest coal-bearing region is located in western Canada, stretching from south Saskatchewan across Alberta into British Columbia (Figure 3.8). The coals that underlie the plains are relatively undisturbed Tertiary and Mesozoic (Late Jurassic–Early Cretaceous) lignites and subbituminous coals, whilst those occurring in the mountains are Mesozoic high volatile to low volatile bituminous coals. The Late Jurassic–Early Cretaceous sediments are up to 2700 m in thickness in the mountainous region of southwest Alberta and southeast British Columbia; numerous coal seams are present of which 14 have been mined with thicknesses of 2 m in Alberta to 14 m in British Columbia. Lower to Upper Cretaceous coals are present in west-central Alberta and northeastern British Columbia, ranging from 2 to 13 m

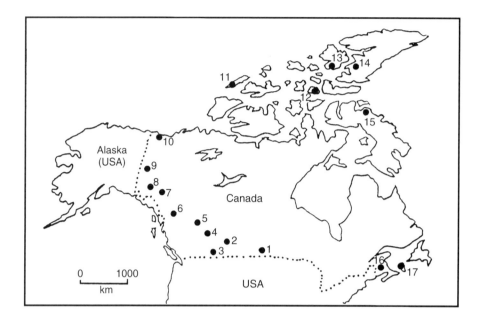

Figure 3.8 Coal deposits of Canada. (1) South Saskatchewan; (2) Central and Eastern Alberta; (3) Southeast British Columbia; (4) West-central Alberta; (5) Northeast Coal Block; (6) Northwest British Columbia; (7) Watson Lake; (8) Whitehorse; (9) Dawson; (10) Mackenzie Bay; (11) Prince Patrick Island; (12) Cornwallis Island; (13) Axel Heiberg Island; (14) Ellesmere Island; (15) Baffin Island; (16) Minto Coalfield; (17) Sydney Coalfield. Reproduced by permission of Dargo Associates Ltd

in thickness. In the central Alberta Plains region occur coals of Upper Cretaceous to Tertiary age; these are subbituminous coals which decrease in rank eastwards to Saskatchewan, where they are preserved as lignites. The majority of the western Canadian coals are low in sulphur content.

In eastern Canada, Carboniferous coals are mined in the Minto Coalfield, New Brunswick and in the Sydney Coalfield, Nova Scotia (Figure 3.7). All are bituminous coals, some with high sulphur contents and coking properties. The Minto Coalfield supplies coal for generating electricity from a single 0.5 m seam. The Sydney Coalfield is now restricted to mining offshore and again supplies power stations, and some coking coal for export.

In northern Canada, coals are found in the Yukon Territory and Northwest Territories. They are of the same age as the coals described for western Canada, comprising Mesozoic high, medium and low volatile bituminous coals, often highly tectonised, and Tertiary subbituminous coals and lignites. The Mesozoic coals are principally from the Yukon Territory, and Tertiary coals are found in both Yukon Territory and the Northwest Territories, including the Canadian Arctic islands.

Older coals (Devonian) are known to occur in the Arctic islands.

3.4.3 Europe

Coal deposits of Paleozoic (Carboniferous), Mesozoic and Tertiary age are developed in a series of basins which stretch from the UK in the west to Turkey in the east. The full range of black and brown coals is present, and all of the most accessible deposits have been extensively worked over the last 150 years.

Those European countries with recorded coal deposits are shown in Figure 3.9.

3.4.3.1 Albania

Albania has small isolated occurrences of subbituminous coal and lignite scattered throughout the country. They produce small amounts of coal for local industrial use, chiefly from Tiranë, Tepelene and Korce.

3.4.3.2 Austria

In Austria there are a number of small coal deposits; all are structurally complex. They include anthracite and

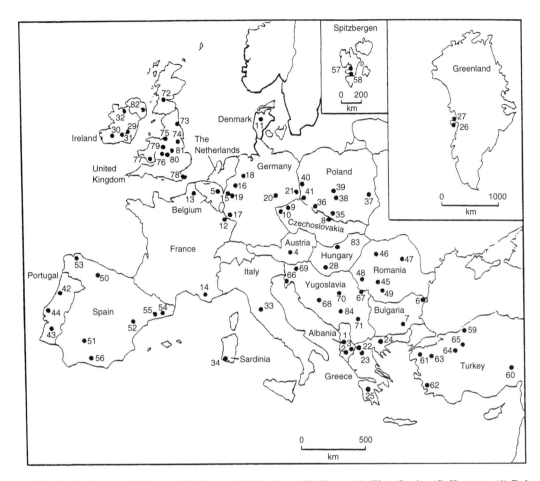

Figure 3.9   Coal deposits of Europe. (1) Tiranë; (2) Tepelene; (3) Korce; (4) West Styria; (5) Kempen; (6) Dobruja; (7) Maritsa; (8) Ostrava–Karviná; (9) North Bohemia; (10) Sokolov; (11) Herning; (12) Lorraine; (13) Nord et Pas de Calais; (14) Provence; (15) Aachen; (16) Ruhr; (17) Saar; (18) Lower Saxony; (19) Rhenish; (20) Halle Leipzig Borna; (21) Lower Lausitz; (22) Florina–Amyndaeon; (23) Ptolemais; (24) Serrae; (25) Megalopolis; (26) Disko; (27) Nugssuaq; (28) Mecsek; (29) Leinster; (30) Kanturk; (31) Slieveardagh; (32) Connaught; (33) Valdarno; (34) Sulsis; (35) Upper Silesia; (36) Lower Silesia; (37) Lublin; (38) Belchatow; (39) Konin; (40) Lausitz; (41) Turoscow; (42) Sao Pedro da Cova; (43) Cabo Mondego; (44) Rio Maior; (45) Jiu; (46) Almas; (47) Comanesti; (48) Banat; (49) Oltenia; (50) Leon; (51) Puertollano; (52) Teruel; (53) Garcia Rodriguez; (54) Calaf; (55) Mequinenza; (56) Arenas del Rey; (57) Longyearbyen; (58) Svea; (59) Zonguldak; (60) Elbistan; (61) Canakkale; (62) Mugla; (63) Bursa; (64) Ankara; (65) Cankiri; (66) Istra; (67) Dobra; (68) Sarajevo–Zenica; (69) Trans-Sava; (70) Kolubara; (71) Kosovo; (72) Scotland; (73) Northeast England; (74) Yorkshire–Nottinghamshire; (75) Lancashire–North Wales; (76) East and West Midlands; (77) South Wales; (78) Kent; (79) North and South Staffordshire; (80) Warwickshire; (81) Leicestershire; (82) Northern Ireland; (83) Visonta-Bukkabrany; (84) Kostolac. Reproduced by permission of Dargo Associates Ltd

bituminous coals, principally in the west of the country. Tertiary lignite and subbituminous coal is present in the West Styria area. Mining is on a small scale only.

3.4.3.3 Belgium

The Carboniferous Coalfield of Kempen in the north of the country produces all of Belgium's coal. The coal

ranges from anthracite to low volatile bituminous, with low sulphur content, some of which is used as coking coal. Seams may be up to 2.0 m thick, and are worked by underground methods.

Smaller coal deposits are known to the south, but mining has long ceased because of thin seams and difficult mining conditions.

3.4.3.4 Bosnia

Tertiary lignite intermontane basins are present in several areas of Bosnia. In the Sarajevo–Zenica area, subbituminous coal and lignite are present in large quantities. The coals have high moisture and ash contents. In the east of Bosnia, at Gacko, lignite seams up to 18 m are mined, and seams of lignite up to 25 m in thickness are present in the Kongora area in west Bosnia, but are as yet unmined. These lignites have variable ash contents due to the presence of numerous thin nonlignite partings, and sulphur contents of around 1%. Current lignite production is used for power generation.

3.4.3.5 Bulgaria

The most extensive coal deposits in Bulgaria are lignites of Tertiary age. Most of the mining areas are situated in the southwest and central parts of the country. The principal producers are the opencast mines at Chukurova, Pernik, Beli Breg and Maritsa East, with underground production from Bobov Dol mine complex and Maritsa West reaching depths of 200 m. The seams are thick, some over 20 m, and have undergone little structural disturbance. They are often low grade with high sulphur content. Lignite is currently used to provide up to 35% of Bulgarian power generation.

High-rank bituminous coals are known from the Dobruja area on the east coast; here the coals are deep and so far have not been mined on a significant scale.

3.4.3.6 Czech Republic

The Czech Republic has numerous deposits of black and brown coals spread widely across the country. The chief black coalfield is that of Ostrava–Karviná on the northeastern border. It is of Upper Carboniferous age and represents a continuation of the Upper Silesian Coalfield in Poland. The lower part of the sequence varies between 1500 and 3000 m in thickness and contains 170 coals with an average thickness of 0.7 m. The upper sequence is 1200 m thick with 90 coal seams of between 5 and 15 m in thickness. The area is structurally complex. The coal produced is low volatile bituminous, which is strongly caking, and anthracite. These coals usually have low ash and sulphur contents. Other smaller bituminous coal deposits are found north and west of Prague.

Tertiary lignites are located in the northwest of the Czech Republic, in North Bohemia (Most), western Bohemia (Sokolov) and in southern Moravia (Hodonin), and have been mined in all these areas. Seams are up to 30.0 m in thickness, with low sulphur contents and little structural disturbance. Other lignite deposits are present in the south of the country but are not currently being exploited.

3.4.3.7 Denmark

Tertiary lignites are found at Herning, but are no longer worked as reserves are small.

3.4.3.8 France

In northern France, two large Carboniferous basins, those of Lorraine and Nord et Pas de Calais, contain high volatile bituminous coal, some suitable as coking coal. Seams are up to 2.0 m thick and have been highly tectonised; all are mined by underground methods. To the south lies the Carboniferous Cevennes Basin, and other smaller scattered coal deposits are known from western and southwestern France; all are structurally complex and mining has ceased in these areas.

The coal-mining industry in France has declined markedly in recent years, due to increasing production costs and competition from other fuel sources.

Lignite is mined in the Provence district; the seams are relatively undisturbed, and supply the local electricity industry.

3.4.3.9 Germany

The black coals of Germany are located in the Carboniferous basins of Aachen, Ruhr and Saar, with small deposits present in the southeast of the country.

The Aachen Basin produces high and low volatile bituminous coal with low ash and sulphur, some of which is coking coal, from seams up to 1.5 m in thickness.

The Ruhr Basin contains 3000 m of Upper Carboniferous sediments containing around 300 coal seams. The area has been subjected to block faulting and folding, and the Ruhr Basin produces high volatile bituminous coal with low ash and sulphur and is a good coking coal. Seam thicknesses range from 0.5 m to 3.0 m. A northern extension of this area is the Lower Saxony Coalfield, which produces low volatile anthracite.

The Saar Basin is less structurally disturbed and produces bituminous coal from seams 0.5 m to 2.0 m in thickness. This coal is not suitable for coking purposes.

The brown coals of Germany are of Tertiary age, and significant deposits are found in three basins, the Rhenish, the Halle Leipzig Borna (the central

German mining area) and Lower Lausitz (the Lusatian mining area).

The Rhenish Basin is situated close to the Ruhr; it contains thick seams (up to 90.0 m) of lignite with low ash and sulphur contents. The coalfield has suffered little structural disturbance, and is mined on a large scale by opencast methods for use in the electricity generation industry.

The Halle Leipzig Borna and Lower Lausitz Coalfields are situated in the east of the country. Again they contain thick seams of lignite which has high volatile, low ash and sulphur contents of 0.3–2.0%. Both areas are mined on a large scale by opencast methods.

Germany is Europe's largest lignite producer.

3.4.3.10 Greece

Lignite deposits occur widely across Greece. The principal deposits are Late Tertiary to Quaternary in age. The principal deposits are situated in the north of Greece at Florina, Amyndaeon, Ptolemais and Elassona. To the south occurs the Megalopolis deposit and, to the northeast, the Drama lignite field. The number and thickness of the lignite seams varies both between and within individual deposits. The Ptolemais Basin contains one major thick lignite seam 60 m in thickness, and in Amyndaeon and Megalopolis, lignite seams are up to 30.0 m thick. The lignite is generally of poor quality due to the presence of numerous nonlignite interbeds. Although the ash and sulphur contents can vary greatly, the current mining areas produce lignite with typically 15–20% ash and sulphur <1%. All is surface mined chiefly for electricity generation.

3.4.3.11 Greenland

Although geographically separate from Europe, Greenland is included here for convenience. Coal occurrences have been reported from several areas in Greenland. The most significant are the subbituminous coals with low sulphur contents that are found on the coast at Disko and Nugssuaq.

3.4.3.12 Holland (Netherlands)

Coal-bearing Carboniferous sequences are known at depth beneath younger sediments; however, seams are thought to be thin. No coal mining is presently carried out in Holland.

3.4.3.13 Hungary

Black coal is mined in the southwest of Hungary, in the area around Mecsek. Here coals of Mesozoic age are present in a structurally complex area, strongly folded and faulted, with associated igneous intrusives. Seams are steeply dipping but can be thick locally (>5.0 m). The coal is weakly caking bituminous with high ash and sulphur contents. The area is a well-established coalfield, mining in difficult conditions.

Brown coal deposits of Mesozoic and Tertiary age are found in a northeast–southwest belt of country in the north of Hungary. Lignite is mined at Oroszlany, Tatabánya, Dorog and Visonta–Bukkabrany. The coals are mined by underground and opencast methods, with ash and sulphur contents of 20% and 1% respectively. All is utilised as fuel for local power plants.

3.4.3.14 Ireland

The coal deposits of Ireland are all of Carboniferous age. Four main coal-bearing areas have been mined, at Leinster, Kanturk and Slieveardagh in the south and at Connaught in the northwest. The coal is anthracite with variable sulphur contents in the south, and medium volatile bituminous coal in the northwest. Coal has been mined on a small scale from thin seams.

3.4.3.15 Italy

Carboniferous coals are present in the structurally complex areas of the Alps and Sardinia.

Tertiary lignites and subbituminous coals are found in the Apennines and Sardinia. The latter coalfield at Sulcis has subbituminous coal as a result of volcanic amelioration. Mining is on a very small scale.

3.4.3.16 Poland

Poland has large reserves of coal and a long-established coal-mining industry. Black coal in Poland is centred around three coalfields, Upper Silesia, Lower Silesia and Lublin. These areas represent large basins which have undergone varying degrees of structural disturbance.

The Upper Silesian Basin contains a thick sequence of Upper Carboniferous sediments, up to 8500 m. The lower part of the sequence contains 250 coal seams while the upper part has 60 coal seams, the thicker seams reaching 6–7 m. The basin is highly tectonised, so that mining operations are complicated by large scale faulting and folding. Coal rank is also affected by igneous intrusions of Permian, Triassic and Miocene age. The coal is primarily high volatile bituminous, with low ash and sulphur contents, and is a major export commodity.

The Lower Silesian Coalfield is a much smaller area containing thinner seams which are highly tectonised. Remaining reserves are deep, but have been an important supplier of coking coal. The number of mining operations has been considerably reduced in this area.

The Lublin Coalfield, only discovered relatively recently, is a very large area with potentially enormous reserves. The seams appear to be less structurally disturbed than in the Silesian Coalfields. Lublin has bituminous coals with low ash and sulphur contents, together with strong coking properties. The Lublin Coalfield is likely to be the principal supplier of black coal in the future.

Tertiary lignite basins are present in central and southwestern Poland, many of which are fault bounded. Production is centred on four areas, namely, Adamov, Belchatow, Konin and Turow. Lignite thickness ranges from 5 m to over 70 m. Ash and sulphur contents are low in the thicker lignites, which are mined to supply local power plants. Poland is the second largest European producer of lignite.

3.4.3.17 Portugal

Anthracite of Carboniferous age is present in the northwest of the country, but is not extensively worked. At Cabo Mondego, bituminous coal, high in ash and sulphur, is produced on a small scale.

A number of basins contain lignite; in particular, that at Rio Maior is a small producer.

3.4.3.18 Romania

The Tertiary lignite reserves of Romania lie in a series of deposits aligned east–west in the south of the country. The principal deposit is at Oltenia, where up to 75% of production is obtained, mining seams up to 30 m in thickness by both opencast and underground methods. Other deposits at Berbesti and Ploiesti are also exploited. The lignite is high in ash and sulphur content and is used exclusively for electricity generation, providing 25% of Romania's power.

A small deposit of higher rank Paleozoic coal is present at Banat in southwest Romania; here, bituminous coal of coking quality is mined for industrial use.

3.4.3.19 Spain

The principal black coal basin is that in the Leon region in the north of the country. The basin contains low volatile bituminous coals with low sulphur contents, some of which are usable as coking coal. Seam thicknesses are up to 1.5 m and the area is structurally complex.

In the south of Spain, southwest of Puertollano, low volatile bituminous coals are found, and Spanish brown coals are located at Teruel, where shallow lignites of Cretaceous age are mined for power generation.

The major lignite production comes from the Puentes de Garcia Rodriguez and Meirama areas in the northwest of the country. Up to 17 lignite seams are exploited in opencast mines to depths of 170 m. the lignite is high in sulphur and is used for local electricity needs. Other lignite deposits are present at Calaf and Mequinenza in the northeast, and Arenas del Rey in the extreme south of Spain. These are also used for local electricity requirements.

3.4.3.20 Spitzbergen

Carboniferous and Tertiary coals are present on the western side of the island. At Longyearbyen and Svea, high volatile bituminous coal with low sulphur is present in seams up to 5.0 m in thickness. Current mining is on a small scale.

3.4.3.21 Turkey

Turkey has considerable reserves of Carboniferous black coal and Tertiary brown coal. The principal black coal deposit is the Zonguldak Coalfield on the northern coast, where numerous seams ranging from 0.7 m to 10.0 m are present. The coal is bituminous with low ash and sulphur contents, and is suitable for use as a coking coal. The coalfield is structurally complex and mining is heavily subsidised.

Tertiary lignite basins are present across west and central Turkey. Older Eocene lignites are found in the north, with seams up to 6 m in thickness. Younger Oligocene and Miocene lignites occur in the northwest and west of the country respectively. The Oligocene is characterised by numerous thin seams, whilst the Miocene lignites form the larger deposits with seams up to 25 m in thickness. In central and eastern Turkey, Pliocene deposits contain large lignite resources, with one or two seams averaging 40 m in thickness. The largest opencast operation at Afsin–Elbistan mines a seam 5–58 m thick. The older lignites have a higher heating value together with high sulphur contents, and the youngest lignites have high ash and moisture values. Current and future use of lignite is planned for electricity generation.

3.4.3.22 United Kingdom

In the United Kingdom, a series of coal-bearing basins is distributed throughout the country. The coals are of Carboniferous age and are principally bituminous with some anthracite, notably in South Wales.

The principal areas are those of Scotland, Northeast England, the Yorkshire–Nottinghamshire region, the Lancashire–North Wales region, the East and West Midlands, South Wales and Kent. Much of the underground mining in the UK has now disappeared and been replaced by opencast operations.

The Yorkshire–Nottinghamshire region is the most important coal-producing area in the UK, supplying coal for electricity generation and coking coal to industry. The region is less structurally affected than other areas, and has a long coal-mining history. The north Nottinghamshire area, together with the Selby area, produces high volatile bituminous coal, some of which is strongly caking. This coal is used chiefly for the electricity generating industry, and declining amounts for the steelmaking industry.

The Lancashire–North Wales region in the past has produced high volatile bituminous coal with low sulphur content. Due to extensive working and difficult mining conditions, mining in the region has now ceased.

In Northeast England, bituminous coals suitable for coking purposes are produced in opencast operations.

The East and West Midlands contain four coal-mining areas: North and South Staffordshire, Warwickshire and Leicestershire. The region has produced high volatile bituminous coal for electricity generation. Warwickshire currently produces from the Warwickshire Thick Seam, which can be up to 8.0 m in thickness and is used for power generation.

In Scotland, high volatile bituminous coals with low sulphur are opencast mined, with only one underground mine in operation at Longannet. This area has been extensively mined in the past.

South Wales, once the principal coalfield in the UK, still produces bituminous coal and anthracite with low sulphur contents; seams are usually between 1.0 m and 3.5 m. Underground mining conditions are difficult and expensive; currently only Tower underground mine is in operation. The bulk of production comes from opencast mining in the coal outcrop areas.

Kent, this small coalfield in the southeast of the UK, contains high-quality bituminous and anthracite coal, but is no longer a producer.

Tertiary lignites are found in southwest England and in Northern Ireland. In Northern Ireland, lignite deposits have been identified for future use in local power stations.

3.4.3.23 Yugoslavia (former)

Former Yugoslavia, in common with many other European countries, has black coal of Paleozoic (Carboniferous) and Mesozoic ages, preserved in the older structurally complex regions of the country, and Tertiary brown coals which are not affected structurally. The latter are more extensive than the older deposits.

The Yugoslav Republic of Serbia has a series of black coal deposits along its northeastern border; most are low volatile bituminous coal with relatively high sulphur contents. Some of the coals have good coking properties but the high sulphur restricts their use. Small occurrences of anthracite are also found in the region. These coals have been mined in the Dobra area and in the Istra area (now in Croatia).

Tertiary lignites are mined at Kolubara, Kosovo, Metohija, Kostolac and Kovin, all providing fuel for electricity generation. The Kosovo Basin contains a lignite seam up to 100 m in thickness, the Kostolac Basin has 45 m of lignite in three seams, and the Kolubara Basin has three lignite seams totalling 18–45 m. The lignites have low ash and sulphur contents, particularly in the Kosovo area.

The Yugoslav Republic of Montenegro has the Pljevlja opencast mine, providing lignite from one thick seam to the local power plant.

3.4.4 Africa

The occurrences of black coal in Africa are, first, those deposits of Carboniferous age found on the northern coast, in Morocco in the west and Egypt in the east, and, second, and more importantly, the widespread Karroo deposits of Late Carboniferous–Permian age which are found throughout central and southern Africa (Haughton 1969). The Karoo sequences were deposited on the Gondwana supercontinent, which split apart in the Mesozoic Period; hence the similarities of African Gondwana coals with those of India and South America.

Brown coals of Tertiary age are present, but in Africa it is the black coals that are of prime interest. The principal coal occurrences are shown in Figure 3.10.

Figure 3.10   Coal deposits of Africa. (1) Lungue–Bungo; (2) Luanda; (3) Moropule; (4) Mmamabula; (5) Bamenda; (6) Al Maghara; (7) Imaloto–Vohibory; (8) Sakoa; (9) Sakamena; (10) Antanifotsy; (11) Livingstonia; (12) North Rukuru; (13) Ngana; (14) Lengwe; (15) Mwabvi; (16) Jerada; (17) Ezzhiliga; (18) Tindouf–Draa; (19) Meknes–Fez; (20) Moatize; (21) Mmambansavu; (22) Chiomo; (23) Itule; (24) Enugu–Ezimo; (25) Orukpa–Okaba–Ogboyoga; (26) Asaba; (27) Karroo Basin; (28) Waterberg; (29) Springbok Flats; (30) Limpopo; (31) Lebombo; (32) Mhlume; (33) Mpaka; (34) Maloma; (35) Ketewaka–Mchuchma; (36) Ngaka; (37) Songwe–Kiwira; (38) Galula; (39) Njuga; (40) Ufipa; (41) Luene; (42) Lukuga; (43) Luangwa; (44) Luano; (45) Maamba; (46) Kahare; (47) Wankie; (48) Lubimbi; (49) Sessami–Kaonga; (50) Tuli; (51) Bubye; (52) Chelga; (53) Wuchalle; (54) Dobre–Brehan; (55) Bourem. Reproduced by permission of Dargo Associates Ltd

3.4.4.1 Angola

Lignites of Tertiary age have been identified in Angola, first in the east around the headwaters of the Lungue–Bungo river, where seams of lignite up to 2.5 m have been recorded, and second in the west around Luanda, where lignites are present in the Tertiary coastal sediments. None of these deposits has been worked.

3.4.4.2 Botswana

Botswana has large reserves of black coal of Karroo age. These coal deposits extend from north to south along the eastern edge of the country. The more important coalfields are those of Morupule and Mmamabula. At Morupule, seams up to 9.5 m, 4.5 m and 2.0 m in thickness are present. At Mmamabula, seam thicknesses average 2.8 m, 5.4 m and 2.0 m. In both these coalfields, the coals are relatively undisturbed, and contain bituminous coal with a high ash and sulphur content; these coals have no coking properties.

Other smaller coalfields are present in close proximity to Morupule and Mmamabula.

Botswana has the potential to be a significant coal producer, but at the present time it is geographically disadvantaged as a coal exporter.

3.4.4.3 Cameroun

In the Bamenda district, lignites are found interbedded with lava flows. They are of Cretaceous–Tertiary age, and locally can be up to 6.0 m in thickness, but are undeveloped.

3.4.4.4 Congo

Small separate basins of coal-bearing Karroo sediments occur in the southeast of the country at Luena and Lukuga. Seams are up to 2.0 m in thickness and are disrupted by faulting. The coals are bituminous with high ash contents and are used locally for electricity generation.

3.4.4.5 Egypt

Carboniferous coal is present in the Sinai Peninsula. Coals are of bituminous and subbituminous rank. Coal has been produced from workings at Al Maghara, at which future development is to be considered.

3.4.4.6 Ethiopia

Tertiary brown coals are known from many localities on the Ethiopian Plateau: beds are up to 15.0 m, and range from lignite to subbituminous coal. They have high ash and low sulphur content. Principal localities are Chelga, Wuchalle and Dobre–Brehan.

3.4.4.7 Malagasy Republic

Black coal is present in the Karroo sediments on the western side of the island, where they overlie the Pre-Cambrian basement. At the southern end of the Karroo outcrop, five coal-bearing areas have been identified.

The northernmost area is the Imaloto Coalfield, which contains seams averaging 1.0 m in thickness. The coal is medium volatile bituminous with high ash and some high sulphur contents. The Vohibory and Ianapera Coalfields have seams up to 2.3 m and 0.6 m respectively; both areas are structurally complex. The Sakoa Coalfield is the best-known area, with seams of 3.0 m and 7.0 m in thickness. The coals are high volatile bituminous with high ash and low sulphur contents; they are noncoking. The Sakamena Coalfield is similar to Sakoa except that the seams are thinner.

Lignite deposits of Tertiary age are present in the region of Antanifotsy, and are thought to cover a large area.

3.4.4.8 Malawi

In Malawi, a series of separate basins contain coal-bearing Karroo sediments; these are located in the extreme north and south of the country.

The main coalfields are those of Livingstonia, Ngana and North Rukuru, with small deposits at Lengwe and Mwabvi in the south.

The Livingstonia coalfield contains seams 1.0 m to 2.0 m in thickness, which are mined to supply fuel to local industry.

At Ngana, one seam is up to 15.0 m in thickness, but seams usually average around 1.0 m, and show rapid vertical and lateral variations in thickness. The southern coalfields have thin seams, and are not well developed.

Malawian coals are subbituminous to high volatile bituminous with high ash and low sulphur contents.

3.4.4.9 Mali

Upper Cretaceous and Tertiary brown coals are recorded from the Mali–Niger Basin in the southeastern part of the country, around Bourem. Seams are thought to reach 2.0 m in thickness, and have moisture values of 24% and ash values of 21%.

3.4.4.10 Morocco

Carboniferous black coals are found in the northeast of Morocco, at Jerada, and have been identified at depth beneath younger sediments at Ezzhiliga and Tindouf–Draa.

At Jerada, four seams of up to 0.7 m in thickness are mined; they are structurally unaffected, and the coal is low volatile anthracite with low ash and high sulphur contents.

Lignites have been identified at Meknes–Fez in northern Morocco.

3.4.4.11 Mozambique

Karroo sediments are preserved in a series of basins in the Pre-Cambrian basement. The coal-bearing Karroo outcrop is a long strip running eastward from the southern tip of Lake Malawi. Four coalfields have been identified; of these the Moatize deposit is the most important.

At Moatize, seams range from 0.4 m to 4.0 m in thickness; structurally the coalfield is heavily faulted. Coal is low volatile bituminous, with high ash and low sulphur contents. The coal deposits at Mmambansavu, Chiomo, and Itule are as yet little known.

3.4.4.12 Namibia

The eastern half of Namibia is covered by post-Karroo sediments of the Kalahari Group. It is possible that Karroo sediments underlie a portion of this area, and may contain coals of similar aspect to those found in Botswana.

3.4.4.13 Niger

The Mali–Niger Basin contains coals of Upper Cretaceous to Tertiary age. Little information is available but these coals may be worked locally.

3.4.4.14 Nigeria

Coal-bearing sediments of Cretaceous and Tertiary age overlie Pre-Cambrian basement in the southeastern part of Nigeria. These sediments dip to the west, where they are overlain by floodplain deposits of the River Niger.

The Nigerian Coalfield is divided into several mining areas, the Enugu, Ezimo, Orukpa, Okaba and Ogboyoga Coalfields. Seams range from less than 1.0 m to 3.0 m, and all the coalfields are affected by faulting and gentle folding. Coals from these coalfields are high volatile subbituminous with high ash and low sulphur contents. Similar coal is reported from the Lafia area situated to the north of the Enugu Coalfields; here the coal is similar but has a higher sulphur content.

Tertiary lignites are found in the Asaba region close to the River Niger; seams are 3.0 m to 7.0 m in thickness. They are high volatile, high ash lignite with low sulphur content.

3.4.4.15 South Africa

The coal deposits of South Africa are found in a series of basins situated in the north and east of the country. The Karroo System contains coal-bearing sediments of Carboniferous–Permian (Gondwana) age.

The main Karroo basin extends for 200 km from Free State Province in the west to south and east Mpumalanga Province (formerly Transvaal), and for 400 km from Mpumalanga in the north to KwaZulu Natal in the south. The Karroo sequence was deposited directly on to basement and the coal seams are shallow and almost horizontal. They have been affected by numerous igneous intrusions which have produced a great variation in rank, often very localised. The western area consists of the Vereeniging–Sasolberg and South Rand coalfields, which contain coals 10–25 m thick. The northern area comprises the Witbank, Eastern Mpumalanga (formerly Eastern Transvaal) and Highveld Coalfields, where up to five seams are present; two of these, up to 10 m thick, are worked. The southern KwaZulu Natal area includes the Vryheid and Utrecht Coalfields, again with five seams, of which two are worked, together with the Kliprivier Coalfield which has two coal seams with a thickness of up to 15 m. The basin as a whole produces high volatile bituminous coal, with high ash and variable sulphur contents; some of the coal is weakly coking, for electricity generation, and to produce liquid fuel. The majority of the mines are underground operations, and the region is a major exporter of steam coals. In eastern Mpumalanga and Natal, anthracite is also produced.

In Cape Province, the Molteno–Indwe coalfield has coals which are of lower rank than those to the northeast.

Other coalfield basins in the northeast of the country are less developed; of these, the Waterberg Coalfield on the Botswana border and the Springbok Flats area appear to have future potential. The Limpopo and Lebombo Coalfields have bituminous coals with high ash contents; these are not worked at this time.

3.4.4.16 Swaziland

Coal-bearing Karroo sediments are located on the eastern side of the country. The seams are thicker in the north and are flat lying; some have been ameliorated by dolerite intrusions. The Mhlume Coalfield in central Swaziland produces anthracite on a small scale; coal is also known from the Maloma area in the south of the country.

3.4.4.17 Tanzania

There are eight coalfields in Tanzania: the Karroo sediments are preserved in depressions in the Pre-Cambrian basement; all are located in the southwest of the country. The Ruhuhu Coalfields have been known for a century but have never been fully developed. Of these, the Ketewaka–Mchuchma and Ngaka coalfields are the most important. In these coalfields, coals occur in two zones, the lower containing the better coals – seams can be as thick as 7.0 m – but this is exceptional. Coals are high to low volatile bituminous with high ash and low sulphur contents.

Other coalfields with similar characteristics are those of Songwe–Kiwira, Galula, Njuga and Ufipa.

3.4.4.18 Zambia

Karroo sediments are preserved in depressions in the Pre-Cambrian basement. A series of such basins are present in the east and southeast of the country, namely the Luangwa, Luano and Maamba areas, and also in the west-central district around Kahare.

The Luangwa coals are up to 1.6 m thick, and are high volatile bituminous high ash coal. The Luano area has fairly thin seams which are high volatile bituminous, with high ash content; some coal has coking properties. The Maamba area in the southeast has seams 2.0 m to 3.0 m in thickness, and is high volatile bituminous with high ash content. The Maamba area produces most of Zambia's coal.

At Kahare, coals are preserved beneath younger sediments, and coal quality is similar to other Zambian coals; this area has not yet been fully investigated.

3.4.4.19 Zimbabwe

The Karroo sequence in Zimbabwe is preserved in the Zambezi Basin in the northwest, and the Limpopo Basin in the southeast. The northwest includes the coalfield districts of Wankie and Lubimbi, with Sessami–Kaonga to the east of these. In these coalfields, the coal is the Wankie Main seam, a medium to high volatile bituminous coal, comprising a lower coking coal up to 4.0 m in thickness, and an upper steam coal up to 8.0 m, all generally with low sulphur contents.

In the southern coalfields of Bubye and Tuli, the coals have variable qualities. Some low sulphur coking coal has been identified in the Tuli Coalfield.

3.4.5 The Indian Subcontinent

The area delineated the Indian Subcontinent extends from Iran in the west to Bangladesh in the east. Black coals are of Paleozoic (Carboniferous–Permian), Meso-zoic and Tertiary age; brown coals are of Tertiary age.

Paleozoic Gondwana coals are found in India, Pakistan and Bangladesh; Mesozoic coals are present in Afghanistan, India, Pakistan and Iran; Tertiary coals are found in all the countries listed in this region. For distribution, see Figure 3.11.

3.4.5.1 Afghanistan

Mesozoic (Jurassic) black coals are present in the northern mountainous region of Afghanistan. The coal is relatively undisturbed with seams up to 1.5 m in thickness. The coal is bituminous with low ash and sulphur contents, with little or no coking properties. Coal is mined at Herat in the northwest and at several other sites in the north. All are small operations and produce for the local market only.

3.4.5.2 Bangladesh

Gondwana coals are found at depth, concealed beneath Tertiary sediments in northwestern and eastern Bangladesh. The Gondwana sediments represent the infilling of depressions in the underlying crystalline basement. These basins have been faulted at the margins, resulting in gently dipping coal seams being preserved in graben structures.

In the northwest, the concealed coal basins of Barapukuria, Khalaspir and Jamalganj contain numerous seams ranging in thickness from less than 1.0 m

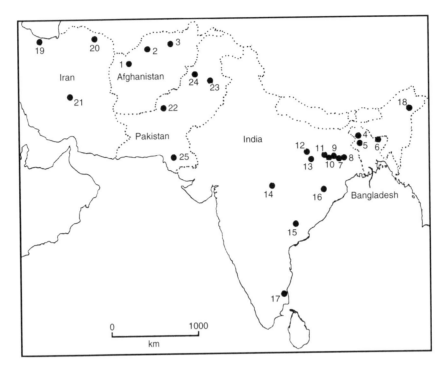

Figure 3.11   Coal deposits of the Indian Subcontinent. (1) Heart; (2) Sari-i-Pul; (3) Dara-i-Suf; (4) Barapukuria–Khala-spir; (5) Jamalganj; (6) Sylhet; (7) Raniganj; (8) Jharia; (9) Bokaro; (10) Ramgarh; (11) Karanpura; (12) Singrauli; (13) Bisrampur; (14) Pench–Kanhan–Tawa; (15) Godavari; (16) Talchir; (17) Neyveli; (18) Makum; (19) Elburz; (20) Khorasan; (21) Kerman; (22) Quetta–Kalat; (23) Salt Range; (24) Makerwal; (25) Hyderabad. Reproduced by permission of Dargo Associates Ltd

to 20.0 m and 30.0 m. The coals are medium to high volatile bituminous, with high ash and low sulphur contents.

In the east of Bangladesh, lower rank coal is located at Sylhet; the coal is subbituminous and lignite.

Coals in Bangladesh have yet to be developed.

3.4.5.3 India

In India coal resources are of Paleozoic (Gondwana) and Tertiary age.

About 98% of India's coal reserves are of Gondwana coal, which also accounts for 95% of production, chiefly for electricity generation and the metallurgical industries. The Gondwana coals are present in over 14 separate basins centred in the northeastern and central eastern parts of peninsular India.

Tertiary brown coals are present in the northeastern and northwestern parts of the country, together with an important lignite deposit in the south at Neyveli.

The principal coalfields containing Gondwana coals are those of Raniganj in West Bengal State; Ramgarh, Jharia, Karanpura and Bokaro in Bihar State; Singrauli, Bisrampur, Pench-Kanhan and Tawa Valley in Madhya Pradesh; Kamptee, Bandar and Wardha Valley in Maharashtra State; Ib river and Talcher Coalfields in Orissa; and Godavari in Andhra Pradesh. In addition, numerous other fields in the same region are producing coal. Seams in these coalfields range in thickness from 1.0 m to 30.0 m, with an exceptionally thick seam of 134.0 m discovered in the Singrauli Coalfield. The coalfields have been faulted but otherwise are not highly tectonised. Coals range from high to low volatile bituminous with high ash and variable sulphur contents. In the Jharia and Raniganj Coalfields, good-quality coking coals are produced.

The Tertiary coals are highly disturbed tectonically and are located in the mountainous regions of northeast India. In the Makum coalfield in Assam, seams are lens-shaped, in places reaching thicknesses of 33.0 m. Coals are subbituminous to high volatile bituminous with high sulphur contents.

In southern India, in the state of Tamil Nadu, Tertiary lignites are found in Neyveli, the thickest seam being up to 20.0 m in thickness. Here the lignite is low volatile, with low ash and sulphur content. Lignite is now also being mined in the northwestern states of Gujarat and Rajasthan. All these areas will increase in importance as the demand for electricity generation increases.

3.4.5.4 Iran

The black coal deposits of Iran are Mesozoic (Jurassic) in age, with some lignites of Tertiary age. The Jurassic coals are bituminous with high ash and sulphur contents, and have coking properties. All are strongly tectonised, with seam thicknesses ranging from 1.0 m to 4.0 m. The coal supplies local needs and the metallurgical industry. Principal coalfields are located at Elburz and Khorasan in the north and at Kerman in central Iran. Lignites are found in northwest Iran but are not worked.

3.4.5.5 Pakistan

All the principal coalfields in Pakistan are of Tertiary age, although Palaeozoic and Mesozoic coals are present. The coalfields of economic importance are situated in the Indus Basin in three distinct coal regions: Hyderabad, Quetta–Kalat and Salt Range–Makerwal. Most of these coalfields have been structurally disturbed.

The Hyderabad Province contains the coalfields of Lakhra, Sonda–Thatta and Meting–Shimpir. Seams are up to 2.0 m in thickness, and the coal is subbituminous, noncoking with high sulphur content. The Quetta–Kalat province contains the coalfields of Sor Range–Daghari, Khost Sharig–Harnai and Duki–Chamalang. Again the coal is subbituminous with high ash and sulphur contents. The Salt Range–Makerwal province comprises the coalfields of eastern, central and western Salt Range, together with the Makerwal coalfield to the west of these. Coals are subbituminous, with high ash and sulphur contents. Overall production is small, the coal being used chiefly for electricity generation.

3.4.6 Central and South America

Coal deposits are distributed throughout Central and South America and make up a significant proportion of world reserves of black coal. The majority of coals are of Tertiary age. Coals of Paleozoic (Gondwana) age are present in eastern South America, in Brazil and Uruguay, and Mesozoic coals are found in discrete deposits throughout the region (see Figure 3.12).

3.4.6.1 Argentina

The coal deposits of Argentina are preserved in a series of basins in the Andean Cordillera and pre-Cordillera and in Austral Patagonia. Coals are of Carboniferous–Triassic, Jurassic and Tertiary ages. Of these the Tertiary coals are the principal deposits of economic interest.

Coals of Carboniferous and Permo-Triassic age are found in the La Rioja–San Juan region. They consist of thin discontinuous seams less than 1.0 m thick and highly tectonised. Coals are low to medium volatile bituminous with some anthracites at Mendoza. Jurassic coals are found south of San Juan in the Neuquen region, preserved in a series of small basins. Seams are thin, normally less than 1.0 m thick and are medium volatile bituminous at Neuquen.

Tertiary coal-bearing sediments are preserved in a large basin which extends from Pico Quemado in the north to Tierra del Fuego in the south. At Pico Quemado coal seams 1.0 m to 2.0 m in thickness are high volatile bituminous with coking properties, their high rank possibly being due to a locally high geothermal gradient related to magmatic phenomena. In the southern part of the basin around Rio Turbio, two coal zones contain seams up to 2.0 m in thickness. They are subbituminous to bituminous with no coking properties. All the Tertiary coals have low sulphur contents and are suitable for the electricity generating industry and, in the case of the Pico Quemado coal, can be used in the metallurgical industry.

3.4.6.2 Bolivia

Two types of coal are known from Bolivia, anthracite of Permian age and lignites of Tertiary age.

Anthracite is located on the Copacabana Peninsula and on the Isla del Sol, Lake Titicaca. Seams are in the form of coal lenses or very thin beds of anthracite with low sulphur content. The Tertiary lignites are found in the Tarija Basin, where seams are thin, under 1.0 m, and have a high sulphur content (6–8%).

3.4.6.3 Brazil

Brazil has five coal-bearing regions which may have potential: the Upper Amazon, the Rio Fresco, Tocantins–Araguaia, Western Piaui and Southern Brazil. Only the Southern Brazil region is currently considered prospective.

The Amazon region contains lignites of Tertiary age; seams are thin (less than 1.5 m) with high ash and

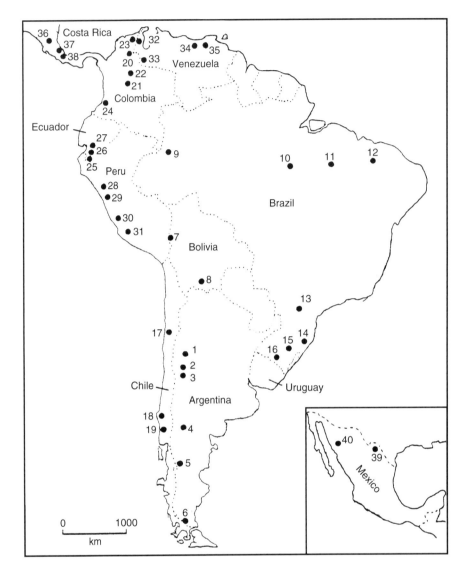

Figure 3.12 Coal deposits of South America. (1) La Rioja; (2) San Juan; (3) Mendoza; (4) Neuquen; (5) Pico Quemado; (6) Rio Turbio; (7) Copacabana Peninsula; (8) Tarija Basin; (9) Amazon; (10) Rio Fresco; (11) Tocantins–Araguaia; (12) Western Piaui; (13) Parana; (14) Santa Catarina; (15) Rio Grande do Sul; (16) Candiota; (17) Copiapo; (18) Arauco; (19) Valdivia; (20) Magallanes; (21) Cundinamarca; (22) Santander; (23) El Cerrejon; (24) Valle del Cauca; (25) Malacatus; (26) Loja; (27) Canar–Azuay; (28) Alto Chicama; (29) Santa; (30) Oyon; (31) Jatunhuasi; (32) Zulia; (33) Lobatera; (34) Caracas–Barcelona; (35) Naricual; (36) Venado; (37) Zent; (38) Uatsi; (39) Coahuila; (40) Sonora. Reproduced by permission of Dargo Associates Ltd

sulphur contents. The Rio Fresco region contains thin seams of anthracite with very high ash contents (40%); seams up to 1.7 m have been reported.

The Tocantins–Araguaia region has very thin coals of Carboniferous age and is not considered of economic importance.

The Western Piaui region also contains Carboniferous coals, which are thin and not significant.

The principal Brazilian coal deposits are situated in the Southern Brazil region. They are of Carboniferous–Permian (Gondwana) age, and are exposed in a lenticular belt which runs from the states of Parana in the north through Santa Catarina to Rio Grande do Sul in the south. Rio Grande do Sul Coalfield contains numerous seams up to 3.0 m in thickness; the Santa Catarina region contains ten coal seams of which the thickest is 2.2 m. In Parana, seams are usually less than 1.0 m thick. The coals are high volatile bituminous with high ash contents. At Candiota in Rio Grande do Sul a 5 m seam is mined, the coal has a high ash content (50%) and a sulphur content of 1%. Santa Catarina and Parana have coals with high sulphur values (3–10%), and in the Parana area, the coals become low volatile bituminous–semi-anthracite due to the intrusion of dolerite dykes into the coals. Some Santa Catarina coals have some coking properties, but the bulk of South Brazilian coal is mined as a thermal coal product.

Some lignites are present in the São Paolo region but their economic potential is unknown.

3.4.6.4 Chile

There are four areas of coal-bearing sediments in Chile: these are, from north to south, the Copiapo region, the Arauco region, the Valdivia region and the Magallanes region.

The Copiapo coals are of Mesozoic (Rhaetic) age; they are strongly folded, with seams occurring as thin lenses of anthracite with high ash and variable sulphur contents.

The Arauco region lies on the Chilean coast just south of Concepcion and the coalfield extends offshore to a distance of 7 km. Dips are steep and faulting common. Seams are of Tertiary age and average 1.0 to 1.5 m in thickness; they are high volatile bituminous coals with low ash and variable sulphur contents, and have poor coking properties. Coals of the Valdivia region are concealed beneath younger sediments. Seams reach 3.0 m in thickness and are subbituminous with low sulphur contents. These coals are used for local purposes.

The Magallanes region forms part of a large sedimentary basin in which over 3800 m of late Cretaceous–Tertiary sediments are preserved. Coal seams up to 7.0 m are present and upwards of 12 coal seams have been identified. All the coals are high volatile subbituminous, noncoking with low ash and sulphur contents. This large coal deposit is geographically remote but is a large resource and may be of future importance.

3.4.6.5 Colombia

Coal deposits of Mesozoic (Cretaceous) and Tertiary age are found in numerous localities in the northern half of Colombia. All have been highly tectonised

and coals range from lignite to anthracite. Cretaceous coals are found just north of Bogota, in the Cundinamarca–Santander region. In this area the coals are bituminous with low ash and sulphur contents, strongly coking, and are suitable for coke production. Tertiary coal deposits are located north of Santander on the Venezuelan border, in the extreme north of Colombia at El Cerrejon, around Cordoba on the north coast and at Valle del Cauca in the west of Colombia. These coalfields produce noncoking, high volatile bituminous coal with generally low ash and sulphur contents. It is ideally suited for use in the electricity generating industry.

The deposit at El Cerrejon is one of the most important in South America. Coals dip gently eastwards and more than 40 seams are greater than 1.0 m in thickness, locally reaching 26.0 m. Because of the high quality of these coals, El Cerrejon is now a significant exporter of steam coal.

3.4.6.6 Costa Rica

Costa Rica contains deposits of Tertiary coals and lignites. Individual seams are up to 1.0 m in thickness. Locally the subbituminous coals and lignites have been ameliorated by igneous intrusions. Sulphur contents range from 1.0% to 4.0%. The three principal coal deposits are at Uatsi and Zent on the southeast coast, and at Venado in the north of Costa Rica.

Areally extensive peat deposits (up to 2.0 m thick) are present in the Talamanca Cordillera and may represent a large resource for future development.

3.4.6.7 Ecuador

Small lignite deposits are present in the Tertiary sequences of the Amazon Basin, the Pacific coast, and in intermontane basins in the Andes. Only the latter are considered to be of significance. The Malacatus Basin contains seams of up to 4.0 m in thickness, disrupted by faulting. To the north the Loja Basin contains seams of up to 2.0 m in thickness and the Canar–Azuay Basin has seams of up to 5.0 m. All the coals are high volatile subbituminous coals with high ash and sulphur contents.

3.4.6.8 Mexico

Coals of Mesozoic (Cretaceous) age are found throughout Mexico; all are highly tectonised and are structurally complex. The principal coalfield is at Coahuila, close to the border with Texas, USA, where shallow, gently dipping seams reach 2.0 m in thickness. The coal is low volatile and bituminous with high ash and low sulphur contents and with no coking properties. Output is used for local industry and power generation. Another location of note is in the northwest of Mexico at Sonora, where anthracites averaging 1.0 m thickness are found.

Numerous other small deposits of bituminous coal are present in Mexico. Seam development is irregular and often ameliorated by volcanic activity.

3.4.6.9 Peru

Mesozoic coals are located within the Andean Cordillera, which extends throughout Peru from north to south.

The northern coalfields are highly tectonised and affected by associated igneous activity, resulting in the formation of anthracite as well as bituminous coal. The principal areas are those of Alto Chicama and Santa. Subbituminous and bituminous coals are found in the southern coalfields of Oyon and Jatunhuasi. All production is for local needs.

3.4.6.10 Uruguay

The northeast of Uruguay contains Carboniferous–Permian (Gondwana) sediments, which represent the southern extension of the South Brazilian coalfields. Coals are found in this area, but no development has yet occurred.

3.4.6.11 Venezuela

All the known coal-bearing sequences in Venezuela are Tertiary in age and occur in a series of basins across the country north of the river Orinoco. The principal areas of interest are Zulia, Lobatera and the Caracas–Barcelona Basin. Other coal occurrences are known in the Lara region and within the eastern Orinoco Basin.

The Zulia deposit is the most important so far identified in Venezuela and is situated in the extreme northwest of the country. Between 25 and 30 seams with thicknesses of between 0.5 m and 15.0 m are present. The coal is high volatile, noncoking bituminous with variable ash and low sulphur contents, suitable as a steam coal for export.

The Lobatera Coalfield is in the west of Venezuela, close to the Columbian border. Here 35 seams over 0.3 m thick are present. The coal is high volatile with low ash and sulphur contents.

The Caracas–Barcelona Coalfield contains the deposits of Naricual and Fila Maestra. Naricual contains 15 seams,

ranging from 1.0 m to 10.0 m in thickness, of high volatile bituminous coal with a low sulphur content; some of these seams have coking properties. The deposit at Fila Maestra is currently being investigated.

In the Lara region thin lenticular seams of low volatile bituminous coal occur with low sulphur content. In the eastern Orinoco Basin seams of lignite occur of up to 1.2 m thick, with high sulphur contents. These have not been considered significant.

3.4.7 CIS (former Soviet Union)

The CIS is the third largest coal producer in the world. It has vast reserves of all ranks of coal stretching across the whole of the country (Figure 3.13). Thick coal-bearing sequences range from Paleozoic (Carboniferous–Permian), Mesozoic (Triassic, Jurassic and Cretaceous) to Tertiary in age. These are preserved in a series of large sedimentary basins, which generally become younger from west to east. Most of the older basins are structurally disturbed, resulting in steeply dipping seams and extensive faulting. The potential for production is enormous; however, geographical position, severe climatic conditions and poor infrastructure may curtail the development of many of these deposits.

3.4.7.1 Kazakhstan

In Kazakhstan, the Karaganda and Ekibastuz Basins are the principal coal producers. The Karaganda Basin contains a thick sequence of carboniferous sediments; numerous coal seams are present varying from <1 to 3.5 m in thickness. The seams range from high volatile bituminous to anthracite, with high ash and medium sulphur contents. The lower seams have good coking properties.

The Ekibastuz Basin contains the same Carboniferous sequence, and the basin is fault bounded. In this area, a number of coal seams have coalesced to form a single seam 130–200 m thick. The coals are of similar quality to those in the Karaganda basin.

3.4.7.2 Russia

Russia has very large coal reserves, of which those in eastern Siberia and the Russian far east remain largely unexploited. Current black coal production is centred on the Kuznetsk and Pechora Basins and the Russian part of the Donetsk Basin.

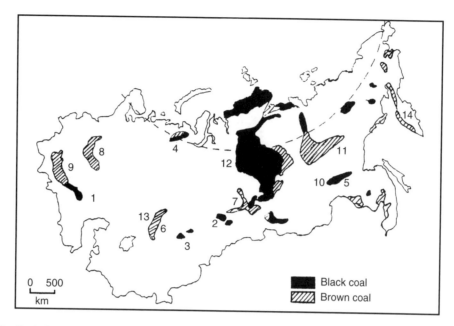

Figure 3.13 Coal Deposits of the Commonwealth of Independent States (formerly USSR). (1) Donetsk Basin; (2) Kuznetsk Basin; (3) Karaganda Basin; (4) Pechora Basin; (5) South Yakutsk Basin; (6) Ekibastuz Basin; (7) Kansk–Achinsk Basin; (8) Moscow Basin; (9) Dnepr Basin; (10) Neryungri; (11) Lena Basin; (12) Tunguska Basin; (13) Turgay; (14) West Kamchatka Area. Reproduced by permission of Dargo Associates Ltd

The Kuznetsk Basin is structurally complex, and the Carboniferous to Jurassic sequence is 7000–8000 m thick. Around 90% of the coal seams (*c*. 300) are found in the Permian, of which 130 are workable with average thicknesses of 1–35 m. Coals range from subbituminous to semi-anthracite. The ash content is variable and sulphur content is generally low. The Pechora Basin also contains Permian coal-bearing sediments which are intensely folded. Up to 5000 m of sediments contain 20–30 workable coal seams of 3–20 m in thickness. Coal rank increases from west to east and with depth. The high volatile bituminous coals have variable ash contents (10–40%), and sulphur is usually less than 1.5%, with occasional high-sulphur coals up to 4%. Many coals are semi-coking.

In the Donetsk Basin, coal rank increases towards the central and eastern parts, and ranges from subbituminous up to anthracite. The greater part of this basin is located in the Ukraine.

The Moscow Basin contains thin seams with difficult mining conditions. This area has traditionally been a large coal producer but in recent years has declined so that little mining now takes place.

In Eastern Siberia, the Kansk–Achinsk and South Yakutsk Basins have coal deposits of Jurassic age. The coals are subbituminous with thicknesses of 40–70 m. Above these occur a number of Lower Cretaceous coals. These basins have simple structures. They supply local power stations and chemical plants, the principal mining area being Neryungri.

3.4.7.3 Ukraine

As for Russia, the Donetsk Basin contains a thick succession of Carboniferous sediments in which numerous coal seams are present, ranging from 0.5 to 2.5 m in thickness. The rank ranges from subbituminous to anthracite; all have variable ash contents and high sulphur contents (average 2–3%). Some seams have good coking properties. The large Dnepr basin produces lignite for the local power stations.

3.4.8 Far East

The Far East region contains 13 countries with known coal deposits (Figure 3.14). By far the largest of these is the People's Republic of China, which has vast resources of all ranks of coal.

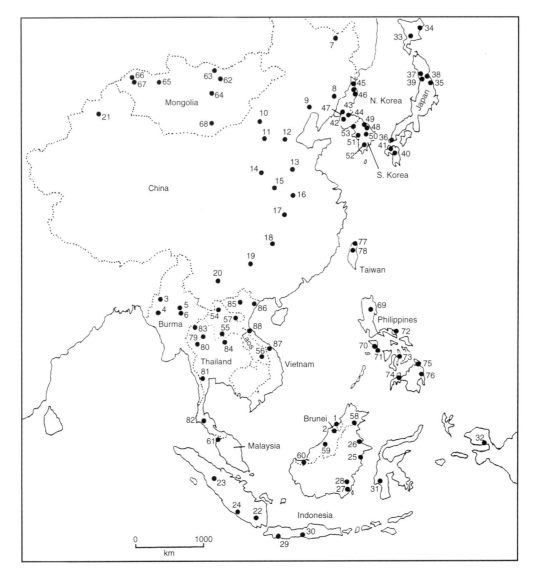

Figure 3.14 Coal deposits of the Far East. (1) Bandar Seri Begawan; (2) Belait Basin; (3) Kalewa; (4) Pakkoku; (5) Panlaung; (6) Henzada; (7) Heilung-kiang; (8) Kirin; (9) Liaoning; (10) Shen Fu–Dong Shen; (11) Shanxi; (12) Hopeh; (13) Shantung; (14) Shaanxi; (15) Honan; (16) Anhwei; (17) Hupeh; (18) Hunan; (19) Kweichow; (20) Yunnan-Guizhou; (21) Singkiang–Uighur; (22) Bukit Asam; (23) Ombilin; (24) Bengkulu; (25) Sangatta; (26) Berau; (27) Senakin–Tanah Grogot; (28) Tanjung; (29) South Java; (30) Central Java; (31) South Sulawesi; (32) Bintuni; (33) Ishikari; (34) Kushiro; (35) Joban; (36) Omine; (37) Mogami; (38) Miyagi; (39) Nishitagawa; (40) Miike; (41) Chikuho; (42) Pyongyang; (43) North Pyongyang; (44) Kowon–Muchon; (45) Kyongsang; (46) Kilchu–Myongchon; (47) Anju; (48) Samcheog; (49) Janseong–Kangnung; (50) Mungyeong; (51) Chungnam; (52) Boeun; (53) Kimpo–Yeongcheon; (54) Phongsaly; (55) Ventiane; (56) Saravan; (57) Muongphan; (58) Silimpopon; (59) Bintulu; (60) Silantek; (61) Bukit Arang; (62) Baganur; (63) Sharin Gol; (64) Nalayh; (65) Mogoyn Gol; (66) Achit Nuur; (67) Khartabagat; (68) Taban Tologoy; (69) Cagayan; (70) Mindoro; (71) Semirara; (72) Catanduanes; (73) Cebu; (74) Zamboanga; (75) Gigaquit; (76) Bislig; (77) Chilung; (78) Hsinchu; (79) Mae Moh-Li; (80) Mae Tun; (81) Nong Ya Plong; (82) Krabi; (83) Vaeng Haeng; (84) Na Duang; (85) Nan Meo–Phan Me–Bo Ha; (86) Quang Yen; (87) Nong Son; (88) Huong Khe. Reproduced by permission of Dargo Associates Ltd

The coals of the Far East range in age from Paleozoic to Tertiary, and all ranks of coal are present.

3.4.8.1 *Brunei*

Coals in Brunei are Tertiary in age, and occur in the northeast of the country close to the capital Bandar Seri Bagawan, and also in the headwaters of the Belait River in the southwest of Brunei. Coal seams are 0.5 m to 5.0 m in thickness and are high volatile bituminous with low ash and variable sulphur content. Those seams close to Bandar Seri Bagawan have been extensively worked in the past, whilst those in the Belait River basin are undeveloped, but are geographically remote.

3.4.8.2 *Burma*

The coal deposits of Burma consist of scattered occurrences of Tertiary lignite, with some Mesozoic black coals, which have been highly tectonised.

Lignites are found in the western and southern parts of the country, notably at Kalewa and Pakokku. The black coals are situated inland in the east-central region of Burma, in the Panlaung and Henzada districts. The coals are reported to be of poor quality and are only worked on a very small scale.

3.4.8.3 *People's Republic of China*

The People's Republic of China (PRC) is the world's largest coal producer. Three-quarters of all proven recoverable coal reserves occur in the northern half of the PRC, and of these, two-thirds are present in the provinces of Inner Mongolia, Shanxi and Shaanxi. These areas provide the bulk of the coal for export and power generation. The majority of mines are underground operations, and range from 70 years old to new (Thomas and Frankland 1999).

China has extensive black coal deposits of Carboniferous, Permian, Triassic, Jurassic and early Cretaceous age, plus lignite reserves of Tertiary age. Carboniferous and Permian coals are found throughout the eastern PRC, whilst Triassic coals are located in the southeast PRC. Coals of Jurassic and early Cretaceous age are located in Inner Mongolia and the northeastern PRC.

The Shenmu–Dengfeng coalfield is located on the Inner Mongolia–Shaanxi border and contains structurally undisturbed coals up to 10 m in thickness. The coal is high volatile bituminous with low sulphur content (0.4%), and is to supply power stations both domestically and for export. Mining is principally by underground methods, with some small opencast operations.

The Datong Coalfield in northern Shanxi Province is also relatively undisturbed structurally. It produces medium volatile bituminous coal with <1% sulphur, again for both domestic use and for export. The Lu'an coal-mining area in south Shanxi Province produces both steam and coking coal for home and export markets.

Shaanxi Province contains five major coalfields; of these, the southern Huang Ling and Tongchuan coal-mining districts have a number of large underground mines in operation. All are producing high and medium volatile bituminous coal with sulphur contents of around 1%.

In Henan Province, the Hebi Coalfield situated in the north of the province is the largest producer, and together with the Gaocheng coal mines in the south, produces a coal range of high volatile bituminous to anthracite coal.

Anhui Province in the east has large coal deposits, all exploited by underground mines. The coalfield areas are structurally complex, seam thicknesses are up to 6 m, and the coals are bituminous with low sulphur contents. Some coals have coking properties.

To the north of Anhui, Shandong Province is an important producer of export-quality coals. Coals are bituminous, low in ash and sulphur and are amongst the best coking coals in the PRC. Seams range in thickness from 1 m to 10 m.

In the northeast of the PRC, the Liaoning region has numerous coal seams up to 100 m in thickness, with little structural disturbance. Coals are high volatile bituminous with low ash and sulphur and with good coking properties. The Heilongjiang and Jilin Coalfields in the far northeast also have thick seams of similar quality.

In Guizhou Province, in the southern PRC, the Pangjiang Coalfield produces medium and low volatile bituminous coal for both power generation and steel production. The coalfield is structurally complex, and anthracite occurs in the more intensely tectonised areas.

To the south of Guizhou, in Yunnan Province, low volatile bituminous coking coals with low sulphur contents are produced.

There are numerous other coal deposits in the PRC, mostly close in location to those listed. The PRC's potential for coal production is enormous, but depends heavily on underground operations and has poor infrastructure in some areas.

3.4.8.4 *Indonesia*

Indonesian coal deposits are Tertiary in age and are situated on the islands of Sumatra, Borneo, Java,

Sulawesi and West Irian. There is a range in rank from lignite to low volatile bituminous, the higher rank coals being affected by local igneous intrusions or, more importantly, by regional heating due to magmatic activity at relatively shallow depths.

On the island of Sumatra, three coalfield areas are currently exploited.

At Bukit Asam at the southeastern end of the island, seams up to 12.0 m in thickness are present. Coals are generally subbituminous with low ash and sulphur contents, but some bituminous coal is present in close proximity to igneous intrusions. The coals are mined by opencast methods and used primarily for electricity generation.

Ombilin, located in central Sumatra, has a few thick seams of high volatile bituminous and subbituminous coal with low ash and sulphur contents. Mining is by both opencast and underground methods, and the coal is used for electricity generation and cement manufacture.

In the Bengkulu region on the southwest coast of Sumatra, small occurrences of mostly subbituminous coals with low sulphur content are worked on a small scale.

On the island of Borneo, the Indonesian territory of Kalimantan has coal deposits situated along the east coast. In East Kalimantan, subbituminous and bituminous coals are found, notably in the Sangatta and Berau areas. These coals are up to 10.0 m in thickness, with extremely low ash and low sulphur contents. Some of these coals are now exported as prime-quality steam coals. In South Kalimantan, in the Senakin, Tanah Grogot and Tanjung areas, subbituminous and bituminous coals with similar characteristics are mined both for export and for local power generation needs. In the northeastern part of Kalimantan, north of Berau, bituminous coals are present at Tarakan, but high sulphur contents have halted the development of these deposits.

In Java, subbituminous coals have been worked on a very small scale in central and western parts. These coals are thin and irregularly developed.

In South Sulawesi, similar subbituminous coals are present which have been mined for local needs.

In West Irian, the western half of the island of New Guinea, subbituminous coals and lignites are present in the Bintuni region at the western end of the island, but have not yet been developed.

Large deposits of recent peat are present in West Kalimantan, but these have not been commercially developed.

3.4.8.5 Japan

Japanese coal deposits are widespread and range from Permian to Tertiary in age. The productive coals are Tertiary, whilst the Permian and Mesozoic coals are of minor importance except for the Omine Coalfield in western Honshu.

The principal Tertiary coalfields are located on the three Japanese islands of Hokkaido, Honshu and Kyushu.

On Hokkaido Island, the structurally complex area of the Ishikari Coalfield provides strongly caking bituminous coal with high ash and low sulphur content. The coals are produced for local use. The Kushiro Coalfield is less disturbed and produces noncoking bituminous and subbituminous coal.

On the island of Honshu, the Joban Coalfield has seams up to 3.0 m in thickness and is thought to extend eastwards offshore. The Omine Coalfield on the southwest coast is important as a source of anthracite for Japanese industry. The Mogami, Nishitagawa and Miyagi Coalfields are situated in the northern half of the island and are the chief lignite producers in Japan.

On Kyushu Island, the Miike Coalfield is structurally undisturbed and is mined offshore. The coal is bituminous with good coking properties. The Chikuho Coalfield has similar coals to Miike, and is a source of coking coal for the metallurgical industry.

Numerous smaller coalfields containing bituminous coals and lignites are worked on a small scale.

3.4.8.6 Democratic Republic of (North) Korea

Coals of Paleozoic, Mesozoic and Tertiary age are present throughout the Korean peninsula.

The principal Paleozoic coalfields are Pyongyang and North Pyongyang in the northwest, and Kowon–Muchon in the east. All have been highly tectonised; consequently seam thicknesses are variable due to intense folding. However, thicknesses of 5.0 m and 15.0 m are reached. The coals are low volatile anthracites with low ash and sulphur contents. All the coal is mined by underground methods and is used for local industry and domestic heating.

Mesozoic coals form small deposits of anthracite; these are also strongly folded, but to a lesser extent than the Paleozoic coals.

The Tertiary coalfields contain subbituminous coal and lignite and are found chiefly in the northeast of the country. The Kyongsang and Kilchu–Myongchon Coalfields contain lignites, and the Tumangang in the extreme northeast contains subbituminous coal which is used for electricity generation. The Anju Coalfield is located north of Pyongyang and is a large deposit of subbituminous coal which is being developed as an opencast operation.

3.4.8.7 Republic of (South) Korea

Coals in South Korea are of similar age and character to those in the north of the peninsula; again, all of the mining operations are underground.

The principal Paleozoic coalfields are Samcheog, Jeongseon, Kangnung, Danyang and Mungyeong. These coalfields are highly tectonised and intensely folded; seam thicknesses vary considerably due to the squeezing of the coals: 1.0 m to 2.0 m is usual. All the coal is anthracite with a low sulphur content and is exclusively used for local industry and domestic heating.

Mesozoic (Jurassic) coal deposits are present at Mungyeong and Chungnam; the latter is structurally complex. Again, all the coal is anthracite.

Small anthracite deposits are found at Boeun and Honam in the south, and at Kimpo and Yeongcheon on the northern border of the country; small workings produce anthracite for local use.

Tertiary deposits containing thin seams of lignite are found in small areas bordering the southeast coast of South Korea.

3.4.8.8 Laos

Paleozoic, Mesozoic and Cainozoic coals are present in Laos. The Paleozoic deposits are chiefly anthracite with a high ash content. The three principal occurrences are at Phongsaly in the north, the Ventiane coal basin in west-central Laos, and the Saravan coal basin in the south of the country. In the Ventiane basin, five seams ranging from 2.6 m to 6.0 m are present; in the other areas, the seams are considerably thinner.

Some Mesozoic (Triassic–Jurassic) coals are found in the Phongsaly region; all are steeply dipping, and seams range in thickness from 0.1 m to 10.0 m. The coals are high volatile bituminous with low ash and low sulphur contents.

Brown coals are present in several Tertiary basins located in the east of the country, chiefly at Muongphan, with other occurrences at Khang Phanieng, Hua Xieng and Bam O. These Tertiary basins are highly faulted and contain subbituminous coals and lignite. At Muongphan, lignite seams are 1.0 m to 6.0 m in thickness and have high volatile and ash contents.

All Laotian coal produced is used for local needs.

3.4.8.9 Malaysia

Malaysian coals are found on the west coast of the West Malaysian peninsula, and on the East Malaysian side of the island of Borneo in the states of Sabah and Sarawak.

All the coals are of Tertiary age; those in Sabah are subbituminous with some coking properties, but often with high sulphur contents. These have been mined at Silimpopon in east Sabah. In Sarawak, higher quality bituminous and subbituminous coals with low sulphur contents have been identified at Bintulu, Balingian and Silantek, and mined on a local scale.

In West Malaysia, at Bukit Arang on the Malaysian/Thailand border, extensive lignite deposits have been identified; another occurrence of lignite is reported north of Kuala Lumpur at Batu Arang.

3.4.8.10 Mongolia

Coal deposits in Mongolia are concentrated in the north of the country. Highly tectonised Paleozoic coals in the form of anthracite and low volatile bituminous are found in small isolated deposits.

Mesozoic (Cretaceous) coals are less deformed and consist of low volatile bituminous coal with low sulphur contents, found principally in the Baganur Coalfield, where seam thicknesses can be up to 25.0 m. In the same region occur the coalfields of Sharin Gol and Nalayh, and in the west of the country are the coalfields of Achit Nuur and Khartarbagat.

In southern Mongolia, at Taban Tologoy, is a large deposit of bituminous coal; however, its geographical location has so far prevented its development.

3.4.8.11 Philippines

Throughout the Philippines archipelago is situated a series of Tertiary basins containing coal-bearing sediments. The coals are predominantly of subbituminous rank, although variations in rank do occur related to local structure and contemporaneous and recent igneous activity.

The northern island of Luzon contains the Cagayan Basin; this area is only partially explored, but is known to contain seams up to 2.0 m in thickness, and is structurally undisturbed. The coals are high volatile subbituminous with low ash and sulphur. The deposit covers a large area and is amenable to opencast mining operations. Such coals would be suitable for local electricity generation.

The island of Mindoro has coal deposits in the south; the seams are up to 2.8 m in thickness, and are subbituminous with variable sulphur contents.

Semirara Island lies to the south of Mindoro and contains coals up to 6.0 m and 12.0 m in thickness. The coals are subbituminous with low ash and sulphur.

Catanduanes Island contains lenticular seams up to 5.0 m in thickness; these are steeply dipping and are ameliorated by igneous intrusions. This has resulted in the formation of bituminous coals with high sulphur and moderate coking properties.

Cebu Island contains several coal deposits; seams are up to 4.0 m in thickness, dip steeply and are high volatile subbituminous coals with low ash and variable sulphur contents.

Mindanao Island has coal deposits at Malangas and Zamboanga in the west, and at Gigaquit and Bislig in the east.

The Malangas–Zamboanga area has ameliorated coals, anthracite and bituminous coking coal. At Gigaquit, low rank coals with high ash contents are characteristic, and at Bislig, some bituminous coal with locally high sulphur content is mined.

Numerous other small coal deposits are worked locally throughout the archipelago.

Small-scale underground mining characterises the bulk of the coal exploration in the Philippines; however, those deposits at Cagayan, Semirara and Zamboanga could be further developed.

3.4.8.12 Taiwan

Coals in Taiwan are Tertiary in age, and the coalfields are grouped into a northern and a central province; of these, the northern province only has economic significance. The Taiwan coalfields have been highly tectonised, and some have been ameliorated by igneous intrusions. The coals are high volatile bituminous and subbituminous with low ash and sulphur contents. At Chilung in the north of Taiwan, ameliorated coals (semi-anthracites) have been mined in small areas. Further south, low volatile bituminous coals, low in sulphur and with good coking properties, have been mined at Hsinchu, Nanchuang, Shuangchi and Mushan. Four seams exceed 1.0 m in thickness and because of the high level of tectonic disturbance there are only underground operations working at increasingly deep levels. This may result in the cessation of mining in these areas.

3.4.8.13 Thailand

In Thailand virtually all the known coal deposits are of Tertiary age, together with some Mesozoic coals found in the northeast of Thailand at Na Duang. The Tertiary sediments are preserved in a series of basins; of these, the Mae Moh basin in northwest Thailand is the most extensive. Other basins in close proximity are Mae Tip, Li, Mae Tun and Vaeng Haeng. Other Tertiary coals are found east of Bangkok at Nong Ya Plong, and at Krabi in the extreme southwest.

Seams in Thailand generally range from 2.0 m to 12.0 m; however, at Mae Moh and Krabi, seams up to 30.0 m are worked. Most coals are relatively undisturbed structurally. These Tertiary coals range from lignite to high volatile subbituminous, with generally low sulphur contents, as found at Mae Moh, Mae Tip, and Li, and some with higher rank, high volatile bituminous as found at Mae Tun and Nong Ya Plong.

The Mesozoic coal at Na Duang is semi-anthracite with low sulphur content.

The bulk of Thailand's coals are mined and supplied to the electricity generating industry.

3.4.8.14 People's Republic of Vietnam

In Vietnam, black coals are of Mesozoic (Triassic) age, and are located first in a broad belt running east to west, situated north and northeast of Hanoi. This belt consists of four sedimentary basins each containing coal-bearing strata; these are the Nan Meo, Phan Me, Bo Ha and Quang Yen basins. Second, they are found in central Vietnam at Nong San, and Huong Khe.

The Nan Meo, Phan Me and Bo Ha basins contain low volatile bituminous coals, some with coking properties. These areas were worked on a small scale in the past.

The most important coal basin is the Quang Yen Basin, the eastern part of which borders the northeast coast. Coals are preserved in a series of folds oriented parallel to the coast, and are bounded by large east–west running faults.

In the east part of the Quang Yen basin, the Hong Gai Coalfield is the chief coal producer in Vietnam; up to six seams with thicknesses 2.0 m to 8.0 m are worked. Coals are low volatile anthracites with low ash and sulphur contents.

In central Vietnam, in the Nong Son area, is a thick seam up to 20.0 m in thickness, and is low volatile bituminous to semi-anthracite with a variable sulphur content. The Huong Khe area is believed to contain several seams of anthracite.

3.4.9 Australasia

Australasia is one of the major coal producers in the world. The bulk of the coal resources are located in the

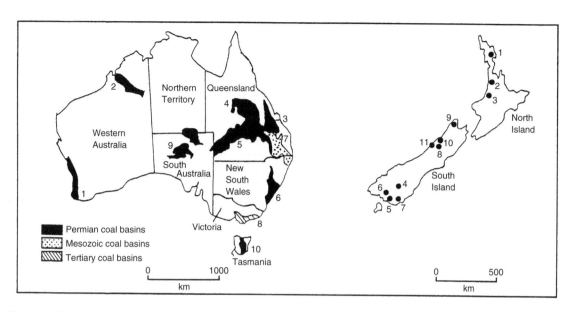

Figure 3.15 Coal deposits of Australasia. Australia: (1) Collie Basin; (2) Fitzroy Basin; (3) Bowen Basin; (4) Galilee Basin; (5) Cooper Basin; (6) Sydney Basin; (7) Brisbane Basin; (8) Gippsland Basin; (9) Ackaringa Basin; (10) Tasmania. New Zealand: (1) Northland Coalfield; (2) Waikato Coalfields; (3) Taranaki Coalfields; (4) Otago; (5) Southland; (6) Ohai; (7) Kaitangata; (8) Reefton; (9) Collingwood; (10) Buller; (11) Pike River–Greymouth. Reproduced by permission of Dargo Associates Ltd

eastern part of Australia, with smaller coal deposits in Western Australia and New Zealand (Figure 3.15).

3.4.9.1 Australia

Australia contains coals of Paleozoic, Mesozoic and Tertiary age. The whole of the black coal resources are of Paleozoic age and are located in Western Australia, Queensland and New South Wales. Mesozoic coal is present in Queensland and an important deposit of Tertiary coal is found in Victoria. The black coals of Queensland and New South Wales are both steam and coking coals, and the bulk of production is for export. In the other coal-producing areas of Australia, coal is primarily used for domestic power generation.

The Paleozoic coals of Australia are Permian (Gondwana) in age and have been generated in a series of basins. The principal ones are the Bowen, Galilee and Cooper Basins in Queensland, the Sydney Basin in New South Wales and the Collie and Fitzroy Basins in Western Australia. Other smaller areas are known from South Australia and Tasmania.

In Queensland, the Bowen Basin has been explored extensively. The eastern side of the basin has subsided more rapidly than the west and has received more sediments, and the lack of structural disturbance has resulted in the preservation of shallow flat-lying coals. The oldest coals are low ash and low sulphur seams which reach thicknesses of up to 30.0 m. The uppermost Permian contains four workable seams. Seam splitting is common, and in the west, igneous intrusions have locally affected the coals. The topmost coal-bearing sediments are the most widespread, with 12 coal seams having thicknesses up to 4 m. The coal is high volatile bituminous, with variable ash and low sulphur content; the coals have good coking properties. The coal is worked by large opencast operations, and large reserves have been identified. In the other coal-bearing areas of Queensland, e.g. the Galilee and Cooper Basins, large reserves of coal have yet to be developed.

In New South Wales, the Sydney Basin is the most important coal-producing area in Australia. Again, as in the Bowen Basin, there is little structural disturbance. Two Permian formations contain coals which are exposed over large areas. The lower part contains up to six seams, and is heavily faulted and intruded. The upper formations have 14–40 coal seams, many of which exhibit splitting, and, in the west, are severely affected by igneous intrusions, which although produce an increase in rank, have also destroyed large reserves of coal. Seam thicknesses reach 10.0 m, and the coal produced is high volatile bituminous with variable ash and low sulphur contents; some of the coals have good coking properties. Coals are mined by underground and opencast methods. The principal mining districts are the Western District, the Burragorang Valley, the Hunter Valley and the Southern District. Much of the coal is exported as steam and coking coal, as well as supplying local needs.

In Western Australia, the Collie Basin contains seams ranging from 1.5 m to 11.2 m; these are structurally undisturbed, and are mined in the Cardiff and Muja areas. The coals are subbituminous with low ash and sulphur contents. Elsewhere in Western Australia, coals have been located but have yet to be developed.

In South Australia, the Ackaringa basin is currently being explored, and in Tasmania some development of the coal deposits may occur in the future.

The Mesozoic coals of southeast Queensland, in the Brisbane area, are subbituminous coals and have not been extensively developed.

The Tertiary coals of the Gippsland Basin in Victoria are thick developments of lignite; the principal seams reach enormous thicknesses of 300.0 m. The seams are shallow and flat-lying, and are high volatile lignite with low ash and low sulphur contents. This basin is worked in a number of separate coalfields, the most important of which is the Latrobe Valley. This lignite is used exclusively for the Victorian electricity industry.

3.4.9.2 New Zealand

With the exception of a few thin uneconomic coals of Jurassic age, all significant New Zealand coals are Cretaceous–Tertiary in age. The coalfields are located in the western part of North Island, and in the northwestern and southeastern districts of South Island.

In North Island, the Waikato Coal region contains New Zealand's major subbituminous coal resource. Coal seams are discontinuous but are thick locally, up to 30.0 m. They are subbituminous with low ash and low sulphur contents. To the north, the Northland area has a few seams up to 2.0 m thick; these are of poorer quality.

To the south of the Waikato Coal region lies the Taranaki Coal area, where seams are usually less than 3.0 m in thickness, and are subbituminous with higher sulphur contents.

In South Island are located the Cretaceous–Tertiary coalfields of Otago, Southland, Ohai and Kaitangata in the far south; on the northwest coast, the Westland Coal region includes the Greymouth, Pike River, Charleston, Buller, Reefton and Collingwood Coalfields.

The southern coalfields have seams up to 6.0 m and 10.0 m, although they are discontinuous and lensoid. The coals vary from lignite to high volatile subbituminous with variable sulphur contents, some of which may be as high as 6%.

In the Westland Coal region, the coals range from subbituminous at Charleston to high volatile bituminous at Pike River, Collingwood and Reefton. In the Buller and Greymouth Coalfields coals range from high volatile to low volatile bituminous.

Production from the New Zealand coalfields is small at the present time.

3.4.9.3 Antarctica

Cretaceous coals of mixed quality have been recorded from James Ross Island, on the southeast flank of the Weddell Sea. Other occurrences have been in the area of the Transantarctic Mountains.

Present legislation will prohibit any development of these possible resources for many years to come.

4

Coal as a Substance

4.1 PHYSICAL DESCRIPTION OF COAL

Coal has been defined by numerous authors; essentially it is a sediment, organoclastic in nature, composed of lithified plant remains, which has the important distinction of being a combustible material.

The composition and character of each coal will be determined first by the nature of the make-up of the original organic and inorganic accumulation, and second by the degree of diagenesis it has undergone.

The inherent constituents of any coal can be divided into 'macerals', the organic equivalent of minerals, and 'mineral matter', the inorganic fraction made up of a variety of primary and secondary minerals. Note that the latter is sometimes erroneously referred to as 'ash' when in fact 'ash' is the mineral residue remaining after combustion of the coal. The composition and ratio of the two fractions reflects the make-up of the original material, and indicates the coal *type*.

The degree of diagenesis or coalification that a coal has undergone by burial and tectonic effect determines the coal *rank*.

The term 'brown coal' is used for low rank coals such as lignite and subbituminous coal, and 'black' or 'hard' coal is used for coals of higher rank, the bituminous, semi-anthracite and anthracite coals.

The majority of coals are composed of discrete layers of organic material. Such layers may possess different physical and chemical properties. It is the relative proportions and petrological characteristics of these layers that determines the character of the coal as a whole, and its usefulness as a mined product.

Coals are divisible into two main groups, the *humic* coals and the *sapropelic* coals.

Humic coals are composed of a diversified mixture of macroscopic plant debris; the coals typically have a banded appearance. Sapropelic coals are composed of a restricted variety of microscopic plant debris; such coals have a homogeneous appearance.

4.1.1 Macroscopic description of coal

4.1.1.1 Humic coals

The use of a simple but distinctive system of description is fundamental to field examination of coals. Several systems to describe the physical character of coal have been proposed and are briefly outlined below.

The term lithotype is applied to the different macroscopically identifiable layers in coal seams.

Stopes (1919) proposed four lithological types (lithotypes) for describing humic coals.

 (i) Vitrain is a black, glassy, vitreous material with a bright lustre, occurring as thin bands; it is brittle. Vitrain breaks into fine angular fragments and is commonly concentrated in the fine fraction of mined coal. Vitrain is found in most humic coals and usually consists of the microlithotype vitrite with some vitrinite-rich clarite.
 (ii) Clarain is bright with a silky lustre between vitrain and durain, and occurs in fine laminations. Clarain comprises alternating thin layers often <1 mm. It can include the microlithotypes vitrite, clarite, durite, fusite and trimacerite.
(iii) Durain is grey to black with a dull lustre; it fractures into rough surfaced fragments. Only lenses thicker than 3–10 mm are referred to as durain. Durain is less common than vitrain and clarain in humic coals, but can occur as extensive layers within a coal seam. Durain is composed of the microlithotypes durite and trimacerite.
(iv) Fusain is black, soft, friable, and easily disintegrates into a black fibrous powder. Fusain occurs in coals as lenses, usually several millimetres thick, often concentrating in discrete layers in the coal. In most coals, fusain is a minor lithotype composed of the microlithotype fusite.

However, difficulties have arisen in using the above terms to describe coals in borehole cores and in exposures. The four lithotypes often occur as thin layers or lenses, often only millimetres in thickness. Strict usage of Stopes's terms would lead to extremely detailed lithological descriptions, whereas in practice only a limited number of lithologically distinct units is required. For practical purposes, alternative terminology has been proposed by various sources which, while essentially retaining the basic classification of Stopes, has a more descriptive lithological bias. The principal types of humic and sapropelic coals are summarised in Table 4.1.

In the USA, Schopf (1960) introduced the term *attrital* coal to include all coal not precisely defined as vitrain or fusain and which can be subdivided into five levels of lustre ranging from bright to dull.

The Australian system is broadly similar in approach, but more descriptive in terminology (Table 4.2). The Australian coal industry defines vitrain and fusain as bright and dull coal respectively, and the five categories of attrital coal are graded according to major and minor constituents of each end member. This is very much a physical description and is eminently more suitable for field recording of coals.

In addition coals with high mineral content contained in discrete bands or as nodules or veins can be best described as *impure* coals. Commonly such mineral matter is in the form of pyrite, calcite, siderite, ankerite, or as clay coatings and infillings. In the USA, coals which contain clay disseminated throughout the coal

Table 4.1 Lithotypes of humic and sapropelic coals. From McCabe (1984), with permission of the International Association of Sedimentologists

Lithotype	Description	Composition
Vitrain	Black, very bright lustre; thin layers break cubically; thick layers have conchoidal fracture	Vitrinite macerals with <20% exinite macerals
Clarain	Finely stratified layers of vitrain, durain and, in some instances, fusain, medium lustre	Variable
Durain	Black or grey, dull, rough fracture surfaces	Mainly inertinite and exinite macerals
Fusain	Black, silky lustre, friable and soft	Mainly fusinite
Cannel coal	Black, dull, lustre 'greasy', breaks with conchoidal fracture	Fine maceral particles usually dominated by sporinite
Boghead coal	Black or brown, dull, homogeneous, breaks with conchoidal fracture, lustre may be 'greasy'	Dominated by alginite

Table 4.2 Macroscopic description of coals in sections and boreholes. From Ward (1984) with permission of Blackwell Scientific Publications

Stopes (1919)	Schopf (1960); ASTM standard (1978)		Australian standard
Banded (humic) coals			
Vitrain	Vitrain		Coal, bright
Clarain		Bright	Coal, bright, dull bands
		Moderately bright	Coal, dull and bright
	Attrital	Mid. lustre	Coal, mainly dull with
Durain	coal	Moderately dull	numerous bright bands
		Dull	Coal, dull minor bright bands
Fusain	Fusain		Coal, dull
Nonbanded (sapropelic) coals			
Cannel coal	Cannel coal		Coal, dull conchoidal (canneloid)
Boghead coal	Boghead coal		
Impure coals	Bone coal		Coal, stony (or shaley)
	Mineralised coal		Coal, heat altered
			Coal, weathered

rather than in layers is termed *bone* coal, and is of dull appearance.

In coals of lower rank, i.e. *brown* coals, the above lithological descriptions are difficult to apply. Brown coals range from *lignite*, which may be anything from soft, dull brown to black in colour, to *subbituminous* coal, which is black, hard and banded. Brown coals are usually described in terms of colour and texture; e.g. they crack and disintegrate when dried out.

Hagemann (1978, 1980) adopted a macroscopic description of lignites and applied it to Saskatchewan lignites and lignite–subbituminous coals. The important criteria in Hagemann's descriptions are the relative proportions of groundmass and woody (xylitic) remains plus the relative abundance of mineral impurities, and the texture or banded characteristics. The groundmass comprises the more finely comminuted particles of varied origin too small to be identified macroscopically. In addition intensity and hue of colour, degree of gelification and presence or absence of inclusions are all incorporated into the system, shown in Table 4.3.

Following the work of Hagemann, the International Committee for Coal and Organic Petrology (ICCP) in 1993 adopted the classification for soft brown coals as shown in Table 4.4 (Taylor *et al* 1998). The classification recognises lithotype groups, lithotypes and lithotype varieties.

The structure and constituents of the lithotype of soft brown coal can be recognised with the naked eye; lithotypes can be distinguished by their degree of gelification and colour. The ICCP classification recognises four coal types as described by Taylor *et al* (1998).

Table 4.3 Macroscopic description of lignites. From Bustin *et al* (1983), based on Hagemann (1980), by permission of Geological Association of Canada

Field observations	Laboratory observations				Additional features
Structure	Texture	Colour	Gelification	Inclusions	
1. Pure coal (nonxylitic)	Unbanded coal	Pale yellow	Gelified groundmass	Resin bodies	Cracking to noncracking
2. Pure coal (xylitic); fibrous/brittle; tree stumps, trunks etc.	Moderately banded coal	Medium light yellow	Gelified tissues	Cuticles	Fracture even
3. Impure coal (nonxylitic) clayey/sandy calcareous coal, iron sulphides etc.	Banded coal	Pale brown	Microgranular humic gel particles	Charcoal	Size breakup coarse to fine
4. Impure coal (xylitic)	Highly banded coal	Medium brown/dark brown/black	–	–	–

Table 4.4 Lithotype classification for soft brown coals. From Taylor *et al* (1998), after International Committee for Coal and Organic Petrology (1993). Reproduced by permission of Gebruder Borntraeger

Lithotype group (constituent elements)	Lithotype (structure)	Lithotype variety (colour; gelification)
Matrix coal	Stratified coal	Brown (weakly gelified) coal Black (gelified) coal
	Unstratified coal	Yellow (ungelified) coal Brown (weakly gelified) coal Black (gelified) coal
Xylite-rich coal Charcoal-rich coal Mineral-rich coal		

Matrix coal consists of a fine detrital groundmass, yellow to dark brown in colour. Plant fragments may be embedded in the groundmass, and matrix coal may be homogeneous in appearance or show some stratification. The homogeneous matrix coals may have originated from peats found in low-lying mires, or from decomposition of swamp forest peats, whereas banded matrix coals are considered to be the product of an open-swamp environment. Matrix coals are common in Tertiary soft brown coals.

Xylite-rich coal includes coals in which xylite (woody tissue) comprises more than 10% of the coal. The groundmass is detrital and may or may not be stratified. Xylite occurs as fibrous tissue and may be mineralised. Inclusions of charcoal or gelified nodules may be present.

Xylite-rich coal occurs in all brown coals and is the dominant lithotype. Its characteristics are thought to be the decomposition of trees and shrubs in the peat-forming mire.

Charcoal-rich coal contains >10% charcoal. The coal can be weakly or strongly stratified, occurring as lenses and occasional more persistent layers. The coal is brownish-black and has a coke-like appearance. It is a minor constituent of soft brown coals. Charcoal-rich coals are considered to be the product of burned forest swamps. Where such coal is stratified, it is indicative of water or wind transported residues in an open-swamp environment.

Mineral-rich coal includes all kinds of mineralisation of the different brown coal lithotype groups, and should be visible to the naked eye. The inorganic materials present include typically quartz, clay, carbonates and sulphides, and other minerals.

In Australia, the State Electricity Commission of Victoria has used a classification of brown coal based on colour and texture. Table 4.5 shows the classification including the additional characteristics of gelification level, weathering character and physical properties. The classification should be assessed on air-dried coal; colour is based on shades of brown, and texture refers to the amount of xylitic material present. The classification does not take into account mineral matter content because of the low ash levels in Victorian brown coals.

4.1.1.2 Sapropelic coals

Sapropelic coals are formed from the biological and physical degradation products of coal peat-forming environments, with the addition of other materials such as plant spores and algae. The resultant sediment is an accumulation of colloidal organic mud in which concentrations of spore remains and/or algae are present. Sapropelic coals are characteristically fine-grained, homogeneous, dark in colour and display a marked conchoidal fracture. They may occur in association with humic coals or as individual coal layers.

Cannel coal is black and dull; it is homogeneous and breaks with a conchoidal fracture. It is composed largely of miospores and organic mud laid down under water, such as in a shallow lake.

Boghead coal is algal coal, and the criterion for the assignment of a coal to a boghead is that the whole mass of that coal originated from algal material without consideration of the state of preservation of the algal colonies, i.e. whether they are well preserved or completely decomposed. Boghead coals may grade laterally or vertically into true oil shales.

Between these two major types of sapropelic coals, transitional or intermediate forms such as cannel–boghead or boghead–cannel are recognised. Essentially all sapropelic coals look similar in hand specimen and can only be readily distinguished microscopically.

Coal descriptions using the above terms result in a considerable amount of data that can be used in conjunction with the laboratory analysis of the coal. Such lithological logs can also provide information on coal quality that will influence the mining and preparation of the coal.

4.1.2 Microscopic description of coal

The organic units or *macerals* that comprise the coal mass can be identified in all ranks of coal. Essentially macerals are divided into three groups:

1. Huminite/vitrinite – woody materials;
2. Exinite (liptinite) – spores, resins and cuticles;
3. Inertinite – oxidised plant material.

The original classification of maceral groups is referred to as the Stopes–Heerlen System, given in Table 4.6. Other detailed descriptions are well summarised by McCabe (1984) in Table 4.7.

However, coals may be made up largely of a single maceral or, more usually, associations of macerals. These associations when studied microscopically are called microlithotypes.

In order to distinguish between the different microlithotypes, the ICCP has agreed that a lithotype can only be recorded if it forms a band >50 μm, and that lithotypes are not composed purely of macerals from one or two maceral groups, it must contain 5% of accessory macerals. All microlithotypes may contain

Table 4.5 Typical characteristics of air-dried brown coal lithotypes from Latrobe Valley, Australia. From Taylor *et al* (1998) modified after George (1975). Reproduced by permission of Gebruder Borntraeger

Lithotype	Abbreviation	Colour	Texture	Gelification	Weathering pattern	Physical properties
Dark	Dk	Dark brown to medium brown	High (20–30%) wood content. Often small (<25 cm) fragments	Gelification, particularly of woody material, common	Cracks wide and deep. Regular pattern	Strong, hard, heavy (high S.G.)
Medium–dark	M–d	Dark brown to medium brown	High to medium (10–20%) wood content. Often large (>25 cm) pieces	Some gelification but not extensive	Cracks wide. Some regularity of pattern	Strength variable, hardness and S.G. above average
Medium–light	M–l	Medium brown to light brown	High to low (0–10%) wood content. Often well preserved	Gelification uncommon. Confined mainly to wood	Cracks shallow. Irregular pattern	Intermediate physical properties
Light	Lt	Light brown	Medium to low wood content	Gelification rare	Cracks generally fine. Random orientation	Generally soft and relatively light (low S.G.)
Pale	Pa	Pale brown to yellow brown	Wood present but uncommon	Gelification very rare	Few extensive cracks	Soft, crumbles readily, very low S.G.

Note: Texture: wood content includes all plant fragments clearly distinguishable from the groundmass. Physical properties: S.G. = specific gravity.

Table 4.6 Stopes–Heerlen classification of maceral groups, macerals and submacerals of hard coals. From Ward (1984), after Stopes (1935), with permission of Blackwell Publishing Ltd

Maceral Group	Maceral	Submaceral
Vitrinite	Telinite Collinite	Telocollinite Gelocollinite Desmocollinite Corpocollinite
Exinite (liptinite)	Sporinite Cutinite Suberinite Resinite Alginite Liptodetrinite Fluorinite Bituminite Exudatinite	
Inertinite	Fusinite Semifusinite Macrinite Micrinite Sclerotinite Inertodetrinite	

amounts of mineral matter; if this reaches 20% then the microlithotype is referred to as a carbominerite (Taylor *et al* 1998).

The composition of the microlithotypes is listed in Table 4.8 (McCabe 1984) and their interrelationship is shown in Figure 4.1 (Bustin *et al* 1983). Taylor *et al* (1998) have used the ICCP recommendations and described the microlithotypes of humic coals; the principal types are shown in Figure 4.2.

Hunt and Hobday (1984) relates the microlithotype composition of Permian coals in eastern Australia to depositional environments, as shown in Figure 4.3. Lower delta plain environments have produced coals relatively vitrinite-rich, whereas upper delta plain and meandering fluvial coals are vitrinite-poor. High subsidence rates prevailed during the accumulation of both, but water tables were high in the early Permian, and low in the late Permian.

Vitrite comprises 95% of the vitrinite macerals, telinite and collinite in bands at least 50 μm thick (Figure 4.2).

Vitrite occurs in coal seams as elongated lenses several millimetres thick. Vitrite originated in anaerobic conditions due to high groundwater table levels in the peat mire.

Table 4.7 Macerals and group macerals recognised in hard coals. From McCabe (1984), with permission of the International Association of Sedimentologists

Maceral group	Maceral	Morphology	Origin
Vitrinite (huminite)	Telinite	Cellular structure	Cell walls of trunks, branches, roots, leaves
	Collinite	Structureless	Reprecipitation of dissolved organic matter in a gel form
	Vitrodetrinite	Fragments of vitrinite	Very early degradation of plant and humic peat particles
Exinite (liptinite)	Sporinite	Fossil form	Mega- and microspores
	Cutinite	Bands which may have appendages	Cuticles – the outer layer of leaves, shoots and thin stems
	Resinite	Cell filling layers or dispersed	Plant resins, waxes and other secretions
	Alginite	Fossil form	Algae
	Liptodetrinite	Fragments of exinite	Degradation residues
Inertinite	Fusinite	Empty or mineral filled cellular structure; cell structure usually well preserved	Oxidised plant material – mostly charcoal from burning of vegetation
	Semifusinite	Cellular structure	Partly oxidised plant material
	Macrinite	Amorphous 'cement'	Oxidised gel material
	Inertodetrinite	Small patches of fusinite, semi-fusinite or macrinite	Redeposited inertinites
	Micrinite	Granular, rounded grains ~1 μm in diameter	Degradation of macerals during coalification
	Sclerotinite	Fossil form	Mainly fungal remains

Table 4.8 Composition of microlithotypes. From McCabe (1984), with permission of the International Association of Sedimentologists

Microlithotype	Composition
Vitrite	Vitrinite >95%
Liptite	Exinite >95%
Inertite	Inertinite >95%
Fusite	Inertite with no macrinite or micrinite
Clarite	Vitrinite and exinite >95%
Durite	Exinite and inertinite >95%
Vitrinertite	Vitrinite and inertinite >95%
Trimacerite	Vitrinite, exinite, inertinite, each >5%

Vitrite makes up 40–50% of the Carboniferous coals in the Northern Hemisphere. In Gondwana coals, however, it rarely exceeds 20–30%. As a group, Late Cretaceous and Tertiary coals are generally rich in vitrinite and comparatively rich in liptinite, usually having >20% inertinite.

Liptite layers form thin lenses or bands a few millimetres thick, and have been deposited in water. Concentrations of up to 95% liptinite group macerals are rare.

Inertite microlithotypes contain >95% inertinite macerals, which include inertodetrite, semifusite and fusite (Figure 4.2). In most coals fusite comprises no more than 5–10% as thin bands and lenses. Fusite-rich coals are thought to be the result of the onset of aerobic conditions in peat formation.

Inertodetrite, which consists of 95% inertodetrinite, is common in Gondwana coals. These coals are composed of numerous inertite layers in which inertodetrinite and semifusinite make up over 95%. Inertodetrite is present in Northern Hemisphere Carboniferous coals and other coals as a minor constituent. The high levels of inertinite in Gondwana coals has been attributed to the peat being oxidised to a high degree during formation (Plumstead 1962). Taylor *et al* (1998) consider the characteristic petrographic composition of many Gondwana coals to be attributable to a climate of wet cool summers and freezing winters, the oxygen content of inertinite having been retained in its structure at an early stage as a result of drying or freeze drying. Similar material in warmer climates would have proceeded toward vitrinization.

Clarite comprises microlithotypes which contain >95% of vitrinite and liptinite (Figure 4.2), each being >5% of the total. Vitrite and clarite are commonly

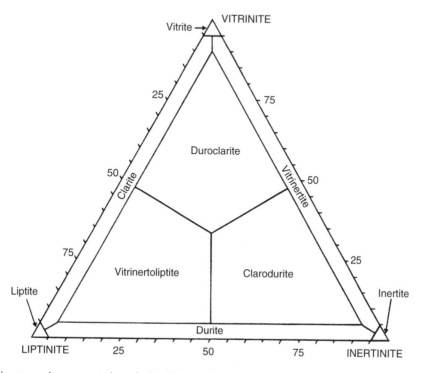

Figure 4.1 Diagrammatic representation of microlithotype classification. From Bustin *et al* (1983), 'Coal Petrology, its Principles, Methods and Applications', Geological Association of Canada. Reproduced with permission

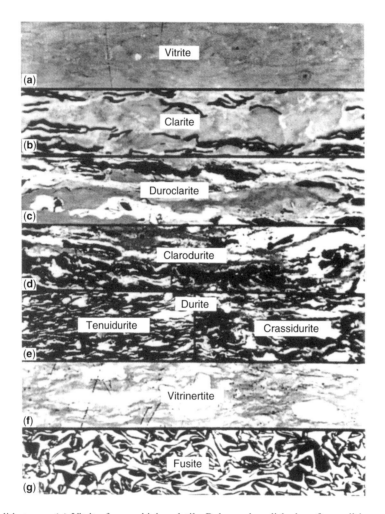

Figure 4.2 Microlithotypes: (**a**) Vitrite from a high volatile Ruhr coal, polished surface, oil imm. × 300; (**b**) Clarite from a Saar coal, polished surface, oil imm. × 300; (**c**) Duroclarite from a high volatile Ruhr coal, polished surface, oil imm. × 300; (**d**) Clarodurite from a high volatile Ruhr coal, polished surface, oil imm. × 300; (**e**) Durite from a high volatile Ruhr coal, showing both tenuidurite and crassidurite, polished surface, oil imm. × 300; (**f**) Vitrinertite from a high volatile Ruhr coal, polished surface, oil imm. × 300; (**g**) Fusite from a high volatile Ruhr coal, polished surface, oil imm. × 300. From Taylor (1998). Reproduced by permission of Gebruder Borntraeger

associated, particularly in Carboniferous coals in the Northern Hemisphere, and in Tertiary hard coals. Liptinite-rich clarites may owe their formation from algae, lipid-rich plants and animal plankton, and as such may grade into sapropelic coals.

Vitrinertite contains 95% vitrinite and inertite. It is rare in Carboniferous coals and common in inertinite-rich Gondwana coals.

Durite is composed of 95% liptinite and inertinite (Figure 4.2). There is a wide variation in the proportion of durite in different coals. Taylor *et al* (1998) suggest

that durite occurs near to the margins of coal basins, as in the case of the Upper Silesian Carboniferous Basin. Some Gondwana coals are particularly durite-rich, as found in South Africa. It is thought that the groundwater tables of the Gondwana peat mires were subject to greater fluctuation than those in the Carboniferous in the Northern Hemisphere.

Trimacerite is the only microlithotype group in which all three maceral groups are present. The trimacerite group is further divided into three microlithotypes: duroclarite, in which vitrinite is more abundant than

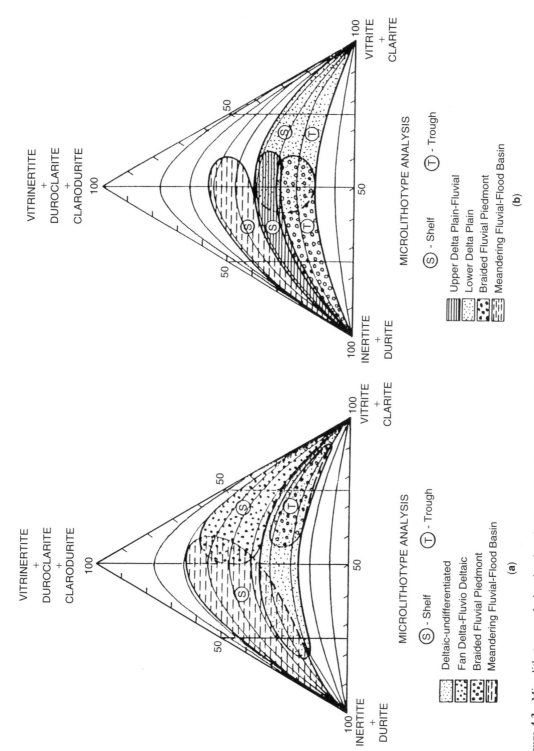

Figure 4.3 Microlithotype analysis related to depositional environment, (**a**) Early and (**b**) Late Permian Coals, Australia. From Hunt & Hobday (1984), with permission of the International Association of Sedimentologists

Table 4.9   Huminite macerals. From Bustin *et al* (1983), by permission of Geological Association of Canada

Maceral	Origin	Petrological features	Equivalents in hard coals
Textinite	Woody tissue	Primary cell wall structure still distinguishable; cell lumina mostly open	Telinite/telocollinite
Ulminite	Woody tissue	Higher degree of humification; texto-ulminite = cell wall structure still visible; eu-ulminite = no visible cell wall structure, cell lumina mostly closed	Telinite/telocollinite
Attrinite	Finely comminuted	Particle size <10 μm, product of degradation of huminite macerals	Desmocollinite
Densinite	Same as attrinite	Tighter packed than attrinite	
Gelinite	Derived from colloidal humic solutions which migrate into existing cavities and precipitate as gels	Secondary cell filling	Gelocollinite
Corpohumite	Condensation products of tannins characteristic of bark tissues	In cross-section globular to tabular shape	Corpocollinite

liptinite; clarodurite, in which the proportion of inertinite is greater than vitrinite and liptinite; and vitrinertoliptite, in which liptinite predominates. In most coals, apart from vitrite, trimacerite occurs most frequently.

In low rank coals, i.e. lignites and subbituminous coals, the *vitrinite* maceral group is referred to as *huminite*, and is regarded as equivalent to, and the precursor of, the vitrinite macerals found in higher-rank coals. The classification of huminite macerals is summarised in Table 4.9, which gives details of their origin and their equivalents in the hard coals. The increase of coal rank leads to the homogenisation of the macerals of the huminite/vitrinite group, the term collinite being used to describe homogeneous structureless vitrite.

The relationship between maceral type and the original plant material has been well documented. The plant materials that make up coal have different chemical compositions, which in turn determine the types of group macerals. Such chemical differences are clear in lower rank coals but it becomes increasingly difficult to distinguish petrographically between the various macerals with increasing coalification. This can be illustrated by an analysis of miospore floras and the petrographic types. Certain relationships have been established based on the investigation of thin layers of coal representing a moment in time during which environmental change was minimal. To illustrate this, in a thick coal seam in a stable area, the ascending miospore sequence and the

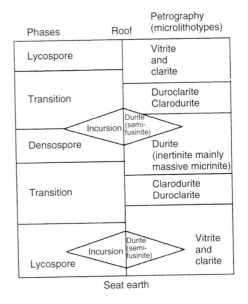

Figure 4.4   Diagrammatic profile of a coal seam showing the sequence of miospore phases and petrographic types. From Smith (1968) with permission

resultant microlithotypes are shown in Figure 4.4. If the coal seam has splits, then the sequence may revert to the early phase of seam development. Above the split the normal sequence of phases may become re-established unless the sequence is again interrupted by splitting.

A microlithotype analysis can give an indication of the texture of a coal. If two coals have equal overall contents of vitrinite and one has a higher vitrite content than the other, this may be due to different thicknesses in the bands of vitrinite, which in turn may influence the preparation of the coal. Similarly the size distribution of masses of inertinite may be important in the coking behaviour of the coal.

4.1.3 Mineral content of coals

The mineral content of coal is the noncombustible inorganic fraction; this is made up of minerals which are either detrital or authigenic in origin, and which are introduced into coal in the first or second phases of coalification. The principal mineral associations are outlined in Table 4.10.

Table 4.10 Minerals identified in coal (not exhaustive), from Taylor *et al* (1998), with permission of Gebruder Borntraeger

Mineral	Occurrence	Mineral	Occurrence
Clay minerals		Chalcopyrite	very rare
Illite–Sericite	common–abundant	Pyrrhotite	very rare
Montmorillonite	rare–common		
Kaolinite	common–abundant	**Phosphates**	
Halloysite	rare	Apatite	rare
		Phosphorite	rare
Iron disulphides		Goyazite	rare
Pyrite	rare–common	Gorceixite	rare
Marcasite	rare–common		
		Sulphates	
Carbonates		Barite	rare
Siderite	common–very common	Gypsum	very rare
Ankerite	common–very common	**Silicates** (other than clays)	
		Zircon	rare
Calcite	common–very common	Biotite	very rare
		Staurolite	very rare
Dolomite	rare–common	Tourmaline	very rare
Aragonite	rare	Garnet	very rare
Witherite	rare	Epidote	very rare
Strontianite	rare	Sanidine	rare
		Orthoclase	very rare
Oxides		Augite	very rare
Hematite	rare	Amphibole	very rare
Quartz	rare–common	Kyanite	very rare
Magnetite	very rare	Chlorite	rare
Rutile	very rare		
		Salts	
Hydroxides		Gypsum	rare
Limonite	rare–common	Bischofite	very rare–common
Goethite	rare	Sylvin (Sylvite)	very rare–common
Diaspore	rare	Halite	very rare–common
		Kieserite	very rare–common
Sulphides (other than iron)		Mirabilite	very rare–rare
Sphalerite	rare	Melanterite	very rare
Galena	rare	Keramohalite	very rare
Millerite	very rare		

Note: minerals classed as abundant to common occur in many coals in significant proportions (5–30% of mineral matter in coal). Minerals classed as rare or very rare commonly in small amounts (<5% of the total mineral matter), but also include some minerals which occur in somewhat larger amounts in only a few coals.

Detrital minerals are those transported into a swamp or bog by air or water. A large variety of minerals can be found in coal; commonly these are dominated by quartz, carbonate, iron and clay minerals with a diverse suite of accessory minerals which may be peculiar to the local source rock.

Water-borne mineral matter is transported into coal swamps along channels which cut through the accumulating organic debris. When such channels are in flood, detritus is laid down on top of the organic material; such events are usually preserved as mineral-rich partings in coals. Mineral-rich materials present in the floor of the peat swamp may be incorporated into the organic layer by differential compaction within the swamp and by bioturbative action.

Wind-borne mineral matter is important, as this can be a significant contributor to the mineral contents of coals because of the slow accumulation rates in peat swamps. Coal swamp areas located in close proximity to active volcanic regions may receive high amounts of mineral matter. Associated lithologies with coals such as flint clays and tonsteins are indicative of such volcanic mineral deposition, and, if the volcanic event was short-lived but widespread, are extremely useful as stratigraphic marker horizons in coal sequences.

Authigenic minerals are those introduced into a peat during or after deposition, or into a coal during coalification. Precipitated minerals may be disseminated through the peat or present as aggregates, whereas mineral-rich fluids present during the later stages of coalification tend to precipitate minerals on joints and any open voids within the coal. Common products of mineralisation are the calcium-iron minerals such as calcite, ankerite and siderite and pyrite, with silica in the form of quartz. The element sulphur is present in almost all coals, it is usually present in the organic fraction of the coal, but inorganic or mineral sulphur is in the form of pyrite. Pyrite may be present as a primary detrital mineral or as secondary pyrite as a result of sulphur reduction of marine waters; thus there is now considered to be a strong correlation between high sulphur coals and marine depositional environments.

Clay minerals on average make up 60–80% of the total mineral matter associated with coal. Their genesis is complex; they can have a detrital origin or be secondary formed from aqueous solutions. Chemical conditions at the site of deposition also influence the type of clay minerals associated with coal. In particular, freshwater swamps with their low pH tend to favour *in situ* alteration of smectites, illite and mixed layer clays to kaolinite. Generally illite is dominant in coals with marine roofs, whereas kaolinite is dominant

in nonmarine-influenced coals. Secondary clays are produced from alteration of primary clays, e.g. chlorite is expected to occur in coals subjected to greater pressure and temperature.

Clay minerals occur in coal in two ways: either in tonsteins or as finely dispersed inclusions in maceral lithotypes.

Tonsteins have been formed by detrital and authigenic processes, and in particular are associated with volcanic activity. They usually contain kaolinite, smectite and mixed layer clays with accessory minerals.

Clay minerals can contaminate all microlithotypes: those with less than 20% (by volume) clay minerals are described as being 'contaminated by clay', for clay mineral contents of 20–60% (by volume) the term 'carbargilite' is used; if higher proportions of clay minerals are present the lithology is no longer a coal but an argillaceous shale.

Clay minerals have the property of swelling in the presence of water. Swelling is accompanied by reduction in strength and disintegration is an end result. This is most significant in mines where coals have clay-rich roofs and floors, which can result in instability, as well as difficulties encountered in drainage and dewatering in both underground and open pit mine operations.

All of the above forms of mineral content in coals can be identified macroscopically in outcrop and borehole core. There are other minerals that may be present in coal which affect its future potential use; these cannot be seen in hand specimens and are detectable only by chemical analysis.

The mineral matter content of coals and the surrounding country rock will influence the properties of the coal roof and floor, and in particular their resistance and response to water. It will also influence the composition of mine dust with a diameter of below $5\,\mu m$, particularly in underground operations. Significant amounts of quartz in dust affects the incidence of silicosis.

The mineral matter in the coal will also affect the washability of the coal and consequently the yield and ash content of the clean coal.

Mineral impurities affect the suitability of a coal as a boiler fuel; the low ash fusion point causes deposition of ash and corrosion in the heating chamber and convection passes of the boiler. Figure 4.5 shows coal ash with low base/acid ratios (<0.25) and an excess of refractory acidic oxides (kaolinite–quartz mineral matter assemblages) that produce high ash fusion temperatures. Coals which have illite–calcite–pyrite mineral matter assemblages have proportionately more of the basic oxides (alkalis and ferric iron), so that ash fusion temperatures are correspondingly reduced. Ash fusion

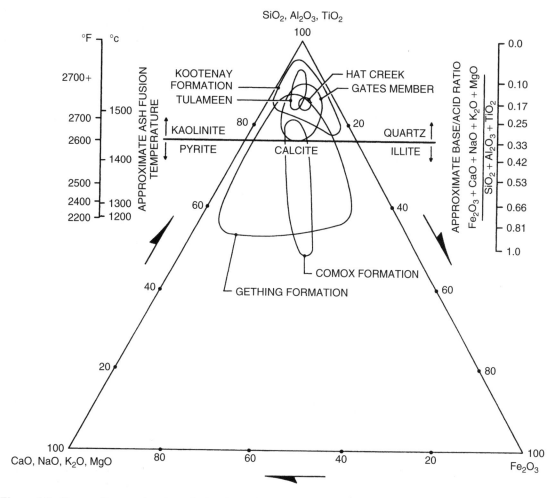

Figure 4.5 Ternary diagram showing ash chemistry of some of Western Canada's coals, together with approximate ash fusion temperatures, and base-acid ratios. Approximate boundary of pyrite–calcite–illite (marine influenced) coal ashes and kaolin–quartz (freshwater dominated) coal ashes is also shown. From Pearson (1985) with permission of the Canadian Institute of Mining, Metallurgy and Petroleum

temperatures are used to predict boiler deposit build-up and slagging performance when used as thermal coals (Pearson 1985). The presence in coal of phosphorous minerals, usually in the form of phosphorite or apatite, causes slagging in certain boilers, and steel produced from such phosphorous-rich coals tends to be brittle.

Halide minerals such as chlorides, sulphates and nitrates are present in coal, usually as infiltration products deposited from brines migrating through the sedimentary sequence. They become significant in mining operations when mine waters enriched with, for example, nitrates create serious corrosion effects on pipework and other metal installations in the mine

workings. In addition, chlorine causes severe corrosion in coal-fired boilers.

The more important trace element minerals found in coals are summarised in Table 4.11. They may originate from the original plant material or be components of other minerals in the coal. Several of them, notably boron, titanium, vanadium and zinc, can have detrimental effects in the metallurgical industry.

The mineral matter in Carboniferous and Gondwana coals is broadly similar. There are, however, differences in the total content and distribution of types of mineral matter. Gondwana coals commonly have higher contents of mineral matter, particularly as well-defined layers

Table 4.11 Contents of trace elements in coals, soils and shales (as ppm), from Taylor *et al* (1998), with permission of Gebruder Borntraeger

Element	Most coals	Soils	Shales
Antimony (Sb)	0.05–10	0.2–10	1.5
Arsenic (As)	0.5–80	1–50	13
Barium (Ba)	20–1000	100–3000	550
Beryllium (Be)	0.1–15	<5–40	3
Boron (B)	5–400	2–100	130
Cadmium (Cd)	0.1–3	0.02–10	0.22
Caesium (Cs)	0.3–5	0.3–20	5
Chlorine (Cl)	50–2000	8–1800	160
Chromium (Cr)	0.5–60	5–1000	90
Cobalt (Co)	0.5–30	1–40	19
Copper (Cu)	0.5–50	2–100	39
Fluorine (F)	20–500	20–700	800
Gallium (Ga)	1–20	5–70	23
Germanium (Ge)	0.5–50	0.1–50	2
Gold (Au)	up to 0.01	0.001–0.02	0.002
Hafnium (Hf)	0.4–5	0.5–34	2.8
Lanthanum (La)	1–40	2–180	4.9
Lead (Pb)	2–80	2–100	23
Lithium (Li)	1–80	5–200	76
Manganese (Mn)	5–300	200–3000	850
Mercury (Hg)	0.02–1	0.01–0.5	0.18
Molybdenum (Mo)	0.1–10	0.2–5	2.6
Nickel (Ni)	0.5–50	5–500	68
Niobium (Nb)	1–20	6–300	18
Phosphorous (P)	10–3000	35–5300	700
Rubidium (Rb)	2–50	20–1000	160
Scandium (Sc)	1–10	<10–25	13
Selenium (Se)	0.2–10	0.1–2	0.5
Silver (Ag)	0.02–2	0.01–8	0.07
Strontium (Sr)	15–500	50–1000	300
Tantalum (Ta)	0.1–1	0.4–6	2
Thallium (Tl)	<0.2–1	0.1–0.8	1.2
Thorium (Th)	0.5–10	1–35	12
Tin (Sn)	1–10	1–20	6
Titanium (Ti)	10–2000	1000–10 000	4600
Tungsten (W)	0.5–5	0.5–80	1.9
Uranium (U)	0.5–10	0.7–9	3.7
Vanadium (V)	2–100	20–500	130
Yttrium (Y)	2–50	10–250	41
Zinc (Zn)	5–300	10–300	120
Zirconium (Zr)	5–200	60–2000	160

of mineral-rich material. Such bands are composed of kaolinite or other clay minerals and quartz. In addition, Gondwana coals tend to have fine clay or other mineral matter dispersed throughout the organic fraction.

The inorganic components of Tertiary coals are strongly affected by the level of rank that the coal will have achieved. Groundwater leaching can lead to precipitation of minerals such as gypsum, barite and other sulphates (Taylor *et al* 1998).

4.1.4 Petrographic applications

The microscopic study of coal has enabled a better understanding of its organic and mineral components and its industrial utilisation. The principal uses of black coals on a worldwide basis are to generate electricity and to produce iron and steel. The latter still depends chiefly on coal, whilst in the electricity generation industry coal has competition from other energy sources. Coal still retains a 40% share of this market, annually generating 4800 TWh (WEC 1998).

The relationship between coal properties and coal usage has been outlined by Taylor and Shibaoka (1976), Pearson (1980, 1985), Callcott and Callcott (1990) and Taylor *et al* (1998).

Coals that are to be used for conventional coke production must have three essential properties.

1. They must be within a specific range in rank for the coking process to occur, i.e. bituminous coal.
2. They must possess a high proportion of fusible macerals (>40% vitrinite) to form a strong, well-fused coke.
3. They must have low levels of certain elements, notably sulphur and phosphorus, and be generally low in mineral matter.

Steam or thermal coals used for electricity generation are required to have a low mineral matter level with a high calorific value. Ash fusion temperatures are preferred to be high and sulphur, nitrogen and trace elements to be low. Local power stations can operate on a wide range of coals, including brown coals, whereas export steam coals are dominated by high volatile bituminous coals with mineral matter contents <15%.

The various macerals and maceral groups react differently to physical stress. Vitrinite is brittle and fractures easily, whilst liptinite–inertinite associations are more durable. Therefore when coal is crushed, a higher percentage of vitrinite will be found in the fine fraction, with inertinite concentrated in the coarser fraction.

In coke production, vitrinite is the maceral group that contributes most to the formation of coke. However, a stronger coke is obtained if the vitrinite is reinforced by inertinite. The liptinite group is characterised by high H/C ratios and therefore produces large amounts of gas on heating, all of which contributes to the fluidity and swelling properties of the coal. However, abundant liptinites are relatively resistant to thermal breakdown and remain after vitrinite has become plastic. In the inertinite maceral group, fusinite and semifusinite do not fuse during carbonisation due to their insufficient hydrogen content. These macerals are characterised by higher O/C ratios. The inertinite maceral group is thought to have little influence during coke-making, although numerous studies on the coking properties of coal suggest that some inertinite is completely fused during the coke-making process. Figure 4.6 shows the petrographic composition expressed as inerts, or the percentage of inertinite macerals plus inert semi-fusinite, plus ash composition calculated by the Parr formula (see Section 4.3.1.1) and rank (expressed as R_o max)

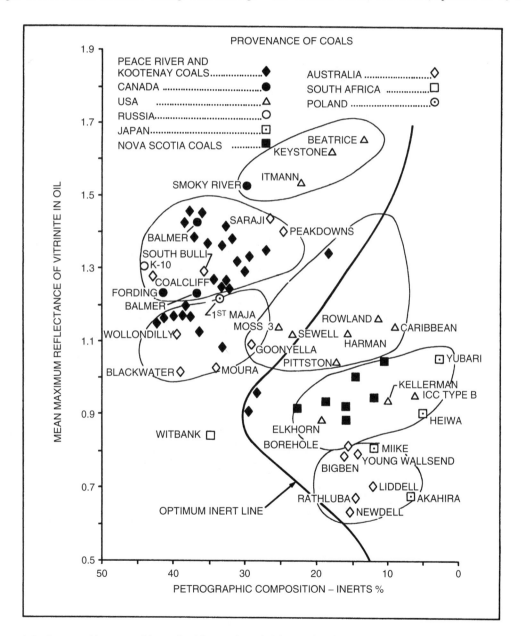

Figure 4.6 Petrographic compositions of coking coals traded internationally. From Pearson (1980) with permission of the Canadian Institute of Mining, Metallurgy and Petroleum

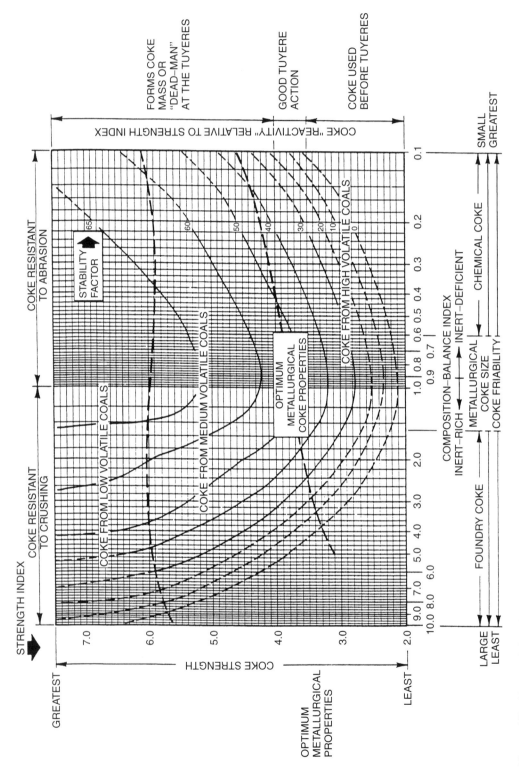

Figure 4.7   Characteristics of coke attainable from bituminous coals. From Zimmerman (1979)

of world traded coking coals. The 'optimum inert' line represents the optimum amount of inert components that would produce the strongest coke for each rank. Coal composition to the left of the line is inertinite-rich and that to the right is reactive-rich (Pearson 1980).

The application of vitrinite reflectance methods (see Section 4.2.1) to reactive macerals has shown that there is a direct relationship between the types of reactive macerals and the amount of inerts or nonreactives in making coke (Zimmerman 1979). By testing coals of different reflectance values in relation to the quantity of inerts in the coal, the relative strength (or stability) of the coal can be determined based on petrographic analyses. It has been shown that reactives in themselves will not make a good coke, but require inert material in proper proportion. The amount of inerts required will vary with the types of reactives present. This ratio of inerts to reactives is used to determine the coke strength, and a balance index is calculated from this ratio and called the Composition Balance Index (CBI). The amount of inerts present and the strength properties of each reactive type can be shown as a series of curves, where each curve peaks at the point at which the optimum amount of inerts is present in the coal. Low reflectance types have low relative strengths. The Strength Index (SI) is the comparative coke strength of the reactive macerals present in the coal, and the SI together with the CBI is required to predict the strength of the coke produced from any coal, and will indicate the ability of the coke to perform in a blast furnace.

Figure 4.7 shows the relationship of SI, CBI to coke strength and coke resistance for bituminous coals (Zimmerman 1979).

Coals used for combustion are less specific in terms of coal rank and type. It is the calorific value of the coal that is of prime interest, i.e. the percentage of combustible matter against noncombustible matter (mineral matter and water). Liptinite with high H/C ratio has the highest calorific value, followed by vitrinite and inertinite; however, vitrinite and inertinite increase in calorific value with increasing rank whereas liptinite declines.

4.2 COALIFICATION (RANK)

4.2.1 Coalification

Coalification is the alteration of vegetation to form peat, succeeded by the transformation of peat through lignite, subbituminous, bituminous, semi-anthracite to anthracite and meta-anthracite coal.

The degree of transformation or coalification is termed the coal rank, and the early identification of the rank of the coal deposit being investigated will determine the future potential and interest in the deposit.

To understand coal rank, a brief examination of the coalification process is given, particularly those conditions under which coals of different rank are produced. A detailed account of coalification and its physical and chemical processes is given by Taylor *et al* (1998). They describe the major stages of coalification from peat to meta-anthracite; these are summarised in Table 4.12. The table outlines not only the denoted rank of coal but also the dominant processes and physico-chemical changes undergone in each stage in order to produce an increase in rank.

The coalification process is essentially an initial bio-chemical phase followed by a geochemical or meta-morphic phase. The biochemical phase includes those processes that occur in the peat swamp following deposition and burial, i.e. during diagenesis. This process is considered to be in operation until the hard brown coal stage is reached. The most intense biochemical changes occur at very shallow depths in the peat swamps. This is chiefly in the form of bacteriological activity which degrades the peat and may be assisted in this by the rate of burial, pH, and levels of groundwater in the swamp. With increased burial, bacteriological activity ceases, and is considered absent at depths greater than 10 m. Carbon-rich components and volatile content of the peat are little affected during the biochemical stage of coalification; however, with increased compaction of the peat, moisture content decreases and calorific value increases.

From the brown coal stage, the alteration of the organic material is severe and can be regarded as metamorphism. Coals react to changes in temperature and pressure much more quickly than do mineral suites in rocks; coals can therefore indicate a degree of metamorphism in sequences which show no mineralogical change.

During the geochemical or metamorphic stage, the progressive changes that occur within coals are an increase in the carbon content and a decrease in the hydrogen and oxygen content, resulting in a loss of volatiles. This, together with continued water loss and compaction, results in the reduction of the coal volume. Products of such coalification are methane, carbon dioxide and water; water is quickly lost and the methane/carbon dioxide ratio increases with rank.

These changes in the physical and chemical properties of the coal are in reality the changes to the inherent coal constituents. During coalification the three maceral groups become enriched in carbon; each maceral group, i.e. exinite, inertinite and huminite (vitrinite) follows a distinct coalification path. Figure 4.8, after van Krevelen

Table 4.12   Major stages of the development from peat to meta-anthracite. From Taylor *et al* (1998), according to Levine (1993), with permission of Gebruder Borntraeger

Coalification stage	Approximate ASTM rank range	Predominant processes	Predominant physico-chemical changes
1. Peatification	Peat	Maceration, humification gelification, fermentation Concentration of resistant substances	Formation of humic substances, increase in aromaticity
2. Dehydration	Lignite to subbituminous	Dehydration, compaction loss of O-bearing groups Expulsion of −COOH, CO_2 and H_2O	Decreased moisture contents and O/C ratio, increased heating value, cleat growth
3. Bituminisation	Upper subbituminous A to high volatile A bituminous	Generation and entrapment of hydrocarbons, depolymerization of matrix, increased hydrogen bonding	Increased vitrinite R_o, increased fluorescence, increased extract yields, decrease in density and sorbate accessibility, increased strength
4. Debituminisation	Uppermost high volatile A to low volatile bituminous	Cracking, expulsion of low molecular weight hydrocarbons, especially methane	Decreased fluorescence, decreased molecular weight of extract, decreased H/C ratio, decreased strength, cleat growth
5. Graphitisation	Semi-anthracite to anthracite to meta-anthracite	Coalescence and ordering of pre-graphitic aromatic lamellae, loss of hydrogen, loss of nitrogen	Decrease in H/C ratio, stronger XRD peaks, increased sorbate accessibility, anisotrophy, strength ring condensation and cleat healing

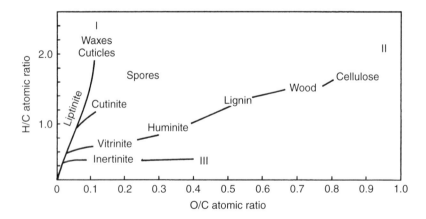

Figure 4.8   Diagram showing the coalification tracks of liptinite, inertite and huminite–vitrinite. Bustin *et al* (1983), based on van Krevelen (1961), by permission of Geological Association of Canada

(1961), illustrates the distinct coalification paths. The petrographic properties of vitrinite change uniformly with increasing rank.

In reflected light the reflectance progressively increases, whereas in transmitted light organic materials become opaque and plant structure becomes difficult to recognise. The optical properties of vitrinite have enabled it to be used as an indicator of rank. Teichmuller and Teichmuller (1982) describe the method used in detail as applied to the medium volatile bituminous to meta-anthracite and semi-graphite range of coals, i.e. coals with less than 30% volatile matter. Also reflectance is considered the best rank parameter for anthracites, and reflectance is nearly comparable to moisture content as a rank indicator in high volatile bituminous coals. It was originally suggested that this is not so for lower rank coals; however, later studies have shown the utility of reflectance in low rank lignitic coals, provided that care is taken in the selection of the component measured. Ward (1984) suggests rank classes in terms of vitrinite reflectance (Table 4.13), and Table 4.14 shows the changing pattern of coal composition with increasing coalification (Diessel 1992). This increase in vitrinite reflectance to increase in coal rank is shown in Figure 4.9(a) for New Zealand coals, these have high proportions of vitrinite and most fall within a restricted band on a volatile matter/calorific value plot. The mean reflectance values given in Figure 4.9(b) are reported to be on the high side, nevertheless the reflectance/rank relationship is a meaningful one (Suggate and Lowery 1982).

Table 4.13 Rank classes in terms of vitrinite reflectance. From Ward (1984), with permission of Blackwell Scientific Publications

Rank	Maximum reflectance (% $R_{o\,max}$)
Subbituminous	<0.47
High volatile bituminous C	0.47–0.57
High volatile bituminous B	0.57–0.71
High volatile bituminous A	0.71–1.10
Medium volatile bituminous	1.10–1.50
Low volatile bituminous	1.50–2.05
Semi-anthracite	2.05–3.00 (approx.)
Anthracite	>3.00

It should be noted that in the case of high volatile South African Gondwana coals, reflectance is a better indicator than moisture due to the presence of higher amounts of inertinite, which has a lower moisture content.

Figure 4.9(c) shows the relationship between R_o max and volatile matter (d.m.m.f. – dry, mineral-matter-free – see Section 4.3.1.1) for nonmarine Canadian Cretaceous coal, Australian nonmarine Gondwana coal, and marine-influenced Pennsylvanian coal from the USA. In general, the nonmarine Cretaceous coals possess lower volatile yields than the nonmarine Gondwana and marine-influenced Pennsylvanian coals. This variation is a consequence of compositional differences, the Cretaceous coals having a higher inertinite content (Pearson 1985).

Table 4.14 Some rank parameters showing the changing pattern of coal composition with increasing coalification. From Diessel (1992), with permission of Springer-Verlag

Rank stages	% volatile matter (d.a.f.)		% in situ moisture		% vitrinite reflectance Random (R_o)	% vitrinite reflectance R max
Wood	50	>65	11.7			
Peat	60	>60	14.7	75	0.20	0.20
Brown coal	71	52	23.0	30	0.40	0.42
Subbituminous coal	80	40	33.5	5	0.60	0.63
High volatile bituminous coal	86	31	35.6	3	0.97	1.03
Medium volatile bituminous coal	90	22	36.0	<1	1.49	1.58
Low volatile bituminous coal	91	14	36.4	1	1.85	1.97
Semi-anthracite	92	8	36.0	1	2.65	2.83
Anthracite	95	2	35.2	2	6.55	7.00
	% carbon (d.a.f.)		Gross CV (MJ/kg)			

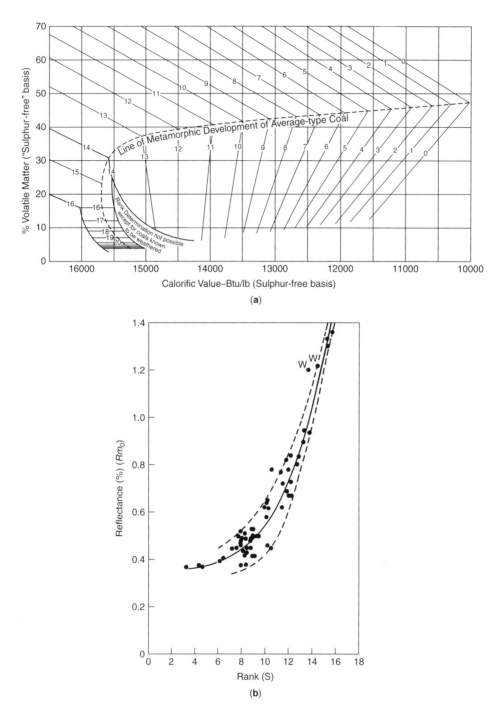

Figure 4.9 (**a**) Rank (S) scale of coal using axes of volatile matter and calorific value (Suggate 1959), with permission of the Royal Society of New Zealand; (**b**) Reflectance/rank relationship (note: w = weathered coal). From Suggate and Lowery (1982), with permission of the Royal Society of New Zealand; (**c**) Relationship between R_o max and volatile matter yield (d.m.m.f.) for nonmarine Canadian Cretaceous coals, and Permian Australian coals, and marine-influenced Pennsylvanian coals from the USA. Pearson (1985) with permission of the Canadian Institute of Mining, Metallurgy and Petroleum

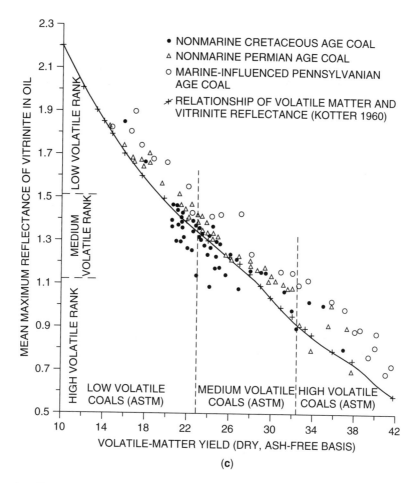

Figure 4.9 *(continued)*

Fluorescence microscopy of the liptinite macerals and the colouration of the liptinite (Thermal Alteration Index) are useful for coals of low rank, but these methods are not as refined as vitrinite reflectance.

During coalification, sapropelic coals undergo alteration similar to that of the liptinite component of humic coals. At the peat stage sapropelic coals are enriched in hydrogen relative to humic coals, but at advanced stages of coalification (90% carbon) the chemical composition of boghead, cannel and humic coals is similar. During coalification, significant amounts of bitumen may be generated from sapropelic coals.

4.2.2 Causes of coalification

The coalification process is governed primarily by rises in temperature and the time during which this occurs.

4.2.2.1 Temperature

Temperature changes can be achieved in two ways:

1. The direct contact of the coal with igneous material, either as minor intrusions or as deep-seated major intrusions. The coals exhibit loss of volatiles, oxygen, methane and water; the surrounding sediments will show evidence of contact metamorphism, e.g. the local development of high rank coal in the Gondwana coals of South Africa and India, and in the Tertiary coals of Sumatra, Indonesia.

2. The rise in temperature associated with the depth of burial. Increasing depth of burial results in a decrease in the oxygen content of the coals, and the increase in the ratio of fixed carbon/volatile matter. Professor Carl Hilt (1873) observed this phenomena and Hilt's Law states:

In a vertical sequence, at any one locality in a coalfield, the rank of the coal seams rises with increasing depth.

The rate of rank increase, known as the rank gradient, is dependent on the geothermal gradient and the heat conductivity of the rocks. Where the geothermal gradient is high (70–80 °C/km depth), bituminous rank can be attained at depths of 1500 m (Upper Rhine Graben, W. Germany), whereas in the same area, the same rank is reached at depths of 2600 m when the geothermal gradient is lower (40 °C/km) (Stach 1982). Similar basinal studies have shown variations in geothermal gradient in different parts of the basin (Teichmuller 1987; Teichmuller and Teichmuller 1982). Studies of the Remus Basin in the Canadian Arctic show differing geothermal gradients of 55 °C/km in the eastern part, and 20 °C/km in the western part. The Remus Basin contains 90 seams of coal, with ranks ranging from lignite to high volatile bituminous with a maximum paleothickness of 4500 m. In South Wales, it is suggested that the coalification that has produced anthracitisation

is due to the proximity of a magmatic heat source. The anthracite field has a present-day geothermal gradient of 25 °C/km.

Figure 4.10 illustrates the manner in which ASTM rank boundaries vary in depth from surface according to the geothermal gradient, as reflected by variations in the moisture and calorific value relationships.

4.2.2.2 Time

Usually coalification temperatures are lower than was once inferred from experimental coalification studies. Stach (1982) quotes temperatures of the order of 100–150 °C sufficient for bituminous coal formation according to geological observations. To attain higher rank, higher temperatures are required with more rapid rates of heating (contact metamorphism) rather than with slower heating rates (subsidence and depth of burial). Therefore it is apparent that the degree of coalification is less where sediments have subsided rapidly and the 'cooking time' was short, and time only has a real

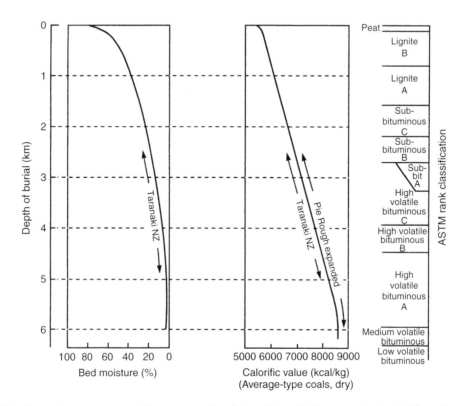

Figure 4.10   Composite sequence providing an example of the relationship between depth, calorific value and ASTM rank. The average geothermal gradient for this sequence is estimated at 26–27 °C/km. Suggate (1982), with permission of the *Journ. of Petroleum Geology*

effect when the temperature is sufficiently high to allow chemical reaction to occur. Where very low temperatures occur over a very long period, little coalification takes place; for example, the Lower Carboniferous lignites in the Moscow Basin. Therefore the influence of time is all the greater, the higher the temperature.

4.2.2.3 Pressure

The influence of pressure is at its greatest during compaction and is most evident from the peat to sub-bituminous coal stages, in the decrease of porosity and the reduction of moisture content with depth. Stach (1982) states that the pressure promotes 'physico-structural coalification', while rise of temperature accelerates 'chemical coalification'. With gradual subsidence of coal, both influences run parallel, but occasionally physico-structural coalification may precede chemical coalification, for example, where relatively low moisture coals have been produced by early folding. Chemical coalification will advance when additional heat is supplied, for example, from intrusive bodies. With increasing chemical coalification, pressure has less influence. Laboratory experiments suggest that the confining pressure may inhibit chemical coalification and retard the process; for example, the removal of gas is more difficult, and the alteration of macerals is postponed by pressure.

Local rises in rank can occur along shear planes; this is probably due to frictional heat.

4.2.2.4 Radioactivity

Increase of rank by radioactivity is rarely observed, and is likely to be only in the form of microscopic contact haloes of higher reflectivity around uranium/thorium concentrations in the coal.

4.3 COAL QUALITY

Coal quality in essence means those chemical and physical properties of a coal that influence its potential use.

It is essential to have an understanding of the chemical and physical properties of coal, especially those properties that will determine whether the coal can be used commercially. Coals need to possess particular qualities for selected usage; should they meet such requirements, then they can be mined and sold as a pure product or, if the quality could be improved, then they can be blended with other selected coals to achieve the saleable product.

The quality of a coal is determined by the make-up of the original maceral and mineral matter content of the coal, and its degree of coalification (rank). In order for this to be understood in analytical terms, procedures for determining chemical and physical properties of coals have been set up (see Karr 1978). A number of countries and organisations have defined standards of procedure which should be consulted (Appendix 1).

A knowledge of the most commonly determined properties of a coal is important, in particular those which are deleterious to the coal. Such coal analyses are essential in the evaluation of a coal deposit, i.e. in awareness of which seams or parts of seams will be unacceptable when mining commences, or conversely, those seams or parts of seams which will yield a premium product for the predetermined market. It is possible that after analysing a coal, hitherto undetected properties may enhance the product or even suggest a different end usage for the coal, e.g. the discovery that a coal has good coking properties when it was originally considered for a steam coal product.

An outline is given of the fundamental chemical and physical properties of coal and what they mean in terms of the coal's usability.

4.3.1 Chemical properties of coal

In simple terms coal can be regarded as being made up of moisture, pure coal and mineral matter.

The moisture consists of surface moisture and chemically bound moisture, the pure coal is the amount of organic matter present, and the mineral matter is the amount of inorganic material present, which when the coal is burnt produces ash. Clearly decomposition during heating of some inorganic minerals means that ash and mineral matter composition cannot be equal.

Coal analyses are often reported as proximate or ultimate analysis. Proximate analysis is a broad analysis which determines the amounts of moisture, volatile matter, fixed carbon and ash. This is the most fundamental of all coal analyses and is of great importance in the practical use of coal. The tests are highly dependent on the procedure used, and different results are obtained using different times and temperatures. It is therefore important to know the procedure used and the reported basis (see Section 4.3.1.1).

Ultimate analysis is the determination of the chemical elements in the coal, i.e. carbon, hydrogen, oxygen, nitrogen and sulphur. In addition, the calculation of the amounts of those elements which have a direct bearing on the usability of the coal is necessary. These may include forms of sulphur, chlorine, and phosphorus, an analysis of those elements making up the mineral matter content of the coal and selected trace elements.

4.3.1.1 Basis of analytical data

Before proceeding to the analysis of the coal, it is important to understand how the moisture, ash, volatile matter and fixed carbon relate to one another, and to the basis on which analytical data are presented.

It is important in evaluating previous coal analyses that the basis on which they are presented is known. It is unfortunately a common problem that analyses are given which do not indicate on what basis they are presented. Indeed, they are often listed together on different bases which are not stated.

Coal analyses may be reported as follows (see Table 4.15):

1. 'As received' basis (a.r.), also 'as sampled'. The data are expressed as percentages of the coal including the total moisture content, i.e. including both the surface and the air-dried moisture content of the coal.
2. 'Air-dried' basis (a.d.b.). The data are expressed as percentages of the air-dried coal; this includes the air-dried moisture but not the surface moisture of the coal.
3. 'Dry' basis (dry). The data are expressed as percentages of the coal after all the moisture has been removed.
4. 'Dry ash-free' basis (d.a.f.). The coal is considered to consist of volatile matter and fixed carbon on the basis of recalculation with moisture and ash removed. It should be noted that this does not allow for the volatile matter derived from minerals present in the air-dried coal. This basis is used as the easiest way to compare organic fractions of coals.
5. 'Dry, mineral-matter-free' basis (d.m.m.f.). Here it is necessary that the total amount of mineral matter rather than ash is determined, so that the volatile matter content in the mineral matter can be removed.

Table 4.16 gives the required formulas for the calculation of results to the above bases (Extract from BS 1016-100 (1994)).

In addition, the following countries have developed equations to calculate the mineral matter content of their coals.

North America:
original Parr formula

$$MM = 1.08A + 0.55S$$

modified Parr formula

$$MM = 1.13A + 0.47Spyr + Cl$$

UK:
BCURA formula

$$MM = 1.10A + 0.53S + 0.74CO_2 - 0.36$$

KMC formula (revised by British Coal)

$$MM = 1.13A + 0.5Spyr + 0.8CO_2 - 2.8SAsh$$
$$+ 2.8SSulph + 0.3Cl$$

Table 4.15 Components of coal reporting to different bases. From Ward (1984), with permission of Blackwell Scientific Publications

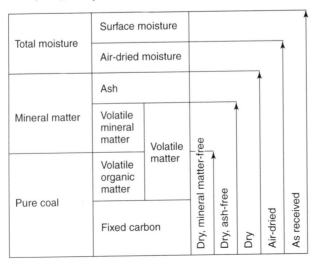

Table 4.16 Formulae for calculation of results to different bases. From BS 1016-100 (1994). Reproduced with permission of BSI under Licence Number 2002 SK/0003

Given result	Wanted result				
	As sampled (as received) (as despatched) (as fired)	Air dried	Dry	Dry, ash-free	Dry, mineral-matter-free
As sampled (as received) (as despatched) (as fired)	–	$\dfrac{100-\text{Mad}}{100-\text{Mar}}$	$\dfrac{100}{100-\text{Mar}}$	$\dfrac{100}{100-(\text{Mar}+\text{Aar})}$	$\dfrac{100}{100-(\text{Mar}+\text{MMar})}$
Air dried (as analysed)	$\dfrac{100-\text{Mar}}{100-\text{Mad}}$	–	$\dfrac{100}{100-\text{Mad}}$	$\dfrac{100}{100-(\text{Mad}+\text{Aad})}$	$\dfrac{100}{100-(\text{Mad}+\text{MMad})}$
Dry	$\dfrac{100-\text{Mar}}{100}$	$\dfrac{100-\text{Mad}}{100}$	–	$\dfrac{100}{100-\text{Ad}}$	$\dfrac{100}{100-\text{MMd}}$
Dry, ash-free	$\dfrac{100-(\text{Mar}+\text{Aar})}{100}$	$\dfrac{100-(\text{Mad}+\text{Aad})}{100}$	$\dfrac{100-\text{Ad}}{100}$	–	$\dfrac{100-\text{Ad}}{100-\text{MMd}}$
Dry, mineral-matter-free	$\dfrac{100-(\text{Mar}+\text{MMar})}{100}$	$\dfrac{100-(\text{Mad}+\text{MMad})}{100}$	$\dfrac{100-\text{MMd}}{100}$	$\dfrac{100-\text{MMd}}{100-\text{Ad}}$	–

M = moisture %; A = ash %; MM = mineral matter %; ar = as received basis; ad = air dried basis; d = dry basis.

Australia:

Standards Association of Australia formula

$$\text{MM} = 1.1\text{A}$$

where MM = mineral matter %, A = ash %, S = total sulphur %, Spyr = pyritic sulphur %, SSulph = sulphate sulphur %, SAsh = sulphur in ash %, Cl = chlorine %, CO_2 = carbon dioxide %.

All values are expressed on an air-dried basis.

4.3.1.2 Proximate analysis

Moisture

The terminology used in describing the moisture content of coals can be confusing and needs to be clarified. The most confusing term is inherent moisture, which has many different definitions and should be avoided if at all possible. If used in any tests it is necessary to ascertain the exact definition that the reference is using.

There is no exact method of determining moisture content. The coal industry has therefore developed a set of empirically determined definitions as follows:

1. Surface moisture. This is adventitious moisture, not naturally occurring with the coal and which can be removed by low temperature air drying (c. 40 °C). This drying step is usually the first in any analysis and the moisture remaining after this step is known as air-dried moisture.

2. As received or as delivered moisture. This is the total moisture of the coal sample when received or delivered to the laboratory. Usually a laboratory will air-dry a coal sample, thereby obtaining the 'loss on air drying'. An aggressive drying step is then carried out which determines the air-dried moisture. These results are added together to give the total/as received/as delivered moisture.

3. Total moisture. This is all the moisture which can be removed by aggressive drying (c. 150 °C in vacuum or nitrogen atmosphere).

4. Air-dried moisture. This is the moisture remaining after air drying and which can be removed by aggressive drying. In addition to this generally used term, the following terms are being increasingly used: moisture holding capacity (MHC); capacity moisture or equilibrium moisture (EQ). It is not within the scope of this book to detail the analytical procedure required but suffice to say that it is lengthy and expensive.

These terms relate to the in-bed or *in situ* moisture of a coal. Numerically the MHC of a bituminous coal will be higher than air-dried moisture and lower than total moisture. Technically it is the MHC which increases with decreasing rank (Figure 4.11).

High moisture is undesirable in coals as it is chemically inert and absorbs heat during combustion, and it creates difficulties in handling and transport. It lowers

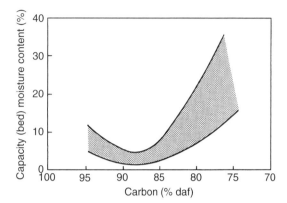

Figure 4.11 Generalised variation of capacity (or air-dried) moisture contents with rank. From Berkowitz (1979), with permission of Academic Press

the calorific value in steam coals and lowers the amount of carbon available in coking coals.

Ash

The ash of a coal is that inorganic residue that remains after combustion. It should be remembered that the determined ash content is not equivalent to the mineral matter content of the coal. It does, however, represent the bulk of the mineral matter in the coal after losing the volatile components such as CO_2, SO_2 and H_2O, which have been driven off from mineral compounds such as carbonates, sulphides and clays.

In a steam coal, a high ash content will effectively reduce its calorific value. Recommended maximum ash contents for steam coals for use as pulverised fuel are around 20% (air-dried), but for some stoker-fired boilers, much lower values are desirable. In coking coals, a maximum of 10–20% (air-dried) is recommended, as higher ash contents reduce the efficiency in the blast furnace.

Volatile matter

Volatile matter represents that component of the coal, except for moisture, that is liberated at high temperature in the absence of air. This material is derived chiefly from the organic fraction of the coal, but minor amounts may also be from the mineral matter present. Correction for the volatile matter derived from the latter may be made in technical works, but is not usually necessary in commercial practice.

In pulverised fuel firing for electricity generation, most boilers are designed for a minimum volatile matter of 20–25% (d.a.f.). In stoker firing for electricity generation, the volatile matter limits recommended are 25–40% (d.a.f.). There is virtually no limit for the volatile matter for coals used in the production of cement. In coke production, high volatile matter content will give a lower coke yield so that the best quality coking coals have a volatile matter range of 20–35% (air-dried) but values of 16–36% can be used.

Fixed carbon

The fixed carbon content of coal is that carbon found in the residue remaining after the volatile matter has been liberated. Fixed carbon is not determined directly, but is the difference, in an air-dried coal, between the total percentages of the other components i.e. moisture, ash and volatile matter, and 100%.

4.3.1.3 Ultimate analysis

Ultimate analysis of coal consists of the determination of carbon and hydrogen as gaseous products of its complete combustion, the determination of sulphur, nitrogen and ash in the material as a whole, and the estimation of oxygen by difference.

Carbon and hydrogen

These are liberated as CO_2 and H_2O when the coal is burned and are most easily determined together. However, CO_2 may be liberated from any carbonate minerals present, and H_2O may be derived from clay minerals or from any inherent moisture in the air-dried coal, or both. Allowances have to be made for these inorganic sources of carbon and hydrogen.

Nitrogen

The nitrogen content of coal is significant particularly in relation to atmospheric pollution. Upon combustion of the coal, nitrogen helps to form oxides which may be released as flue gases and thereby pollute the atmosphere; as a result, coals which are low in nitrogen are preferred in industry.

Coals should not as a rule have nitrogen contents of more than 1.5–2.0% (d.a.f.) because of these NO_X emissions.

Sulphur

As in the case of nitrogen, the sulphur content of coals presents problems with utilisation and resultant pollution. Sulphur causes corrosion and fouling of boiler tubes, and atmospheric pollution when released in flue gases.

Sulphur can be present in coal in three forms:

1. Organic sulphur, present in the organic compounds of the coal.
2. Pyritic sulphur, present as sulphide minerals in the coal, principally iron pyrite.
3. Sulphate minerals, usually hydrous iron or calcium sulphates, produced by oxidation of the sulphide fraction of the coal.

In the ultimate analysis of the coal, only the total sulphur content is determined; however, in many instances, the relative amount of sulphur in each form is required. This is carried out as a separate analysis.

The total sulphur content in steam coals used for electricity generation should not exceed 0.8–1.0% (air-dried); the maximum value will depend upon local emission regulations. In the cement industry, a total sulphur content of up to 2.0% (air-dried) is acceptable, but a maximum of 0.8% (air-dried) is required in coking coals, because higher values affect the quality of steel.

Oxygen

Oxygen is a component of many of the organic and inorganic compounds in coal as well as the moisture content. When the coal is oxidised, oxygen may be present in oxides, hydroxides and sulphate minerals, as well as oxidised organic material.

It should be remembered that oxygen is an important indicator of rank in coal.

Oxygen is traditionally determined by subtracting the amount of the other elements, carbon, hydrogen, nitrogen and sulphur from 100%.

4.3.1.4 Other analysis

Forms of sulphur

The proportions of organic, inorganic and sulphate forms of sulphur are important when considering the commercial usefulness of a coal. Coal preparation can reduce the inorganic (pyritic) and sulphate fractions, but will not reduce the organic sulphur content. Therefore if a coal has a high sulphur content, it is essential to know if this can be reduced by coal preparation methods; if not, then it may mean that the coal is unusable, or at best used in a blend with a low sulphur product. Also, pyritic sulphur can be linked to liability to spontaneous combustion.

Carbon dioxide

Carbon dioxide in coal occurs in the carbonate mineral matter fraction. The carbonates liberate CO_2 on combustion, and contribute to the total carbon content of the coal; this reaction, however, reduces the amount of energy available from the coal.

Chlorine

The chlorine content of coal is low, usually occurring as the inorganic salts of sodium, potassium and calcium chloride. The presence of relatively high amounts of chlorine in coal is detrimental to its use. In boilers chlorine causes corrosion and fouling, and when present in flue gas, it contributes to atmospheric pollution.

Steam coals should have a maximum chlorine content of 0.2–0.3% (air-dried), and for coals used in the production of cement, a maximum of 0.1% (air-dried) is recommended.

Phosphorus

Phosphorus may be present in coal, usually concentrated in the mineral apatite. It is undesirable for large amounts of phosphorus to be present in coking coals to be used in the metallurgical industry, as it contributes to producing brittle steel. It is also undesirable in stoker firing coal as it causes fouling in the boiler.

Coking coals should have a maximum phosphorus content of 0.1% (air-dried).

Ash analysis

The ash in coal represents the residue of the combusted mineral matter, and it can be broken down and expressed as the series of metal oxides which make up the lithosphere. These are SiO_2, Al_2O_3, TiO_2, CaO, MgO, K_2O, Na_2O, P_2O_5, Fe_2O_3, and SO_3. These data are important in determining how a coal will behave, such as steam coal in boilers where slagging and fouling can result, because the presence of large amounts of the oxides of iron, calcium, sodium and/or potassium can result in ashes with low ash fusion temperatures. In coking coals, sodium and potassium oxide content should be a maximum of 3% in ash, as high alkalis cause high coke reactivity.

Trace element analysis

Coals contain diverse amounts of trace elements in their overall composition. Those predominantly associated with the organic fraction are boron, beryllium and germanium; those predominantly associated with the inorganic fraction include arsenic, cadmium, mercury, manganese, molybdenum, lead, zinc and zirconium. Other trace elements have varying associations with the organic and inorganic fractions: those usually associated with the organic fraction are gallium, phosphorus, antimony, titanium and vanadium; those with the inorganic fraction are cobalt, chromium, nickel and selenium.

Boron can be a useful index in indicating the paleosalinity of the coal's depositional conditions.

Certain trace elements such as lead, arsenic, cadmium, chromium and mercury, if present in high amounts, could preclude the coal from being used in environmentally sensitive situations. Others have detrimental effects on the metallurgical industry; these include boron, titanium, vanadium and zinc.

As a result of the high tonnages of coal used in industry, significant amounts of trace elements may be concentrated in residues after combustion. Therefore trace element determinations are carried out before the coal is accepted for industrial usage.

4.3.2 Combustion properties of coal

The determination of the effects of combustion on coal will influence the selection of coals for particular industrial uses.

Tests are carried out to determine a coal's performance in a furnace, i.e. its calorific value and its ash fusion temperatures. In addition the caking and coking properties of coals need to be determined if the coal is intended for use in the metallurgical industry.

These parameters are particularly significant as they form the basis for the classification of coals (see Section 4.4).

4.3.2.1 Calorific value

The calorific value (CV) of a coal is the amount of heat per unit mass of coal when combusted. Calorific value is often referred to as specific energy (SE), particularly in Australia.

The CV of a coal is expressed in two ways:

1. The gross calorific or higher heating value. This is the amount of heat liberated during testing in a laboratory, when a coal is combusted under standardised conditions at constant volume, so that all of the water in products remains in liquid form.
2. The net calorific or lower heating value. During actual combustion in furnaces, the gross calorific value is never achieved because some products,

especially water, are lost with their associated latent heat of vapourisation. The maximum achievable calorific value under these conditions is the net calorific value at constant pressure. This can be calculated and expressed in absolute joules, calories per gram, or Btu per pound. The simplified equations for these are as follows:

In MJ/kg: Net CV = Gross CV
$$- 0.212H - 0.024M$$

In kcal/kg: Net CV = Gross CV
$$- 50.7H - 5.83M$$

In Btu/lb: Net CV = Gross CV
$$- 91.2H - 10.5M$$

where H = Hydrogen % and M = Moisture %.

As an approximate value, in bituminous coals gross as received calorific value can be converted to net as received calorific value by subtracting the following values: 1.09 MJ/kg; 260 kcal/kg; 470 Btu/lb.

It should be noted that in practice, the USA uses Btu/lb, the UK has used Btu/lb, although the British coal industry uses gigajoule per tonne (this is not used elsewhere). South Africa and Australia use megajoule per kilogram, whilst the rest of the world usually use kilocalories per kilogram. A conversion chart is given in Appendix 3.

4.3.2.2 Ash fusion temperatures

How the coal's ash residue reacts at high temperatures can be critical in selecting coals for combustion, i.e. how it will behave in a furnace or boiler.

A laboratory-prepared and moulded-ash sample (either in the shape of a cone, cube or cylinder) is heated in a mildly reducing or oxidising atmosphere, usually to about 1000–1600 °C.

Four critical temperature points are recognised:

1. Initial deformation temperature (IT), the temperature at which the first rounding of the apex or corners of the sample occurs.
2. Softening (sphere) temperature (ST), the temperature at which the moulded sample has fused down to a lump, the width of which equals its height.
3. Hemisphere temperature (HT), the temperature at which the mould sample has fused down to a lump the height of which is half of its width.
4. Fluid temperature (FT), the temperature at which the mould has collapsed as a flattened layer.

Temperatures recorded under a reducing atmosphere are lower than or equal to those recorded under oxidising atmosphere. The IT and FT temperatures are the most difficult to reproduce.

The behaviour of ash at high temperatures is a direct response to its chemical composition. Oxides of iron, calcium and potassium act as fluxes and reduce the temperature at which fusion occurs; high aluminium is the most refractory. In stoker boilers a minimum IT of 1200 °C is recommended as lower values lead to excessive clinker formation. In Pf (pulverised fuel) combustion, in dry bottom boilers a minimum IT of 1200 °C and in wet bottom boilers a maximum of 1300 °C are recommended.

4.3.2.3 Caking tests

Free swelling index

The free swelling index (FSI) in BSI nomenclature (the crucible swelling number (CSN) in ISO nomenclature) is a measure of increase in the volume of coal when heated, with the exclusion of air. The test is useful in evaluating coals for coking and combustion.

The coal sample is heated for a specific time. When all the volatiles have been liberated, a small coke 'button' remains. The cross-section of the button is then compared to a series of standard profiles (Figure 4.12).

Coals with a low swelling index (0–2) are not suitable for coke manufacture. Coals with high swelling numbers (8+) cannot be used by themselves to produce coke, as the resultant coke is usually weak and will not support the loads imposed within the blast furnace. However, they are often blended to produce strong coke.

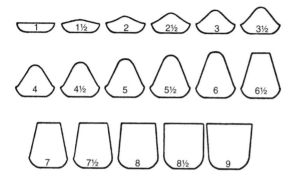

Figure 4.12 Characteristic profiles of coke buttons for different values of the crucible swelling number (free swelling index). From BS 1016-107.1 1991. Reproduced with permission of BSI under Licence Number 2002SK/0003

Roga index test

The Roga index test again indicates the caking properties of coals. A sample of coal is combined with a standard measure of anthracite and then heated. The resultant button is then tested for mechanical strength rather than the change in dimensions, by being rotated in a drum for a specific time.

There is a correlation between Roga index values and free swelling index values. For example, the Roga index values 0–5 are equivalent to the free swelling index values 0–1/2, values 5–20 = 1–2, values 20–45 = 21/2–4 and values >45 = >4.

4.3.2.4 Coking tests

Gray–King coke type

Finely crushed coal is heated slowly in a sealed tube, and the appearance and texture of the coke residue is compared to standards and assigned a letter, the Gray–King coke type.

Values range from A, no coking properties at all, to G, where the coal has retained its volume and forms a well-fused product. If it swells beyond its volume, it is said to have superior coking properties and is further tested and designated coke type G1–8. Table 4.17 outlines the characteristics of the Gray–King coke types.

Gray–King coke types approximate to free swelling indexes as follows: Gray–King coke type A–B is equivalent to the free swelling index value 0–1/2, C–G2 = 1–4, F–G4 = 41/2–6, G3–G9 = 61/2–8, and G7 or above = 81/2–9.

Fischer assay

This test is most widely used for testing low rank coals to low temperature carbonisation. The percentages of coke, tar and water driven off by the dry coal are determined; gas is calculated by subtraction.

Gieseler plastometer

To form coke, coal passes from a solid form through a fluid or plastic state to become a fused porous solid. The temperature range of the fluid phase and the viscosity of the fluid are important features when blending coals for coke manufacture. These parameters are measured by the Gieseler plastometer, in which a coal sample is pressed around a spindle under torque. As the coal reaches its fluid state, the spindle begins to revolve; the rate at which it turns is measured in 'dial divisions

Table 4.17 Characteristics for classification of Gray–King coke type. From BS 1016-107.2 1991. Reproduced with permission of BSI under Licence Number 2002 SK/0003

A, B & C Retains initial cross section Examine for strength			D, E & F Shrunken Examine for strength		
Noncoherent	Badly coherent	Coherent	Moderately hard and shrunken	Hard and very shrunken	Hard, strong and shrunken
Usually in powder form but may contain some pieces which, however, cannot be handled without breaking	In several pieces and some loose powder. Pieces can be picked up but break into powder on handling	Usually in one piece but easily broken; may be in two or three pieces with practically no loose powder; very friable and dull	May be fissured but can be scratched with fingernail and stains the fingers on rubbing the curved surface vigorously; usually dull and black & appearing fritted rather than fused	Usually very fissured, moderate metallic ring when tapped on a hard wooden surface, does not stain the fingers on rubbing; grey or black with slight lustre	May be fissured; moderate metallic ring when tapped on a hard wooden surface; does not stain the fingers on rubbing. Cross section well fused and greyish.
A	B	C	D	E	F

G Retains initial volume Examine for strength			G1 to Gx Swollen Examine for degree of swelling		
Hard and strong			Slightly swollen	Moderately swollen	Highly swollen
Well fused with a good metallic ring when tapped on a hard wooden surface					G3 & higher. Guided by swelling number, blend with minimum number of parts of electrode carbon to give a standard G-type coke
G			G1	G2	Gx

per minute' (d.d.m.), which are then plotted against temperature.

Coals with high and low fluidity may be blended to obtain improved coking properties.

Audibert–Arnu dilatometer

Coals shrink during carbonisation; such volume changes that accompany the heating of a coking coal are measured with a dilatometer. Several have been developed for this purpose, the most widely used are the Audibert–Arnu and Ruhr dilatometers.

Dimensional changes in a coal can be measured as functions of time. While the temperature of the coal is being raised at a constant rate, curves record the length of a coal sample to define the extent of contraction and dilatation, and the temperatures at which these changes begin or end.

These properties are significant in determining the volume of coal that can be fed into a coke oven, and also in blending different coals for coke production.

The resultant coke is itself subjected to rigorous testing to confirm its strength and quality for use in commercial operations.

4.3.3 Physical properties of coal

In addition to the chemical and combustion properties of a coal, its evaluation for commercial usage requires the determination of several physical properties. These are the coal's density, hardness, grindability, abrasiveness, size distribution and float–sink tests.

4.3.3.1 Density

The density of a coal will depend on its rank and mineral matter content. It is an essential factor in converting coal units of volume into units of mass for coal reserves calculation.

Density is determined by the loss of weight incurred when immersed in water. The testing of field samples and core samples in this way gives 'apparent density', because air remains trapped within the coal. True density is determined by crushing the coal and using a standard density bottle or pycnometer. The ease with which apparent densities can be determined in the field is an important facility available to the geologist when describing coal types whose mineral matter contents may fluctuate up to levels where the coal could become uneconomic on quality grounds.

It should be noted that density is not synonymous with specific gravity or relative density; the former is defined as the weight per unit volume given as grams per cubic centimetre, whereas specific gravity or relative density is its density with reference to water at 4 °C.

4.3.3.2 Hardness and grindability

In modern commercial operations, coals are required to be crushed to a fine powder (pulverised) before being fed into a boiler. The relative ease with which a coal can be pulverised depends on the strength of the coal and is measured by the Hardgrove grindability index (HGI). This is an index of how easily a coal can be pulverised in comparison with coals chosen as standards.

Coals with a high HGI are relatively soft and easy to grind. Those coals with low HGI values (less than 50) are hard and difficult to grind into a pulverised product.

HGI varies with coal rank, as shown in Figure 4.13.

4.3.3.3 Abrasion index

Coarse mineral matter in coal, particularly quartz, can cause serious abrasion of machinery used to pulverise coal. Coal samples are tested in a mill equipped with four metal blades. The loss in mass of these blades determines the 'abrasion index', and is expressed as milligrams of metal per kilogram of coal used.

4.3.3.4 Particle size distribution

Size distribution in a coal depends on the mining and handling it undergoes, together with its hardness, strength and its inherent degree of fracturing.

The size of coal particles affects coal preparation plant design, which in turn is related to the sized product to be sold. Tests are based on sieve analysis as for other geological materials, and the results expressed in various size-distribution parameters, such as mean particle size and cumulative size percentages.

4.3.3.5 Float–sink tests

The particles in coal are of different relative densities. The densities represent the varying amounts of mineral

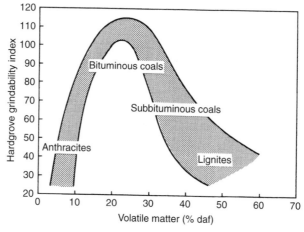

Figure 4.13 Generalised variation of the Hardgrove grindability index with rank. From Berkowitz (1979) with permission of Academic Press

matter present. Consequently the coal preparation process is designed to remove these, so that the ash level of the coal is reduced, and the product to be used or sold, improved.

Coal particles are separated into density fractions by immersion in a series of liquids of known relative density, usually ranging from 1.30 to 2.00. Commencing with the lowest relative density, the sinking fraction is transferred to the next liquid in the series and so on. An example of a float–sink analysis is shown in Table 4.18. Using the results given in Table 4.18, these may be plotted graphically as a series of 'washability curves'. These are used to calculate the amount of coal which can be obtained at a particular quality, the density required to effect such a separation and the quality of the discard left behind.

The curves shown in Figure 4.14 are the classical washability curves, i.e. the cumulative floats curve which plots column I values against column G, the densimetric curve which plots column G against column C, the cumulative sinks curve which plots column L against column J, and the elementary ash curve which plots column G against column E (see Table 4.18).

Quantitatively, an examination of the cumulative floats curve will give yield values for a given quality, and the densimetric curve will indicate the density at which to wash (i.e. washing density) in order to obtain that yield and quality. This can also be calculated in reverse.

The curves can also be used on a more qualitative basis, e.g. if the density value that is required is on the

steep part of the densimetric curve, then it will be more difficult to maintain a consistent quality.

The significance of this is that the amounts of coal and mineral matter or discard can be determined for a specific relative density, so enabling a product of specified ash content to be produced using liquids of known relative density.

For example, in Figure 4.14, a coal with an ash content of 5% will give a yield of 68.6%, and a density of 1.47 will be needed to achieve this. The ash of the sinks (reject) will be 76% and the percentage of those ash particles in the floats will be 16.6%. The latter figure is useful to the coal preparation engineer for coal-blending calculations.

Sometimes the coal is cleaned to produce two products, a prime product and a lower quality product (the so-called 'middlings'), plus a discard. Classical washability curves cannot be used to calculate yield or quality of middlings; in order to determine these values, an M curve is used (M = Mayer, middlings or mean value curve), as shown in Figure 4.15. The M curve is produced by plotting column G against column I (see Table 4.18). The angle of lines drawn from point A to intersect the abscissa represents the ash value, and the value of the ordinate represents the yield.

For example, in Figure 4.15, to calculate the yield of a prime product of 4.5% ash, a line is drawn from A to intersect the ash axis at 4.5% (F); where this line crosses the M curve at B, a yield of 65.17% can be read off from the yield axis. To calculate the yield of a 25% ash middling, a line is drawn from A to intersect the ash

Table 4.18   Washability data. Reproduced by courtesy of S.C. Frankland

Relative density			Fractional			Cumulative floats			Cumulative sinks		
A	B	C	D	E	F	G	H	I	J	K	L
Sink RD	Float RD	Midpoint RD	Weight %	Ash %	Ash points	Weight %	Ash points	Ash %	Weight %	Ash points	Ash %
		$\frac{A+B}{2}$			$\frac{D \times E}{100}$	$\sum D \downarrow$	$\sum F \downarrow$	$\frac{H}{G} \times 100$	$\sum D \uparrow$	$\sum F \uparrow$	$\frac{K}{J} \times 100$
–	1.30	–	43.31	3.10	1.34	43.31	1.34	3.10	100.00	27.66	27.66
1.30	1.35	1.325	18.47	6.61	1.22	61.78	2.56	4.15	56.69	26.31	46.42
1.35	1.40	1.375	4.91	11.11	0.55	66.69	3.11	4.66	38.22	25.09	65.66
1.40	1.45	1.425	1.41	15.26	0.22	68.10	3.32	4.88	33.31	24.55	73.70
1.45	1.50	1.475	1.45	19.04	0.28	69.55	3.60	5.18	31.90	24.33	76.28
1.50	1.55	1.525	1.04	21.69	0.23	70.59	3.83	5.42	30.45	24.06	79.00
1.55	1.60	1.575	0.77	28.08	0.22	71.36	4.04	5.66	29.41	23.83	81.03
1.60	1.70	1.650	1.07	34.70	0.37	72.43	4.41	6.09	28.64	23.62	82.46
1.70	1.80	1.750	0.68	45.85	0.31	73.11	4.73	6.46	27.57	23.24	84.31
1.80	2.00	1.900	1.02	55.38	0.56	74.13	5.29	7.14	26.89	22.93	85.28
2.00	–	–	25.87	86.46	22.37	100.00	27.66	27.66	25.87	22.37	86.46

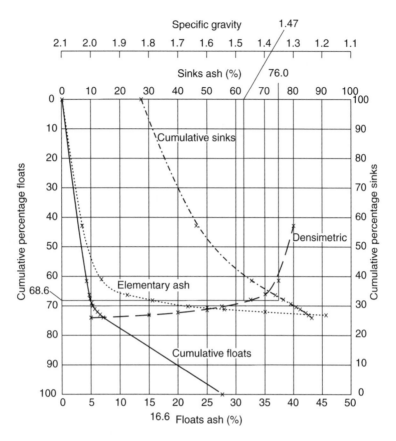

Figure 4.14   Washability curves based on data given in Table 4.18. Reproduced by courtesy of S.C. Frankland

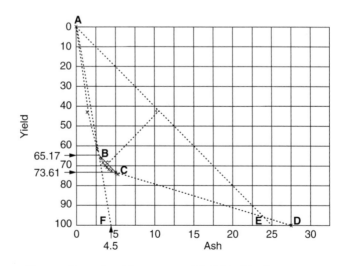

Figure 4.15   Middlings or M curve. Reproduced by courtesy of S.C. Frankland

axis at 25% (E). A line is drawn from B parallel to the 25% ash line A–E to intersect the M curve at C. This gives a total yield of 4.5% ash prime product and 25% ash middlings of 73.61%. Therefore, the yield of 25% ash middlings is 73.61% − 65.17% = 8.44%.

The densimetric curve may also be drawn on the M curve and used in an identical way to the classical washability curves. Intersects of the densimetric curve with all lines drawn from the yield axis give densities of separation.

4.3.4 Coal oxidation

Exposure of coals to weathering in the atmosphere, or by oxygenated groundwaters, results in the oxidation of the organic and inorganic constituents of the coal. Oxidation reduces the coal quality by altering the chemical and physical properties of coal. In particular, the calorific value is lowered, and caking is eliminated. There is also a loss of floatability during washing of the coal.

The weathering of coal results in its physical breakdown to fine particles, which enhances hydration and hydrolysis. If the coal is structurally fractured, the extent of oxidation will be greater.

The degree of oxidation is determined by the maceral and mineral matter content. Vitrinite is considered by some to be the most readily oxidised maceral; however, Gondwana coals high in inertinite have a high propensity for spontaneous combustion, which would indicate a rapid oxidation of the inertinite. In addition, pyrite and other sulphides readily oxidise to sulphates.

All ranks of coal are affected by oxidation, and the degree to which this may occur is influenced by the coal rank, pyrite content, climate, hydrology, and by the surface area within the coal accessible to oxidation.

It is extremely important to establish how much of a coal deposit has been oxidised. The oxidised coal may well be excluded from the tonnage produced.

One direct side-effect of oxidation is that of spontaneous combustion. This occurs when the rate of heat generation by oxidation exceeds the rate of heat dissipation. All coals have the propensity to heat spontaneously, but lower rank coals have a greater tendency to self-heat. When the temperature of the coal is raised, the rate of oxidation is also increased; it is suggested that the oxidation rate doubles for every 10 °C rise in temperature at least up to 100 °C. It has also been demonstrated that low rank coal produces heat when wetted, and that if dispersed pyrite is present, reactivity is increased tenfold.

Where coals possessing some or all of these properties are stockpiled or loaded into vessels, tests and monitoring are rigorously carried out. Procedures carried out to

lessen heating effects include compaction of the coal, which reduces the oxidation rate, and protection of the coal from heat sources such as solar radiation.

Spontaneous combustion is also a hazard to underground mining. Oxidation of *in situ* coals and coal dust particles presents a potential danger. The following factors contribute to the possibility of combustion: if the coal is thicker than its mined section, steep dips, faulting and coal outbursts. Where workings are deep, the natural strata temperature is higher and therefore so will be the base temperature of the *in situ* coal. Care in mine design and careful monitoring is needed in these circumstances to minimise heating effects. Potential fires or explosions are costly in terms of labour, materials and time, with a corresponding loss in production. This is particularly true if an area of mine has to be abandoned and sealed off through spontaneous combustion, so losing the potential reserves of coal in that area.

4.4 CLASSIFICATION OF COALS

Coals have usually been classified according to the coal's chemical properties in relation to their industrial usage.

Several classifications are in common usage, which classify both humic and brown coals, and refer to particular parameters; these range from the percentage of fixed carbon and volatile matter (on a dry mineral-matter-free basis), calorific value (on a moist, mineral-matter-free basis), the caking properties of coal (free swelling index and Roga index), and the coking properties of coal (dilatometer and Gray–King tests).

Coals have been classified either for 'scientific' purposes or for coal use. The scientific classifications use carbon/oxygen or carbon/hydrogen correlations; of these, the best known is that of Seyler (Figure 4.16). This classification is applicable only to British Carboniferous coals, however, and takes little account of lower rank coals. It uses the terms 'perhydrous' for hydrogen-rich material and 'subhydrous' for hydrogen-poor samples; the terms for each rank are given in Table 4.19.

The principal commercial classifications of coal in current use are given below.

The ASTM (American Society for Testing and Materials) classification (D 388–99) is used on a worldwide basis. This is based on two coal properties: the fixed carbon values and the calorific values (on d.m.m.f. basis). The higher rank coals are classified according to fixed carbon on the dry basis; the lower rank coals are classified according to gross calorific value on the moist basis. A correlation of the rank's property and volatile

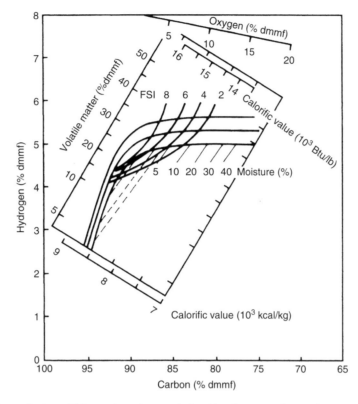

Figure 4.16 Seyler's coal chart. This version shows relationships between elemental composition, volatile matter contents, moisture contents, and caking properties. From Berkowitz (1979), based on Seyler (1899, 1931) with permission of Academic Press

matter with the mean maximum reflectance group of macerals (D 2798–99) is used as supplemental information. Further classification is given for those coals with agglomerating or coking properties; see Table 4.20. The classification is applicable to coals composed mainly of vitrinite, so that coals rich in inertinite or liptinite (exinite) cannot be properly classified because the properties that determine rank differ greatly between these maceral groups. In North America, such coals are mostly non-banded varieties containing only a small proportion of vitrinite and consist mainly of attrital materials.

The British classification system was devised by British Coal before privatisation and is shown in Table 4.21. It uses a three-figure numeral code to classify bituminous and anthracite coals. The first two digits are based on the amount of volatile matter in the coal (on d.m.m.f. basis) and the third digit is based on the Gray–King assay value. Coals with less than 19.6% volatile matter (d.m.m.f.) are classified by this property alone. It should be noted that coals with ash contents greater than 10% must be cleaned prior to analysis.

Coals that have been thermally altered by igneous intrusions have the suffix H added to the coal code, and coals that have been oxidised by weathering may be distinguished by adding the suffix W to the coal code.

In Europe, in 1988, the Codification System for medium and high rank coals was published by the United Nations Economic Commission for Europe (UNECE) and approved by the International Organisation for Standardisation (ISO). The Codification System uses a series of numbers to illustrate the chemical and physical characteristics which determine the usage of the coal. It does not include low rank coals, which are defined as coals with a gross calorific value <24.0 MJ/kg (5700 kcal/kg) and a mean random reflectance (R_r) <0.6% (Figure 4.17).

The Codification System is applicable to both run-of-mine and washed coals. The coals are characterised by using a 14-digit code number based on eight property-related parameters providing information concerning rank, type and grade of a coal (Table 4.22). In addition, a list of 'Supplementary Parameters' is

Table 4.19 Parameters used in Seyler's coal classification. From Ward (1984), with permission of Blackwell Scientific Publications

Genus	Hydrogen (%)	Class (% carbon)						
		Anthracite (>93.3)	Carbonaceous (93.3–91.2)	Bituminous			Lignitous	
				Meta- (91.2–89.0)	Ortho- (89.0–87.0)	Para- (87.0–84.0)	Meta- (84–85)	Ortho- (80–75)
Per-bituminous	>5.8	—	—	Per-bituminous (per-meta-bituminous)	Per-bituminous (per-ortho-bituminous)	Per-bituminous (per-para-bituminous)	Per-lignitous	
Bituminous	5–5.8	—	Pseudo-bituminous species	Meta-bituminous	Ortho-bituminous	Para-bituminous	Lignitous (meta, ortho)	
Semi-bituminous	4.5–5.0	—	Semi-bituminous species (ortho-semi-bituminous)	Subbituminous (sub-meta-bituminous)	Subbituminous (sub-ortho-bituminous)	Subbituminous	Sub-lignitous (meta, ortho)	
Carbonaceous	4.0–4.5	Semi-anthracitic species Dry steam coal	Carbonaceous species (ortho-carbonaceous)	Pseudo-carbonaceous (sub-meta-bituminous)	Pseudo-carbonaceous (sub-ortho-bituminous)	Pseudo-carbonaceous (sub-para-bituminous)		
Anthracitic	<4	Ortho-anthracite True anthracite	Pseudo-anthracite (sub-carbonaceous)	Pseudo-anthracite (sub-meta-bituminous)	Pseudo-anthracite (sub-ortho-bituminous)	Pseudo-anthracite (sub-para-bituminous)		

Table 4.20 ASTM classification of coals by rank.[a] Reprinted with permission from ASTM D388–1999

Class/Group	Fixed carbon limits (dry, mineral-matter-free basis), %		Volatile matter limits (dry, mineral-matter-free basis), %		Gross calorific value limits (moist,[b] mineral-matter-free basis)				Agglomerating character
					Btu/lb		Mj/kg[c]		
	Equal or greater than	Less than	Greater than	Equal or less than	Equal or greater than	Less than	Equal or greater than	Less than	
Anthracitic:									
Meta-anthracite	98	—	—	2	—	—	—	—	nonagglomerating
Anthracite	92	98	2	8	—	—	—	—	
Semianthracite[d]	86	92	8	14	—	—	—	—	
Bituminous:									
Low volatile bituminous coal	78	86	14	22	—	—	—	—	
Medium volatile bituminous coal	69	78	22	31	—	—	—	—	
High volatile A bituminous coal	—	69	31	—	14 000[e]	—	32.6	—	commonly agglomerating[f]
High volatile B bituminous coal	—	—	—	—	13 000[e]	14 000	30.2	32.6	
High volatile C bituminous coal	—	—	—	—	11 500	13 000	26.7	30.2	agglomerating
	—	—	—	—	10 500	11 500	24.4	26.7	
Subbituminous:									
Subbituminous A coal	—	—	—	—	10 500	11 500	24.4	26.7	nonagglomerating
Subbituminous B coal	—	—	—	—	9 500	10 500	22.1	24.4	
Subbituminous C coal	—	—	—	—	8 300	9 500	19.3	22.1	
Lignitic:									
Lignite A	—	—	—	—	6 300[g]	8 300	14.7	19.3	
Lignite B	—	—	—	—	—	6 300	—	14.7	

[a] This classification does not apply to certain coals.
[b] Moist refers to coal containing its natural inherent moisture but not including visible water on the surface of the coal.
[c] Megajoules per kilogram. To convert British thermal units per pound to megajoules per kilogram, multiply by 0.002326.
[d] If agglomerating, classify in low volatile group of the bituminous class.
[e] Coals having 69% or more fixed carbon on the dry, mineral-matter-free basis shall be classified according to fixed carbon, regardless of gross calorific value.
[f] It is recognized that there may be nonagglomerating varieties in these groups of the bituminous class, and that there are notable exceptions in the high volatile[c] bituminous group.
[g] Editorially corrected.

Table 4.21 Coal classification system used by British Coal (revision of 1964). Reproduced by permission of the British Coal Corporation

Coal rank code			Volatile matter (d.m.m.f.) (%)	Gray–King coke type[a]	General description
Main class(es)	Class	Subclass			
100			Under 9.1	A	
	101[b]		Under 6.1	} A	} Anthracites
	102[b]		6.1–9.0		
200			9.1–19.5	A–G8	Low Volatile steam coals
	201		9.1–13.5	A–C	
		201a	9.1–11.5	A–B	} Dry steam coals
		201b	11.6–13.5	B–C	
	202		13.6–15.0	B–G	
	203		15.1–17.0	E–G4	} Coking steam coals
	204		17.1–19.5	G1–G8	
300			19.6–32.0	A–G9 and over	Medium volatile coals
	301		19.6–32.0	G4 and over	
		301a	19.6–27.5	} G4 and over	} Prime coking coals
		301b	27.6–32.0		
	302		19.6–32.0	G–G3	Medium volatile, medium caking or weakly caking coals
	303		19.6–32.0	A–F	Medium volatile, weakly caking to noncaking coals
400–900			Over 32.0	A–G9 and over	High volatile coals
400			Over 32.0	G9 and over	
	401		32.1–36.0	} G9 and over	} High volatile, very strongly caking coals
	402		Over 36.0		
500			Over 32.0	G5–G8	
	501		32.1–36.0	} G5–G8	} High volatile, strongly caking coals
	502		Over 36.0		
600			Over 32.0	G1–G4	
	601		32.1–36.0	} G1–G4	} High volatile, medium caking coals
	602		Over 36.0		
700			Over 32.0	E–G	
	701		32.1–36.0	} E–G	} High volatile, weakly caking coals
	702		Over 36.0		
800			Over 32.0	C–D	
	801		32.1–36.0	} C–D	} High volatile, very weakly caking coals
	802		Over 36.0		
900			Over 32.0	A–B	
	901		32.1–36.0	} A–B	} High volatile, noncaking coals
	902		Over 36.0		

[a] Coals with volatile matter of less than 19.6% are classified by using the parameter of volatile matter alone; the Gray–King coke types quoted for these coals indicate the general ranges found in practice and are not criteria for classification.

[b] To divide anthracites into two classes, it is sometimes convenient to use a hydrogen content of 3.35% (d.m.m.f.) instead of a volatile matter of 6.0% as the limiting criterion. In the original Coal Survey rank coding system the anthracites were divided into four classes then designated 101, 102, 103 and 104. Although the present division into two classes satisfies most requirements, it may sometimes be necessary to recognise more than two classes.

Notes:
(1) Coals that have been affected by igneous intrusions ('heat-altered' coals) occur mainly in classes 100, 200 and 300, and when recognised should be distinguished by adding the suffix H to the coal rank code, e.g. 102H, 201bH.
(2) Coals that have been oxidised by weathering may occur in any class and when recognised should be distinguished by adding the suffix W to the coal rank code, e.g. 801W.

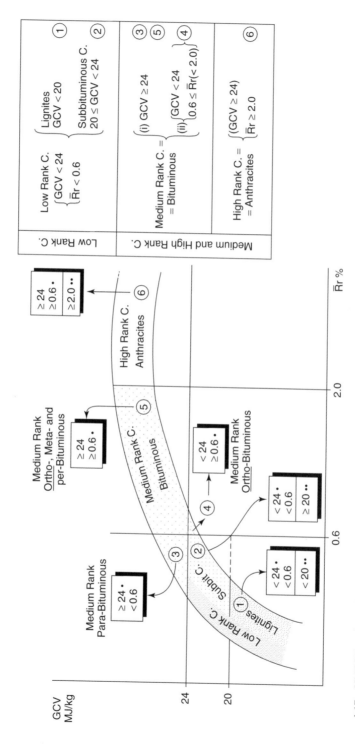

Figure 4.17 UNECE main coal classification categories, defined by gross calorific value (MJ/kg m, af) and vitrinite reflectance % in oil (R_{ro}%). Reproduced with permission of UNECE

Table 4.22 Codification system used by UNECE for medium and high rank coals (1988), reproduced with permission of UNECE

Digit	1:2	3	4	5	6	7;8	9;10	11;12	13;14
	Random reflectance of vitrinite %	Characteristics of reflectogram	Maceral group composition index % by vol. (mmF) 4 = Inertinite	5 = Liptinite	Swelling crucible number	Volatile matter mass % (d.a.f.)	Ash mass % (db)	Total sulphur mass % (db)	Gross calorific value MJ/kg (d.a.f.)
Code No.	02 0.20–0.29	0	0 0–<10	0 exempted	0 0–½	48 ≥48	00 0–<1	00 0.0–<0.1	21 <22
	03 0.30–0.39	1	1 10–<20	1 0–<5	1 1–1½	46 46–<48	01 1–<2	01 0.1–<0.2	22 22–<23
	04 0.40–0.49	2	2 20–<30	2 5–<10	2 2–2½	44 44–<46	02 2–<3	02 0.2–<0.3	23 23–<24
	05 0.50–0.59	3	3 30–<40	3 10–<15	3 3–3½	42 42–<44	03 3–<4	03 0.3–<0.4	24 24–<25
	06 0.60–0.69	4	4 40–<50	4 15–<20	4 4–4½	40 40–<42	04 4–<5	04 0.4–<0.5	25 25–<26
	07 0.70–0.79	5	5 50–<60	5 20–<25	5 5–5½	38 38–<40	05 5–<6	05 0.5–<0.6	26 26–<27
	08 0.80–0.89		6 60–<70	6 25–<30	6 6–6½	36 36–<38	06 6–<7	06 0.6–<0.7	27 27–<28
	09 0.90–0.99		7 70–<80	7 30–<35	7 7–7½	34 34–<36	07 7–<8	07 0.7–<0.8	28 28–<29
	10 1.00–1.09		8 80–<90	8 35–<40	8 8–8½	32 32–<34	08 8–<9	08 0.8–<0.9	29 29–<30
	11 1.10–1.19		9 ≥90	9 ≥40	9 9–9½	30 30–<32	09 9–<10	09 0.9–<1.0	30 30–<31
	12 1.20–1.29					28 28–<30	10 10–<11	10 1.0–<1.1	31 31–<32
	13 1.30–1.39					26 26–<28	11 11–<12	11 1.1–<1.2	32 32–<33
	14 1.40–1.49					24 24–<26	12 12–<13	12 1.2–<1.3	33 33–<34
	15 1.50–1.59					22 22–<24	13 13–<14	13 1.3–<1.4	34 34–<35
	16 1.60–1.69					20 20–<22	14 14–<15	14 1.4–<1.5	35 35–<36
	17 1.70–1.79					18 18–<20	15 15–<16	15 1.5–<1.6	36 36–<37
	18 1.80–1.89					16 16–<18	16 16–<17	16 1.6–<1.7	37 37–<38
	19 1.90–1.99					14 14–<16	17 17–<18	17 1.7–<1.8	38 38–<39
	20 2.00–2.09					12 12–<14	18 18–<19	18 1.8–<1.9	39 ≥39
	21 2.10–2.19					10 10–<12	19 19–<20	19 1.9–<2.0	
	22 2.20–2.29					09 9–<10	20 20–<21	20 2.0–<2.1	
	23 2.30–2.39					08 8–<9		21 2.1–<2.2	
	24 2.40–2.49					07 7–<8		22 2.2–<2.3	
	25 2.50–2.59					06 6–<7		23 2.3–<2.4	
	26 2.60–2.69					05 5–<6		24 2.4–<2.5	
	27 2.70–2.79					04 4–<5		25 2.5–<2.6	
	28 2.80–2.89					03 3–<4		26 2.6–<2.7	
	29 2.90–2.99					02 2–<3		27 2.7–<2.8	
	30 3.00–3.09					01 1–<2		28 2.8–<2.9	
	50 >5.00								

Maceral group composition index % by vol. (mmF) 4 = Inertinite; 5 = Liptinite

used where appropriate, e.g. chlorine content, ash fusion temperature.

The code is based on:

1. Mean random reflectance (R_r) as two digits. Codes 2–50 cover R_r values from 0.20 to >5.00.
2. Description of a reflectogram, the third digit covering codes 0–5, dependent on whether the coal is a single seam or a blend.
3. Maceral group composition; the fourth digit provides the lower limit of a 10% range of the inertinite content, and the fifth digit indicates the upper limit of a 5% range of the liptinite content.
4. The sixth digit indicates the crucible swelling number in terms of the lower limit of two half-step numbers.
5. The seventh and eighth digits indicate the lower limit of a 2% range of volatile matter down to

Table 4.23 Australian classification of hard coal. From Ward (1984), based on AS 2096-1987, with permission of Blackwell Scientific Publications

	Value	Volatile matter (d.m.m.f.) (%)	Gross calorific value (d.a.f.) (MJ/kg^{-1})
1st digit (coal class)	1	<10.0	
	2	10.1–14.0	
	3	14.1–20.0	
	4A	20.1–24.0	
	4B	24.1–28.0	
	5	28.1–33.0	
	6	33–41[a]	>33.82
	7	33–44[a]	32.02–33.82
	8	35–50[a]	28.43–32.02
	9	42–50[a]	27.08–28.42
	Value	**Crucible swelling no.**	
2nd digit (coal group)	0	0–$\frac{1}{2}$	
	1	1–2	
	2	$2\frac{1}{2}$–4	
	3	$4\frac{1}{2}$–6	
	4	$6\frac{1}{2}$–9	
	Value	**Gray–King coke type**	
3rd digit (coal sub-group)	0	A	
	1	B–D	
	2	E–G	
	3	G1–G4	
	4	G5–G8	
	5	G9–	
	Value	**Ash (dry basis %)**	
4th digit (ash number)	(0)	<4.0	
	(1)	4.1–8.0	
	(2)	8.1–12.0	
	(3)	12.1–16.0	
	(4)	16.1–20.0	
	(5)	20.1–24.0	
	(6)	24.1–28.0	
	(7)	28.1–32.0	
	(8)	>32.0	

[a] Values for information only.

10% (d.a.f.) and a 1% range when volatile matter <10%.

6. The ninth and tenth digits indicate the lower limit of 1% range of the ash (dry basis – db).
7. The eleventh and twelfth digits indicate the lower limit of a 0.10% range of the total sulphur content (db) multiplied by 10.
8. The thirteenth and fourteenth digits indicate the lower limit of 1 MJ/kg range of the gross calorific value (d.a.f.).

All of the testing is in accordance with ISO Standards as listed in Appendix 1.
An example of the Codification System when applied to an Australian coal is:

R_r	1.25	Digit no. 12
Reflectogram	s = 0.14, no gap	1
Maceral composition		
Inertinite	51	
Liptinite	3	51
Free swelling index	6½	6
Volatile matter % (d.a.f.)	24.3	24
Ash content % (db)	5.53	05
Total sulphur content % (db)	0.42	04
Gross CV MJ/kg (d.a.f.)	35.9	35

CODE NO. 12 1 51 6 24 05 04 35

The Australian Standard Coal Classification for hard coals again assigns a multi-digit number to determine coal type. The first digit represents volatile matter for coals with less than 33% volatile matter (d.m.m.f.) and gross calorific value (d.a.f.) for other coals. The second digit is the free swelling index of the coal, the third digit is the Gray–King assay value and the fourth digit (given in brackets) is based on the ash content (dry basis) of the coal; see Table 4.23.

In South Africa, coals are divided for commercial purposes into three broad classes on the basis of volatile matter (d.a.f.). These are South African anthracite, semi-anthracite and steam coal. Coals of each class are graded on the basis of calorific value (ad), ash (ad) and ash fusibility.

Because of the wide spectrum of criteria used to define the boundary between 'brown' coal and high rank coals, plus the variation in brown colour, the United

Table 4.24 Classification of low rank coals. UNECE (2000). Reproduced with permission of UNECE

Coal type	Gross CV MJ/kg (m, af)	Moisture (% ar)	R_{ro} %
Low rank C (Ortholignites)	<15.0	<75	–
Low rank B (Meta-lignites)	15–20	–	–
Low rank A (Subbituminous)	20–24	–	<0.6

Table 4.25 Codification system used by UNECE for low rank coals (2000). Reproduced with permission of UNECE

Digit 1, 2 Code no.	Gross calorific value MJ/kg (d.a.f.)	Digit 3, 4 Code no.	Total moisture % (mass) (ar)	Digit 5, 6 Code no.	Ash content % (db)	Digit 7, 8 Code no.	Total sulphur content % (db)
15	15.00–15.98 incl.			00	0.0–0.9 incl.	00	0.00–0.09 incl.
16	16.00–16.98 incl.			01	1.0–1.9 incl.	01	0.10–0.19 incl.
17	17.00–17.98 incl.			02	2.0–2.9 incl.	02	0.20–0.29 incl.
		20	20.0–20.9 incl.				
		21	21.0–21.9 incl.				
		22	22.0–22.9 incl.			11	1.10–1.19 incl.
						12	1.20–1.29 incl.
						13	1.30–1.39 incl.
		38	38.0–38.9 incl.	28	28.0–28.9 incl.		
		39	39.0–39.9 incl.	29	29.0–29.9 incl.		
		40	40.0–49.9 incl.	30	30.0–30.9 incl.		
						20	2.00–2.09 incl.
		50	50.0–50.9 incl.				
		56	56.0–56.9 incl.	49	49.0–49.9 incl.		
						32	3.20–3.29 incl.
		65	65.0–65.9 incl.				

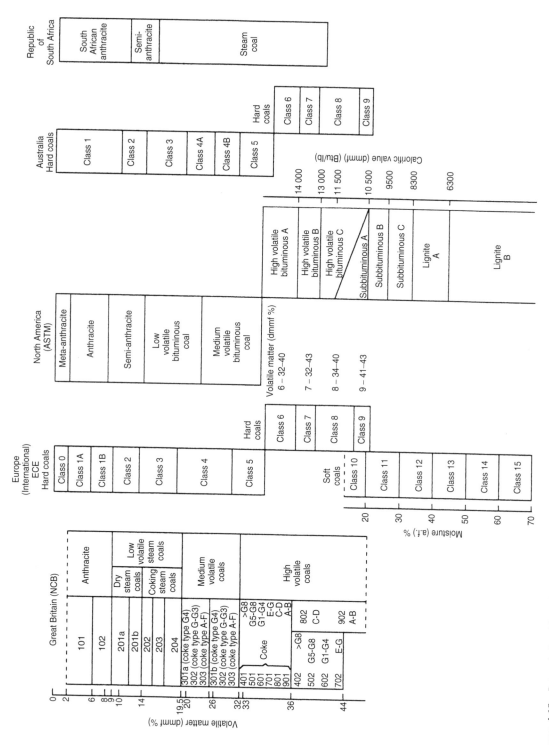

Figure 4.18 Interrelationships of coal classification systems used in various countries. Unpublished data, reproduced by permission of BP Coal Ltd

Nations Economic Commission for Europe decided to abandon the term 'brown' coal, and devised a new classification and codification system for low rank coals (2000); see Table 4.24.

As stated above, the threshold between low rank and medium and high rank coals is defined as follows: low rank coals are those coals with a gross calorific value less than 24.0 MJ/kg (moist, af or m, af), and a vitrinite reflectance % in oil (R_{ro}%) less than 0.6%. The gross calorific value calculation uses total moisture according to ISO 1015-1975, and low rank coals are classified as shown in Figure 4.17.

As for medium and high rank coals, the UNECE has devised a codification system for low rank coals based on an eight-digit number.

1. Digits one and two, the gross calorific value in MJ/kg (m, af).
2. Digits three and four, the total moisture % recalculated to as received basis (ar).
3. Digits five and six, the ash content % recalculated to dry basis (db).
4. Digits seven and eight, the total sulphur content % recalculated to dry basis (db).

The coded parameters are always coded in this order, and the make-up of the code numbers is shown in Table 4.25.

Before a final selection of a low rank coal for a particular purpose, a set of relevant analyses is carried out. These supplementary parameters consist of the numerous chemical and physical tests outlined above and which are also used for the medium and high rank coals.

The interrelationships of the various coal classifications are shown in Figure 4.18; it should be noted, however, that the UNECE high and low rank coal classifications shown in the figure have been superseded as described above.

5

Coal Sampling and Analysis

The sampling of coal can be a difficult task in that coal is a heterogeneous material.

Samples are the representative fractions of a body of material, acquired for testing and analysis, in order to assess the nature and composition of the parent body, collected by approved methods and protected from contamination and chemical change.

Such samples should be differentiated from those materials collected in ways that may not be truly representative of the coal from which they have been collected. These materials may still be useful but should be regarded as specimens rather than samples (Pryor 1965).

Coal samples may be required as part of a greenfields exploration programme to determine whether the coal is suitable for further investigation, or as part of a mine development programme, or as routine samples in opencast and underground mines to ensure that the quality of the coal to be mined will provide the specified run-of-mine product.

In situ coal samples are taken from surface exposures, exposed coal seams in opencast and underground workings, and from drill cores and cuttings.

Non *in situ* samples are taken from run-of-mine coal streams, coal transport containers and coal stockpiles.

Such coal samples may have to be taken under widely differing conditions, particularly those of climate and topography. It is essential that the sample taken is truly representative as it will provide the basic quality data on which decisions to carry out further investigation, development, or to make changes to the mine output will be made.

It is important to avoid weathered coal sections, coals contaminated by extraneous clay or other such materials, coals containing a bias of mineralisation, and coals in close contact with major faults and igneous intrusions.

5.1 *IN SITU* COAL SAMPLING

Several types of *in situ* samples can be taken, dependent upon the analysis required.

5.1.1 Grab samples

Generally this is a most unsatisfactory method of obtaining coal for analysis, as there are no controls on whether the coal is representative, and can easily lead to a bias in selection, e.g. the bright coal sections attract attention. However, grab samples can be used to determine vitrinite reflectance measurements as an indicator of coal rank.

5.1.2 Channel samples

Channel samples are representative of the coal from which they are taken. If the coal to be sampled is a surface exposure, the outcrop must be cleaned and cut back to expose as fresh a section as possible. Ideally the full seam section should be exposed, but in the case of thick coals (especially in stream sections), sections of the roof and coal or coal and floor only may be seen. To obtain a full seam section under these circumstances, two or more channels will need to be cut and the overlap carefully recorded. The resultant samples will consist of broken coal and will not preserve the lithological sequence.

In opencast workings, the complete seam section should be exposed, and is less likely to be weathered than natural surface exposures.

In underground workings, the seam will be unweathered, but the whole seam section may not always be seen, due to the workings only exposing the selected mining section of the seam.

To carry out a channel sample, the coal is normally sampled perpendicular to the bedding. A channel of uniform cross-section is cut manually into the coal seam, and all the coal within the cut section is collected on a plastic sheet placed at the base of the channel (Figure 5.1(a)). Most channels are around 1.0 m across and samples should not be less than 15 kg per metre of coal thickness. Such channel samples will provide a composite quality analysis for the seam, i.e. an analysis

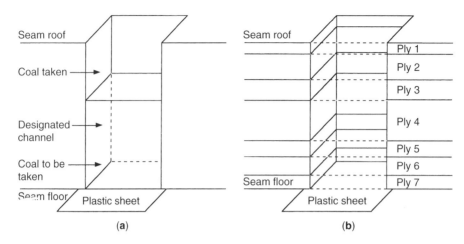

Figure 5.1 Channel sampling procedure. (**a**) Whole seam channel sampling; (**b**) Coal seam ply channel sampling. Reproduced by permission of Dargo Associates Ltd

of all the coal and mineral matter present in the seam as a whole.

Whilst this is suitable for general seam quality assessment, more detailed analysis of the seam from top to bottom may be required. To achieve this, a channel ply sample is taken; this entails a similar procedure as for the whole seam channel sample except that the seam is divided into plies or subsections, as shown in Figure 5.1(b).

Coal seams are rarely homogeneous throughout their thickness; most are divisible into distinct lithological sections. Plies are lithological subdivisions of the seam, each of which has a uniform character. When the lithology changes, such as at a clay parting in the seam, a separate ply is designated.

Where the roof and floor of a seam are exposed, ply samples of at least 0.25 m of roof material immediately above the seam and 0.25 m of floor underlying the seam should be included in the samples. This will allow the effects of dilution on coal quality to be assessed. In general the thickness of coal plies should be a minimum of 0.1 m and a maximum of 1.0 m. In the case of banded coals containing alternating thin (<0.1 m) layers of bright coal/dull coal/clay, the seams may be sampled as a series of composite plies, with the details of the individual layers shown on the record sheet. An interbedded noncoal ply greater than 0.25 m in thickness may be regarded as a seam split and recorded as such. Ply samples should be at least 2.0 kg where possible, it may be that the sample will be split into two fractions and one stored for later use.

Once the outcrop or face is cleaned, a shallow box cut is made for the total thickness of exposed coal seam.

Once this is completed, the seam is divided into plies, each of which is measured and recorded on a record sheet similar to that shown in Figure 5.2.

The channel sample record sheet should show the following information:

1. Record card number.
2. Map or air photograph number on which locality is located.
3. Location of sample point, grid reference or reference number.
4. Description of the locality, stream section, working face etc. including dip, strike, coal seam roof and floor contacts.
5. Extent of weathering, fracturing, mineralisation etc.
6. Lithological description of each ply interval.
7. Thickness of each ply interval.
8. Designated sample number of each ply interval.

Space can also be allocated on the record card for analytical details, i.e. proximate analysis, to be added later to complete the record.

The fresh surface is then sampled as a channel cut from top to bottom (Figure 5.1(b)), cutting and collecting all material from each ply section in turn. Each ply sample should be sealed in a strong plastic bag immediately after collection to prevent moisture loss and oxidation. All sample bags must be clearly labelled with a designated number, a copy of which should be placed in a small plastic bag inside the sample bag, and another attached to the outside of the sample bag. This number must be recorded on the channel sample record sheet.

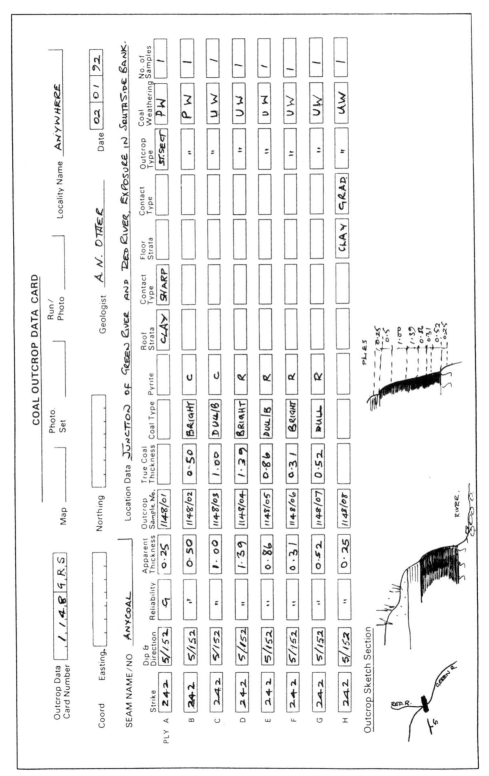

Figure 5.2 Coal outcrop data card. Reproduced by permission of BP Coal Ltd

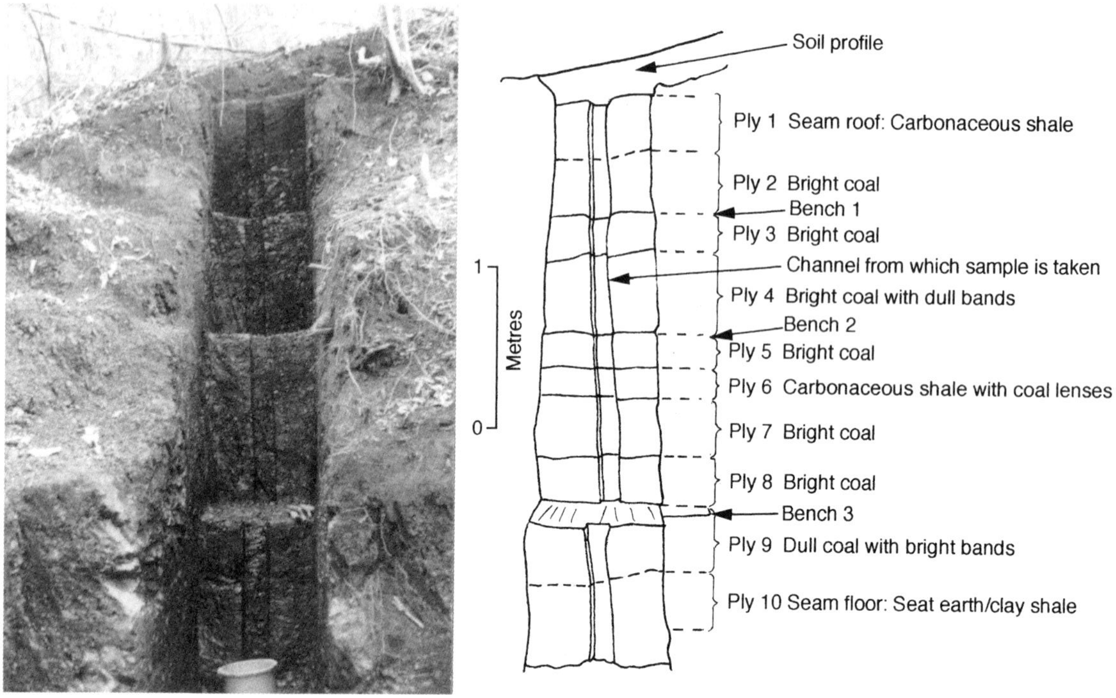
Soil profile
Ply 1 Seam roof: Carbonaceous shale
Ply 2 Bright coal
Bench 1
Ply 3 Bright coal
Channel from which sample is taken
Ply 4 Bright coal with dull bands
Bench 2
Ply 5 Bright coal
Ply 6 Carbonaceous shale with coal lenses
Ply 7 Bright coal
Ply 8 Bright coal
Bench 3
Ply 9 Dull coal with bright bands
Ply 10 Seam floor: Seat earth/clay shale
1
Metres
0

Figure 5.3 Surface coal ply channel sample taken in shallow dipping seam. Central narrow channel taken for ply sample analysis, including coal seam roof and floor. Photograph by LPT, reproduced by permission of Dargo Associates Ltd

Because this task is invariably a dirty one, labels get wet, blackened and unreadable very easily; care must therefore be taken to ensure that the sample numbers do not get lost or obliterated during transit to the laboratory as unidentifiable samples are useless and an expensive waste of time.

The advantage of channel ply sampling is that not only can the analysis of the individual plies be obtained, but that by combining a fraction of each ply sample, a whole seam composite analysis can also be made. An example of a channel ply sample from a surface exposure is illustrated in Figure 5.3, which shows a channel cut to expose fresh coal, and then a thinner channel (*c.* 0.25 m wide) cut from the fresh coal from which ply samples are collected for analysis.

5.1.3 Pillar samples

In underground coal mining, samples of large blocks of undisturbed coal are taken to provide technical information on the strength and quality of the coal. These pillar samples are taken when a specific problem may have arisen or is anticipated. Such samples are taken in much the same way as whole seam channel samples except that extra care is required not to disturb the cut-out section of coal during removal. Samples are then boxed and taken to the laboratory. Pillar sampling is a long and arduous business and is only undertaken in special circumstance, such as when mining becomes difficult or new roadways or faces are planned.

5.1.4 Core samples

Core sampling is an integral part of coal exploration and mine development. It has the advantage of producing nonweathered coal, including the coal seam floor and roof; unlike channel samples, core samples preserve the lithological sequence within the coal seam.

First, the borehole core has to be cleaned if drilling fluids have been used, and then lithologically logged. Following this, the lithological log should be compared with the geophysical log of the borehole to select ply intervals and to check for core losses and any other length discrepancies.

Once the core has been reconciled to the geophysical logs and the ply intervals have been selected, sampling can begin. Core ply samples are taken in the same way as for surface channel ply samples (as shown in

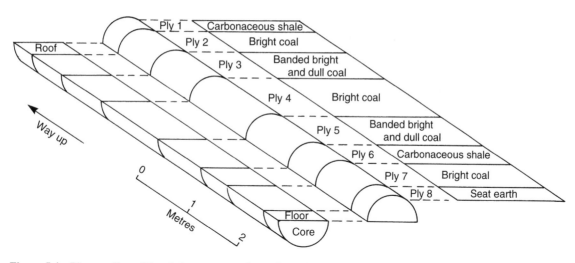

Figure 5.4 Ply sampling of borehole core; run of samples to include all the coal seam and roof and floor. Core may be split in this fashion and one half kept for future examination and analysis. Reproduced by permission of Dargo Associates Ltd

Figure 5.4); again a ply sample of the coal seam roof and floor (up to 0.25 m) is taken to determine dilution effects. Then the individual plies are sampled, making sure no core is discarded. As in the case of surface samples, bright coal tends to fragment and make up the finer particles that may easily be left in the core tray.

The samples are bagged and labelled as for surface ply samples, and the sample numbers recorded on the core logging sheet in the manner shown in Figure 6.20.

Large diameter cores may be split lengthways with a bolster chisel and then one half-ply sampled, the other being retained for future analysis.

5.1.5 Cuttings samples

This method of sampling is considerably less accurate than that of core sampling. As with core samples, cuttings are unweathered and are a useful indicator as to the general nature of the seam. Air flush and mud flush noncore drilling is a quicker operation than core drilling and will produce cuttings for each horizon encountered in drilling. In the case of mud flush cuttings, they will need to be washed to remove any drilling fluid before sampling.

Cuttings are usually produced for every metre drilled; those cuttings returns which are all coal may be collected, bagged and numbered in the same way as channel samples. The depth to the top and bottom of the seam sampled should be determined from the geophysical log. The drawback with using cuttings samples is that only a general analysis of the seam can be made, and even this

is unlikely to be truly representative. Also contamination from strata above the coal may be included; a close study of the geology and geophysical log will determine whether this is so.

5.1.6 Specimen samples

Orientated specimens of coal may be collected so that their precise orientation can be re-created in the laboratory. The dip and strike of the coal is marked on the specimen before removal. This method is commonly used for the studies of the optical fabric of the coal, or for the structural features in the coal.

5.1.7 Bulk samples

Bulk samples are taken from outcrops, small pits or minishafts (i.e. 2 m diameter shaft excavations). A bulk sample is normally 5–25 tonnes and is taken as a whole seam channel sample on a large scale. Such a bulk sample is taken in order to carry out test work on a larger scale; this is designed to indicate the coal's likely performance under actual conditions of usage.

Steam coals are taken for small combustion tests in a pulverised fuel (pf) rig, to simulate conditions in a pf boiler. Pulverised coal firing is the combustion of powdered coal suspended as a cloud of small particles in the combustion air. Substantially more heat is released per unit volume in pf boilers than in stoker type boilers.

Coking coals are taken to carry out moving wall oven tests, i.e. to determine that when the coal is combusted, how much the coal swells and thus puts pressure on the oven walls, which are constructed of uncemented brickwork. High pressure coals are undesirable, and are normally blended with low pressure coals to reduce the problem. In the USA, low volatile coking coals (VM = 20–25%, SI = 9) are high pressure coals, whereas in general, high volatile coals do not have such high pressures. It is significant that Gondwana coking coals are low pressure coals, an important factor in Australia being able to export coking coals.

Bulk samples are collected from a site already channel-sampled, loaded into drums, numbered and shipped to the selected test centre.

5.1.8 Sample storage

In the majority of cases, the channel and core samples will be required immediately for laboratory analysis. However, there are circumstances where duplicate coal samples for future reference are taken. Usually the channel plies are divided into two or the cores are split and one half retained.

If the duplicate samples are to be put into storage, this presents a problem; the exposure of the coal to air will allow oxidation to take place during storage and will result in anomalous quality results when analysed at a later date.

The usual procedure to prevent oxidation of samples is to store them under nitrogen or in water.

To store in nitrogen, place a tube connected to a pressurised cylinder containing nitrogen in a plastic sample bag, then add the coal sample, flush the sample with nitrogen, regulating the flow by means of a flow meter. The nitrogen has to fill the spaces between the coal fragments, so flushing with nitrogen is required for several minutes. One difficulty with this method is that nitrogen is lighter than air, so inevitably some is lost in the process. Once the bag has been thoroughly flushed, it should be heat-sealed; no other form of sealing is anywhere near as effective. The coal samples can be as received or air-dried and can be in the form of lump or crushed coal. It should be noted that for all *in situ* and non *in situ* samples, the top size to which any sample is crushed to is important in determining the weight of the sample required. The size of the sample is calculated as:

$$5.24 \times \text{mean particle size} = x\,\text{kg (where mean}$$

$$\text{particle size is top size} \times \text{bottom size).}$$

5.24 is an empirically determined number quoted in BS 1017-1.

A cheaper method of storage is by immersing the channel or core sample in the form of lump coal in water. This method has the advantage over storing in nitrogen in that it preserves fluidity of the coal, but it does present handling problems when the sample is required. The sample will have to be air-dried before analysis can begin.

Samples can be kept by these methods for one to two years before analysis.

5.2 NON *IN SITU* SAMPLING

The object of collecting coal samples after mining is to determine the quality of coal actually being produced. This coal may differ significantly from the *in situ* seam analysis in that not all of the seam may be included in the mining section, or in that more than one seam may be worked and fed to the mine mouth and mixed with coal from other seams. In addition there may be dilution from seam roof and/or floor which becomes part of the mined coal product.

The mined coal is broken up and therefore contains fragments which vary a great deal in size and shape. Representative samples are collected by taking a definite number of portions known as increments distributed throughout the total quantity of coal being sampled. Such increments represent a sample or portion of coal obtained by using a specified sampling procedure, either manually or using some sampling apparatus.

The various practices used in collecting non *in situ* samples and the mathematical analysis of the representativeness of samples, i.e. quality control, is reviewed in Laurila and Corriveau (1995).

Increments are taken by three methods:

1. Systematic sampling, where increments are spaced evenly in time or in position over the unit.
2. Random sampling, where increments are spaced at random but a prerequisite number are taken.
3. Stratified random sampling, where the unit is divided by time or quantity into a number of equal strata and one or more increments are taken at random from each.

It is good practice that, whatever the method used, duplicate sampling should be employed to verify that the required precision has been attained.

Non *in situ* coal sampling is carried out on moving streams of coal, from rail wagons, trucks, barges, grabs or conveyors unloading ships, from the holds of ships and from coal stockpiles.

1. Hand sampling from streams is carried out using ladles or scoops; the width of the sampler should be 2.5 times the size of the largest lump likely to be encountered. However, this type of sampling is not suitable for coal larger than 80 mm. For larger samples mechanical sampling equipment is used; moving streams of coal (conveyors) can be sampled by using:
   (a) falling stream samplers; these make either a linear traverse across the coal conveyor in a straight-line path perpendicular to the direction of flow, or opposite to the direction of flow or in the same direction of flow, or they make a rotational traverse by moving in an arc such that the entire stream is within the radius of the arc;
   (b) cross-belt samplers, which move across the belt pushing a section of coal to the side while the belt runs;
   (c) stop-belt method, whereby the conveyor is stopped and all coal occurring within a selected interval, usually a couple of metres, is collected (Figure 5.5).

Laurila and Corriveau (1995) state that the correct increment selection occurs when all the elements of the transversal cross-section are intercepted by the sampling cutter during the same length of time. This should avoid any increase in error. These sampling systems are checked for bias by using a reference sampling method as recommended by ISO 13909 or ASTM D2234.

2. Wagons and trucks are sampled by taking samples from their tops by means of probes, or by sampling from bottom or side-door wagons during discharge, or sampling from the exposed face of coal as the wagons or trucks are tipping into bunkers or ships, or wagons being emptied via tipplers.

3. Ships are sampled either from conveyors loading and unloading coal, at a point where bias can be avoided, or from the hold of the ship. Samples from the hold are taken every 4 m of the depth of the coal within the hold. It is important to estimate the proportion of fine and lump coal in the consignment. It should be noted that free moisture, if present, will tend to settle towards the bottom of the hold. This increase of moisture with depth makes it difficult to collect samples for moisture content determination.

4. Sampling from barges is the same as for ships except that if the depth of coal is less than 4 m it should be sampled in onstage during unloading, once the bottom of the barge is partially uncovered.

5. Sampling from stockpiles: where the preferred procedure of sampling from a conveyor belt during stocking and unstocking cannot be used, then the stockpile is sampled based on collecting increments spaced as evenly as possible over the surface and layers of the stockpile. Sampling is by means of probes or by digging holes. If the stockpile is known to consist of different coals piled in separate areas of the total pile, a separate gross sample must be taken from each such area. The stockpile should be divided into a number of portions, each 1000 tonnes or less, from which a separate sample with a specified number of increments is taken. This normally takes a long time to accomplish, but can be speeded up if automated auger units are employed. It is important that all levels in the stockpile are sampled.

Table 5.1 indicates the minimum number of increments required for gross samples of single load consignment up to 1000 tonnes, from all of the above.

5.3 COAL ANALYSIS

The marketability of coals depends on their quality. This will determine whether they are to be sold as steam or coking coals, prime or lower-grade coals. Customer

Figure 5.5 The collection of a stop belt sample from a main conveyor (Mazzone 1998) reproduced by permission of World Coal, Palladian Publications Ltd

Table 5.1 Minimum number of increments required for gross samples of a single coal consignment up to 1000 tonnes. From Osborne (1988) 'Coal Preparation Technology' published by Graham & Trotman, with kind permission of Kluwer Academic Publishers

Sampling situation	Common sample for total moisture and general analysis			Total moisture sample		General analysis sample		Size analysis sample
	Sized coals; dry cleaned or washed	Washed smalls (50 mm)	Blended part treated untreated, run-of-mine & 'unknown' coals	Sized coals dry cleaned or washed, unwashed dry coals	Washed smalls (50 mm) blended, part treated, untreated, run-of-mine & 'unknown' coals	Sized coals: dry cleaned, or washed or unwashed dry coals	Blended part treated, untreated, run-of-mine & 'unknown' coals	All coals
Moving streams	20	35	35	20	35	20	35	40
Wagons & trucks, barges, grabs or conveyors unloading ships	25	35	50	20	35	25	50	40
Holds of ships, stockpiles	35	35	65	20	35	35	65	40

requirements vary considerably from those who will accept a broad spectrum of coal qualities to those who require coal for a specialised purpose and have set restricted specifications for the coal.

The coal producer, i.e. the mining company, will have assessed the potential market before developing any coal deposit, i.e. whether to mine coal for export or local use. The mining company will also need to know the quality limitations of the coal that can be produced from the deposit.

The quality of the coal has to be determined at an early stage of exploration and monitored during all later phases of development.

All coals should be sampled using the procedures outlined in 5.1.2, 5.1.4 and 5.1.6, and sent to the laboratory where they are weighed, crushed and split for analysis.

5.3.1 Outcrop/core samples

The procedure for weighing, crushing and splitting outcrop/core samples is shown in Figure 5.6 for steam (thermal) coals, and Figure 5.7 for coking (metallurgical) coals. However, there is no universal set procedure and differences do occur.

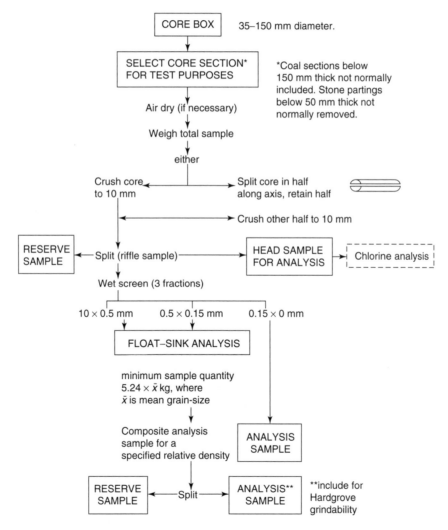

Figure 5.6   Sample preparation diagram for drill core samples from a steam (thermal) coal deposit. From Osborne (1988) 'Coal Preparation Technology' published by Graham & Trotman, with kind permission of Kluwer Academic Publishers

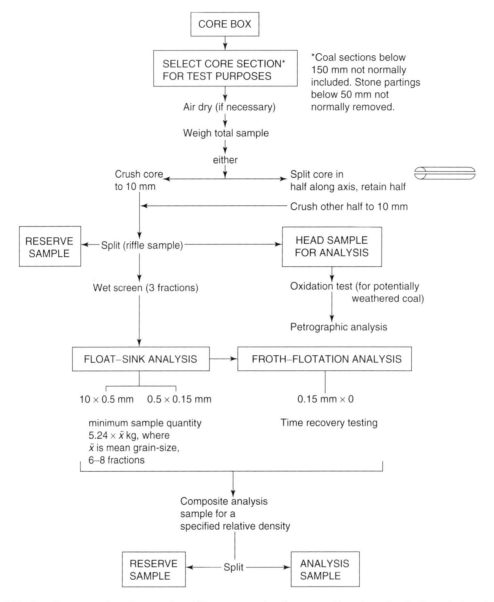

Figure 5.7 Sample preparation diagram for drill core samples from a coking (metallurgical) coal deposit. From Osborne (1988) 'Coal Preparation Technology' published by Graham & Trotman, with kind permission of Kluwer Academic Publishers

In Australia, a standard similar to that shown in the tables is set, with the difference that the samples are crushed to 11.2 mm. Large diameter core samples are preferred when sampling Gondwana coals in order to be more confident of yield values obtained during analysis. There should be a correlation in properties between outcrop/small diameter core, large diameter core and bulk samples.

The analysis undertaken for each float–sink fraction is proximate analysis, plus total sulphur and calorific value. This is intended to produce a simulated product by combining several float–sink fractions. This product is then analysed for ultimate analysis, ash analysis, ash fusion temperatures, hardgrove grindability, swelling index and Geiseler plastometer test, the last only if the coal has coking properties.

In South Africa, the coal is generally crushed to only 25 mm, then analysed for proximate analysis, total sulphur and calorific value. Float–sink tests are done but no simulated product is made.

In Australia and South Africa, the fine fraction (0.5 mm or 0.1 mm) is screened out before analysis, dependent upon expectations for coal preparation.

In the UK, a similar procedure is used except that for outcrop/core samples no float–sink analysis is carried out, only proximate analysis, total sulphur and calorific value.

In the USA, there are no defined crushed parameters; proximate analysis, total sulphur and calorific value are determined. Float–sink analysis is done with the results reported to zero; this often means that the fines have to be screened out. Because in the USA there is no hard and fast procedure for outcrop/core analysis, the individual procedures have to be verified in order to correctly assess the reliability of the analytical results.

5.3.2 Bulk samples

The procedure for sample preparation of bulk samples of both steam and coking coals is shown in Figure 5.8 for steam (thermal) coals and Figure 5.9 for coking

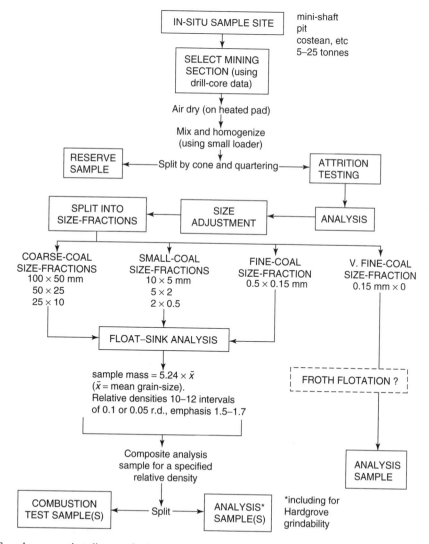

Figure 5.8 Sample preparation diagram for bulk sample(s) from a steam (thermal) coal deposit. From Osborne (1988) 'Coal Preparation Technology' published by Graham & Trotman, with kind permission of Kluwer Academic Publishers

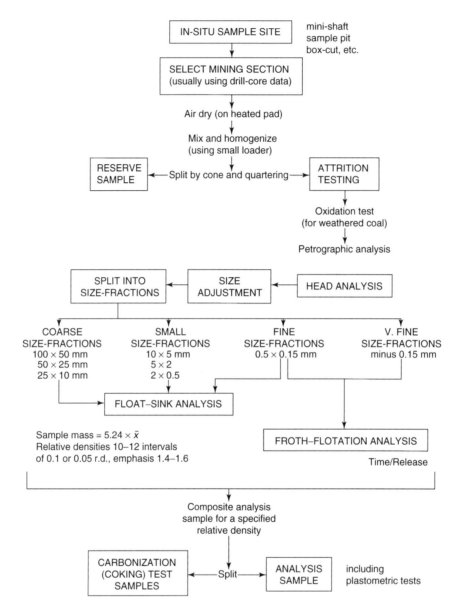

Figure 5.9 Sample preparation diagram for bulk sample(s) from a coking (metallurgical) coal deposit. From Osborne (1988) 'Coal Preparation Technology' published by Graham & Trotman, with kind permission of Kluwer Academic Publishers

(metallurgical) coals. The coals are analysed as for outcrop/core samples with the additional tests for combustion and coking properties.

5.3.3 Non *in situ* samples

The analysis of non *in situ* coal samples is undertaken to ascertain the quality of the coal leaving the mine, leaving the coal preparation plant, if one is installed,

and in stockpiles prior to shipment, to ensure that the agreed specification of the coal is maintained.

Stream samples will be crushed and analysed for proximate analysis, total sulphur and calorific value, plus other properties if requested; normally no float–sink analysis is done. In addition, other stream samples such as stop belt samples will be used for size analysis and float–sink analysis. Samples from small stockpiles are rarely taken, and although in large stockpiles augured samples may be taken, it is difficult to obtain a truly representative sample.

Various types of on-line analysers are currently used at coal-fired power stations, coal mines and coal-handling facilities. These are usually ash and moisture monitors or elemental analysers. Modern on-line analysers can provide high-precision ash, moisture, sulphur and energy monitoring, and are increasingly used on coal conveyor systems and for coal shipments. These are designed to withstand tropical weather conditions such as high humidity and heavy rainfall.

On-line analysers monitor dual-energy gamma-ray transmission for ash and microwave moisture measurement, and there are more expensive methods for measuring sulphur using gamma neutron activation analysis. Between these, cost-wise, X-ray fluorescence can be used, but this only penetrates a thin layer of material. These types of measurement can be influenced by the composition of the noncoal material; for example, if the shale fraction included in the coal stream is of marine origin it will have a higher radioactivity, and if pyrite is present, it can influence the background scatter by having high fluorescence.

6

Coal Exploration and Data Collection

6.1 INTRODUCTION

The principal objective in the exploration for coal is to determine the location, extent and quality of the resources available in a particular area, and to identify those geological factors which will facilitate or constrain mine development.

Such a role encompasses the evaluation of existing data, geological mapping and sampling, the use of geophysics and drilling. Once adequate resources of coal of suitable quality have been identified, the geological input will be concentrated on supporting the engineers in the design and development of the mine; this will include additional drilling and sampling succeeded by geotechnical studies.

The emphasis of geological input will gradually change from exploration to development without a break in continuity. Figure 6.1 illustrates the various stages in this process from exploration mapping and sampling through to reserve calculations, coal quality results and geotechnical investigations.

In the last 20 years, the use of microelectronics-based technology has resulted in significant changes in how coal exploration is carried out, in particular, how data are collected, analysed and presented. This technology has the ability to handle large amounts of information and maintain consistency, which has resulted in higher standards of data acquisition than was possible using only traditional exploration techniques. Nevertheless, the basic geological practice of observation and data recording is still widely used and forms the geological database on which computerised studies are developed.

6.2 FIELD TECHNIQUES

The field examination of coal-bearing sequences is an essential component of any exploration programme, particularly the identification and assessment of a new potential coal-bearing area. Field examination of surface exposures of coal is the precursor to formulating a drilling programme to identify coal in the subsurface.

The first step in carrying out a geological study of a selected area is to collate all available information on that area. This may include published geological maps, topographic and cadastral maps, scientific papers and reports, land records, air photographs and satellite imagery. If such information exists for the selected area, then the geological setting can be ascertained as well as topography, water supply and land access, and the availability of base maps. If no surveyed maps exist, photogrammetric maps constructed from available air photographs will be required to carry out ground surveys.

The bulk of this information is usually available from the Geological Survey or Land Survey Departments of the country in question. Additional data can be obtained from universities, libraries and local government departments, plus writers such as Knutson (1983), who outlined planning and implementation of coal exploration programmes in reconnaissance geology for coal exploration.

The study should make special note of any previous history of coal exploration or mining, and details of which companies were involved. In particular attention should be given to any coal quality data, and the basis on which they are quoted, reserves calculations and production figures.

The scale of maps is important; in order to carry out a reconnaissance survey, base maps of 1:20 000 and 1:50 000 will be adequate. However, for further detailed mapping and sampling, and for planning drilling sites at set intervals, a scale of 1:10 000 is necessary. In the case of mine operations, large-scale plans are required; those of 1:2000 and 1:5000 are most commonly used. If the area has not previously been surveyed, air photographs will be used for map compilation (see Section 6.2.2). Modern advances in remote sensing and global positioning systems (GPS) together with enormous improvements in communications have changed the 'pioneering' aspect of field coal exploration. However, fieldwork will be necessary at all stages of coal mine development, with

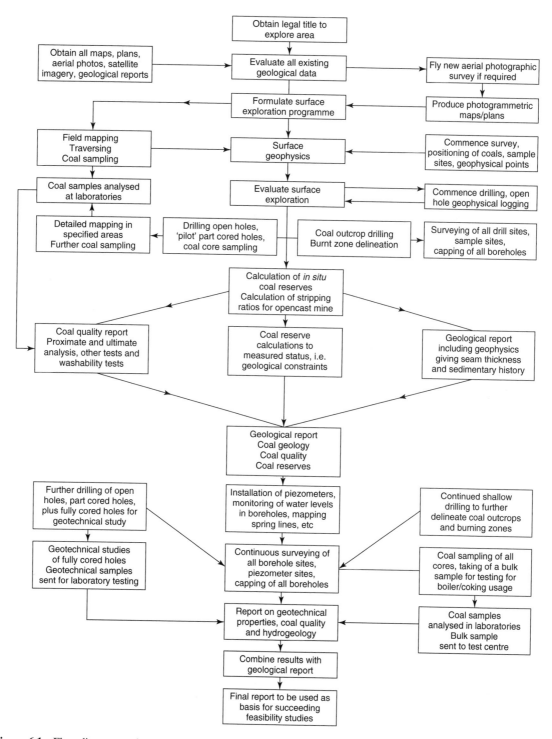

Figure 6.1 Flow diagram to show principal activities requiring a geological input during the exploration phase of a coal mine development project. Reproduced by permission of Dargo Associates Ltd

the concentration of field activities in the reconnaissance and exploration stages of the project.

The style of field surveys varies greatly; in tropical rainforest traverses are made along watercourses either on foot or by using some type of boat. Alternatively local agriculture or industry may have constructed roads or tracks with exposures of bedrock. In hill or mountain country, outcrops may be frequent but require a helicopter for access and observation, whilst in low-relief arid or scrub country, outcrop may be almost nonexistent, but allows unrestricted access by 4-wheel-drive vehicles and trail bikes.

Safety during fieldwork is of prime importance; suitable clothing should be worn dependent on the climatic region, e.g. tropical jungle boots for wet jungle conditions, together with strong cotton shirts and trousers with as many pockets as possible. When working in temporary pits or sampling coal, protective headwear and footwear are essential.

The principles of working in hostile terrain are a combination of common sense and experience (or learning from an experienced colleague). The adoption of routines to check one's own health and hygiene are essential at the end of a day's fieldwork. This greatly facilitates the next day's work and monitors any potential problems that might arise, e.g. the daily foot inspection is important as the feet take the brunt of fieldwork, particularly in mountainous and jungle areas. Carelessness about one's health can result in discomfort and an inability to carry out fieldwork, and in extreme cases, give rise to the necessity of leaving a remote field area, which may be difficult and expensive. Although such situations are more likely to occur in geographically difficult areas, such routines are still good practice in logistically accessible countries. In addition to health, at the end of each day's fieldwork all equipment should be checked for damage or losses, and all notebooks and field record sheets and maps should be clearly labelled and kept as dry and secure as possible.

Temporary and permanent campsites must be kept clean; food and equipment should not be left lying about. Kitchen, washing and toilet areas should be clearly separated and adhered to; toilet areas should never be in an upstream position to the campsite.

Campfires must be carefully controlled, especially in areas in which the vegetation is likely to ignite easily. Carelessness in this respect can result in destruction of habitat and/or property, and even loss of life to plants and animals in the area.

Each day's work should be planned in sequence, with time allowed for travelling, sampling and data recording. It is important to remember that the field mapping is the

only method apart from drilling by which basic information on the geology of an area is obtained. All later studies are only as good as the original information collected.

6.2.1 Outcrop mapping

The basic elements of geological mapping, rock identification and structural measurements have been described by Barnes (1981) and Berkman and Ryall (1987). Mapping is ideally suited to areas where coal-bearing sequences are exposed due to erosion, folding and faulting. However, it is common to have to evaluate areas with a scarcity of outcrops to provide at least some basic geological data.

In order to carry out fieldwork, the following items of equipment will be needed: topographical, geological maps and site plans (if any), map case, air photographs and pocket stereoscope, 'Chinagraph' pencils, notebook/field record sheets, marker pens, geological hammer, chisel, trowel/fold-up spade, polythene bags, sample bags and labels, clinometer/compass, hand lens, small and large tape measures, penknife and camera. Some of these items are shown in Figure 6.2.

The use of personal and laptop computers has speeded up the field input of data, and readily available software programs provide the ability to construct maps in the field for any desired purpose, e.g. a coal sampling programme.

The first aim of field exploration will be to determine the location, structural attitude and extent of coals and associated strata, together with structural features such as faults and fold axes and igneous intrusions, all of which if present influence future mining conditions. If for the area of interest geological information already exists, then further fieldwork may only involve verification of coal seam locations, taking fresh samples and filling in any gaps in the previous data.

In the absence of published base maps, the traditional method has been to produce plans by field traverses, usually in the form of tape and compass traverses. A long plastic tape measure (30 m, graduated in cm) is used along the traverse in conjunction with a compass bearing at the beginning of each measurement. All such traverses must be connected in closed loops and the closing errors between the surveys must be corrected before any geological information is plotted. The latter is usually done at base camp. The beginning and end of each traverse must be clearly marked, together with all distinct physical features such as hills, river bends, waterfalls, road crossings and buildings. Geological features such as thick sandstones, coal seam outcrops, sample locations, faults and fold axes will also be put on to the plan. Standard symbols used to portray geological elements on plans together with mining symbols are shown in Figure 6.3, and the graphic portrayal of the principal lithotypes found in coal-bearing sequences is shown in Figure 6.4. Figure 6.5 shows the results of a typical traverse survey in dissected terrain using these methods. It is important that these identified features can be revisited at a later date to survey the area and plot elevations accurately using a theodolite.

Figure 6.2 Field equipment used by coal geologists. Reproduced by permission of Dargo Associates Ltd

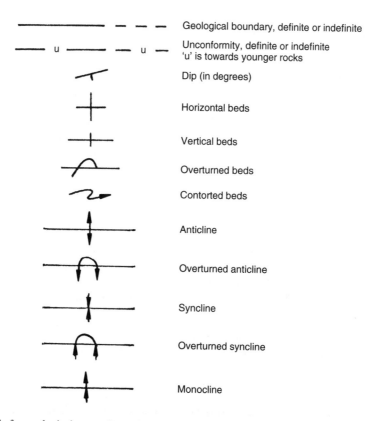

Figure 6.3 Symbols for geological maps. Reproduced by permission of Dargo Associates Ltd

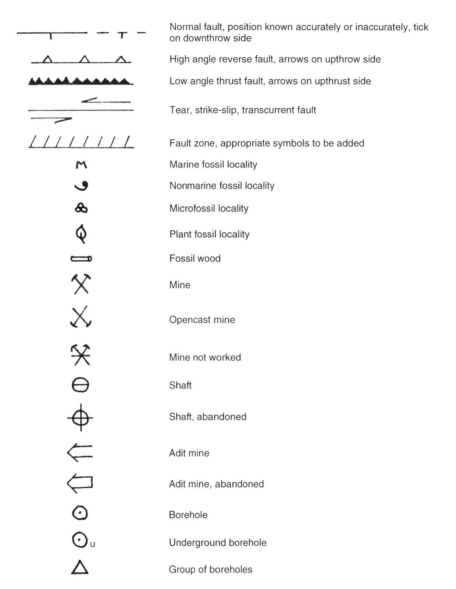

Figure 6.3   (*continued*)

Field traverses should record all geological features seen; when lithological associations have been recognised they should be linked up wherever possible between traverses. Extrapolation across country from one traverse to another should take into account effects of dip and topography. Where lines obviously do not tie up, then this may be the effect of faulting and should be checked on the ground and on the air photographs.

Modern methods of positioning in the field are with the use of GPS; these are digitally based surveying systems together with digital topography. The use of these systems has greatly increased the accuracy of positioning outcrops etc. in the field. These techniques allow the rapid location and installation of control points for the development of digitally based topographic maps.

Significant coal-bearing outcrops and the surrounding strata should be mapped in detail; stratigraphic sections should be measured where exposure permits. Individual units such as coal beds and marker beds should be traced laterally to ascertain their lateral correlation. Where possible, the environments of deposition should be interpreted during section measuring and coal seam

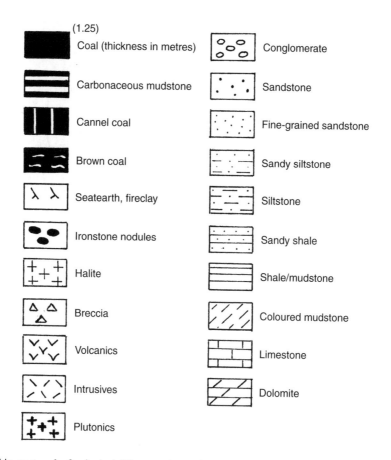

Figure 6.4 Graphic portrayal of principal lithotypes in coal-bearing sequences. Reproduced by permission of Dargo Associates Ltd

correlation. Care must be taken in measuring sections of strata which may be faulted; low angle or bedding-plane thrust faults can pass undetected through a coal section, which can result in exaggerated thickness or missing intervals. Gentle folding can change along strike to isoclinal and recumbent folding.

Recognition and tracing of marker horizons is important; these may be beds of distinctive lithology such as volcanic deposits or limestone, or beds containing fauna such as foraminifera, bivalves and other organisms, or beds containing floral assemblages in the form of plant remains, spores and pollens. Such distinctive horizons can serve to correlate coal seams, identify structural dislocations and establish facies patterns.

National Geological Surveys adopt this approach when mapping coalfield areas. Figure 6.6 shows a typical field map produced by the British Geological Survey for a coalfield area where a large amount of data has been compiled to produce the final geological map.

It is essential to be able to recognise all the lithological types associated with coal and coal-bearing sequences; the ability to do this undoubtedly improves with experience. In the USA, handbooks have been produced illustrating in colour all the lithologies found in coal-bearing sequences in the eastern USA (Ferm and Smith 1980; Ferm and Weisenfluh 1991). These are invaluable to the inexperienced geologist, and, for the most part, can be used on a global basis, allowing for the occasional local lithological term. The descriptions of the varieties of sandstone, siltstone, shales, mudstones and coals and their intimate interrelationships are a fundamental part of the data recording stage of fieldwork.

The lithological description of coal itself is essential to the understanding of the physical subdivisions (plies) of the coal. Most coals have components of bright and dull coal, ranging from all bright to all dull with combinations of both in between. Brightness in coal is

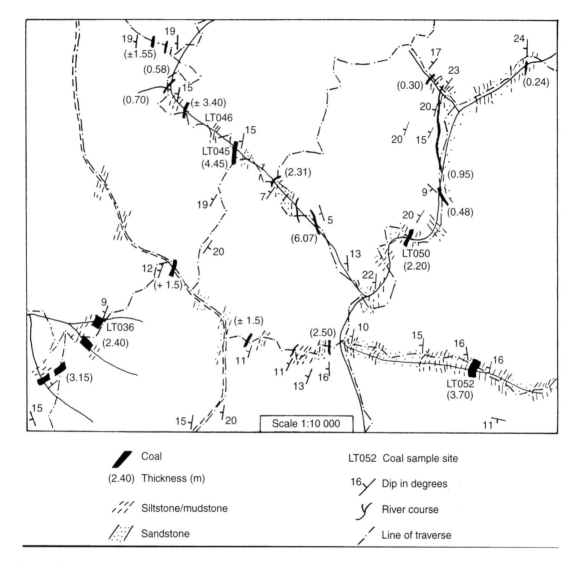

Figure 6.5   Typical traverse survey showing coal outcrops, sample sites and structure. Reproduced by permission of Dargo Associates Ltd

indicative of the ash content: the brighter the coal, the lower the ash content.

Figure 6.7 shows examples of bright coals: a banded bituminous coal showing pyrite mineralisation on the cleat face from the Carboniferous of the UK; a bright nonbanded high volatile coal of Tertiary age from Indonesia; a bright high rank anthracite from the Jurassic, Republic of Korea; and a banded, high ash Gondwana coal from India.

Figure 6.8 gives the graphic representation of coal based on but not identical to Australian Standard

ASK183-1970. A shading, ruling and letter system is used, beginning with bright coal (>90% bright coal) ranging to dull coal (<1% bright coal); also illustrated are symbols for cannel coal, weathered and heat-altered coals.

The system used in South Africa closely follows the Australian Standard, and this is also the case for countries influenced by these areas, e.g. Indonesia.

In the UK, British Coal has used a system of graphic representation for coals in its underground operations. These range from bright coal to banded coal to dull

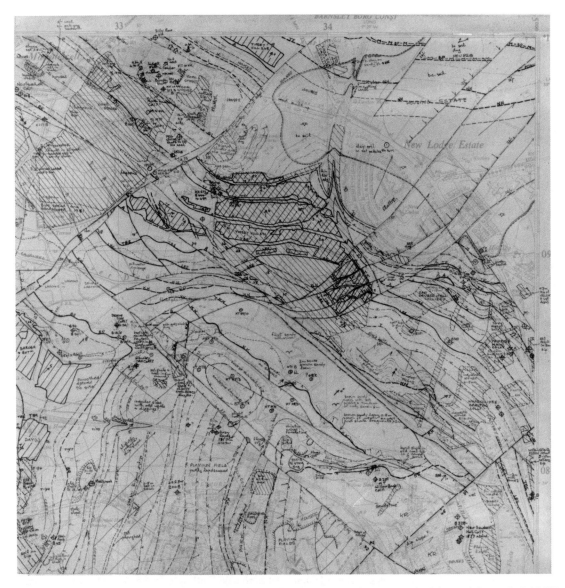

Figure 6.6   Field map of a UK coalfield area, showing geology and past and present coal mining activity; scale 1:10 000. 1PR/25-6C British Geological Survey. © NERC. All rights reserved

coal; Figure 6.9 shows the symbols used, together with those for cannel and dirty coal.

In the USA, there is no standardised system for the graphic representation of coals; however, an example produced for the ASTM is shown in Figure 6.10. Coals are described as ranging from bright to intermediate bright, intermediate dull and dull. Coals with high mineral content (bone coal) are also shown.

In addition to the coal itself, a careful description of the roof and floor of the coal seam is necessary to provide a useful framework for later geotechnical and mining studies.

A description of a coal seam section should include the following:

1. Composition of coal seam roof/floor: siltstone, sand-stone, carbonaceous shale etc.
2. Structure of coal seam and immediate roof/floor: faults, strike/dip, stratigraphic displacement.
3. Coal cleat: face cleat, end cleat, strike.

(a)

(b)

Figure 6.7　Varieties in black (hard) coal. (**a**) Banded bituminous coal from Northumberland, UK with pyrite mineralisation on the cleat face; (**b**) high volatile, low-ash bituminous coal, nonbanded, from East Kalimantan, Indonesia; (**c**) bright, high-rank anthracite, highly tectonised, from Samcheog Coalfield, Republic of Korea; (**d**) banded, high-ash bituminous coal from Talcher Coalfield, India. Photographs by M.C. Coultas and LPT. Reproduced by permission of Dargo Associates Ltd

(c)

(d)

Figure 6.7   (*continued*)

4. Slickensides: frequency, continuity.
5. Joints: strike/dip, frequency, continuity.
6. Chemical structures: nodules, ironstone, pyrite, concretionary structures.
7. Soft sediment structures: slumping, folding, liquefaction structures.
8. Degree of weathering: from fresh to completely weathered.
9. Roof/floor and/or seam structure: flat, rolling, discontinuities of bedding, splitting.
10. Mineralisation.

6.2.2 Remote sensing

6.2.2.1 Satellite imagery

The most widely used sources of satellite remote sensing imagery have been from the Landsat series of the

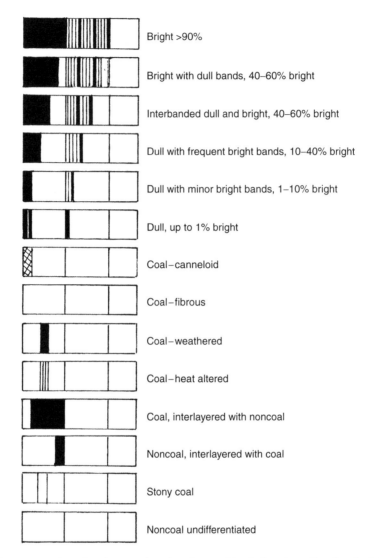

Figure 6.8   Graphic representation of coal seams: based on but not identical to Australian Standard ASK183-1970

Earth Observation Satellite Co. (EOSTAT), and from the French SPOT series of spacecraft. Both have worldwide coverage from which data can be obtained in computer-compatible tape (CCT) and photographic form for ground scenes, each 34 225 km^2 in area. Landsat data are a direct record of the sensor data and can be processed into images which can be enhanced to highlight selected features and enlarged to scales of 1:50 000 for map construction or 1:24 000 for geological field maps.

Landsat Imagery illustrates regional fault patterns and the different geological 'imprint' of a variety of lithological successions within delineated areas. Landsat Imagery is provided in four separate bands of the visible spectrum, from blue to infrared. In tropical terrain with thick vegetation cover, the red and infrared spectrum is most useful for interpreting geological features.

Landsat interpretation is usually used in conjunction with geological and geophysical maps, if available. As well as highlighting geological structures, variations in rocks, soils and vegetation is provided in each spectral band. Comparison of data from a number of bands at their different wavelengths allows types of lithology and ground surface cover to be recognised.

SPOT collects data for ground scenes, each of 3600 km^2 in area, and are also available as computer tapes and derived photographic images.

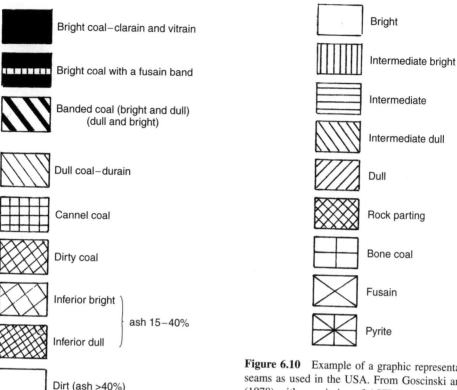

Figure 6.9 Graphic representation of coal seams as used in the UK by kind permission of the British Coal Corporation

Figure 6.10 Example of a graphic representation of coal seams as used in the USA. From Goscinski and Robinson (1978) with permission of ASTM

Other sources of satellite imagery include data from a number of international satellites together with photography from manned spacecraft. Both shuttle multispectral infrared radiometer (SMIRR) and shuttle imaging radar (SIR) have geological applications.

Satellite-borne radar has the great advantage of penetrating the persistent cloud cover and forest cover that characterise large areas, particularly those in the tropics.

6.2.2.2 Airborne imagery

Air photographs have been available for the last 80 years, although not for all areas of the world. Photogeological mapping is an established technique in reconnaissance, as it provides an economical and effective map on which geological data are located to ground features, particularly the topography and structural style of an area. Such a framework can then be used for more detailed coalfield exploration in the form of ground surveys.

The use of overlapping air photographs to provide stereoscopic (three-dimensional) interpretation of areas will give a more detailed geological picture. The scale of the photographic coverage, usually 1:50 000 but can be 1:25 000, provides a basis for making photo-interpretation maps of prospective coal deposits by defining regional geological features such as fault lines and zones, persistent lithological horizons, major fold patterns and amounts and changes of dip and strike. On plans the symbols used to portray such geological features are shown in Figure 6.11.

When no, or inadequate, air photograph coverage is available, it is necessary to prepare a new set of photographs to work from. This enables a ground crew to mark out 'targets' on the ground which can later be accurately surveyed and used as permanent reference points. The scale of the photographs should be in the same order of magnitude as the scale required for the ground plans. For the compilation of maps and plans from air photographs see Barnes (1981) and Berkman and Ryall (1987).

Dense vegetation cover can mask the ground surface, but can still reflect closely the underlying geology. When combined with fieldwork, subtle lithological differences become apparent. Such photographic interpretation is essential when used in reconnaissance and detailed exploration fieldwork, in identifying traverses and pinpointing physical features on those traverses, e.g. river junctions and road intersections, coal outcrops and for locating preliminary drilling sites.

The drawbacks with air photographs are that in areas of high rainfall, cloud-free conditions are infrequent;

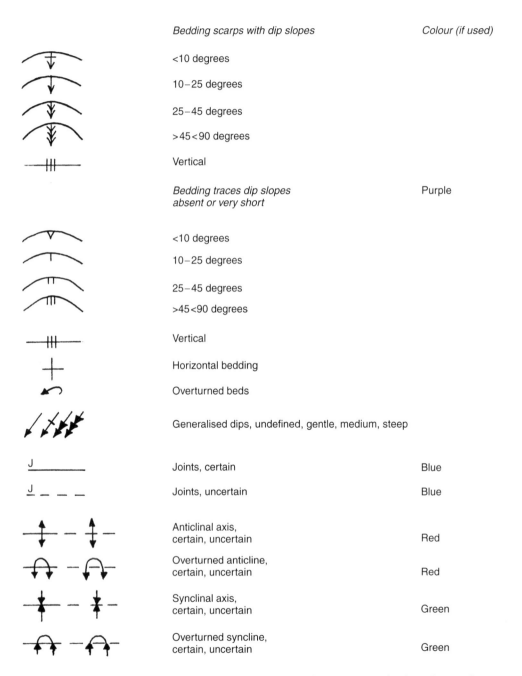

	Bedding scarps with dip slopes	*Colour (if used)*
	<10 degrees	
	10–25 degrees	
	25–45 degrees	
	>45<90 degrees	
	Vertical	
	Bedding traces dip slopes absent or very short	Purple
	<10 degrees	
	10–25 degrees	
	25–45 degrees	
	>45<90 degrees	
	Vertical	
	Horizontal bedding	
	Overturned beds	
	Generalised dips, undefined, gentle, medium, steep	
	Joints, certain	Blue
	Joints, uncertain	Blue
	Anticlinal axis, certain, uncertain	Red
	Overturned anticline, certain, uncertain	Red
	Synclinal axis, certain, uncertain	Green
	Overturned syncline, certain, uncertain	Green

Figure 6.11   Photogeological symbols for use on aerial photographs and photogrammetric plans (from various sources). Reproduced by permission of Dargo Associates Ltd

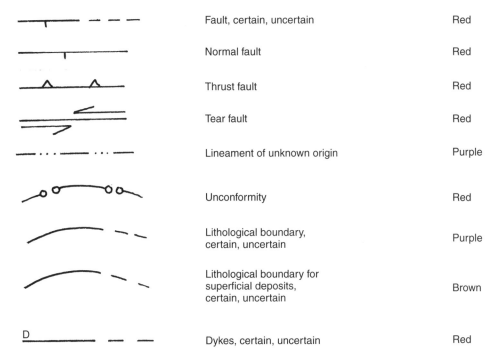

Fault, certain, uncertain		Red
Normal fault		Red
Thrust fault		Red
Tear fault		Red
Lineament of unknown origin		Purple
Unconformity		Red
Lithological boundary, certain, uncertain		Purple
Lithological boundary for superficial deposits, certain, uncertain		Brown
Dykes, certain, uncertain		Red

Figure 6.11 (*continued*)

this prohibits good photographic coverage of the area. The topography may be severely dissected so that the photographs are too strongly distorted for photogrammetric mapping. Colour infrared photography may also be used.

Side Looking Airborne Radar (SLAR) is also used in exploration, but has poorer resolution than photography, particularly where vegetation and cloud cover limit the use of air photographs. However, linear patterns do show up better than actual geological features. SLAR can be used to compensate for the lack of air photograph cover, as it can highlight structural features not seen at ground level, accurately locate structural elements in the field, and provide a regional structural framework that can be used for planning field traverses and exploration drilling grids. SLAR has been used extensively in the USA and Australia (Hartman 1992).

6.3 DRILLING

Once the field survey has been completed, the position and attitude of all coal seams will be plotted on the base plan. If the dip, structure and initial quality results indicate that the coal is of economic interest, then the next stage in exploration will be the locating of drill sites to provide data in those areas

between known coal outcrops and in areas where no outcrops have been located but in which coal is thought to occur.

It is important that the geologist maintains a good and close relationship with the drilling supervisor, drillers and mechanics during the drilling programme. In order to maintain proper records of the drilling operation, the geologist should liaise with the drilling supervisor to ensure that the driller records the following information for each shift drilled: site details, borehole number, details of openhole drilling, details of hole diameter, details of casing sizes and depths (if used), details of each core run, length of core recovered, details of water encountered, strata description, details of timing of each operation, details of flush losses and bit changes, details of core barrel and core bit type, and details of drilling crew.

If the geologist is supervising junior or less experienced geological staff who may be assigned to a particular drilling rig, the procedure that should be followed is outlined in Figure 6.12. This will ensure that the borehole will be completed in the correct fashion by recording the presence or absence of all coal seams targeted, and that the borehole is drilled to at least 5.0 m below the lowest coal seam to allow geophysical logging tools to record the total coal section.

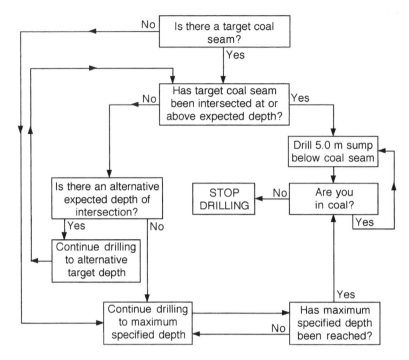

Figure 6.12   Drilling procedure to correctly complete a borehole containing one or more targeted coal seams. Reproduced by courtesy of M.C. Coultas

6.3.1 Openhole drilling

Exploration drilling will be carried out to determine coal seam depth, thickness and quality at any given point within an area, and details of the strata associated with the coal.

For the majority of boreholes drilled in the main area of interest, rotary drilling rigs are used. These rigs have good penetration rates, relatively low cost and are mobile. They provide the most economical means of shallow (down to 400 m) openhole and core drilling of coal deposits. Exploration boreholes are usually vertical, but in areas of high dips inclined boreholes may be drilled, particularly in underground workings.

Rotary rigs are usually truck mounted but can be adapted to fit on a bulldozer (Figure 6.13) or on skids. The rigs can use high-pressure air circulation (air flush), water or drilling fluid such as bentonite mud (fluid flush) to cut the rock with tungsten carbide bits. The introduction of high density polymer foam to facilitate cuttings removal and stabilise hole sidewalls is now replacing bentonite mud. The use of foam has the added advantage of allowing boreholes to be later used as water observation or abstraction wells. In principle, a string of metal rods is rotated axially, and a bit at the base of the string is forced downward under controlled pressure, cutting into the sediments, therefore advancing the depth of the hole. Rock cuttings are circulated away from the bit and lifted to the surface by means of the pumped fluid or compressed air. Several types of drill bits are available; the blade bit gives a high penetration rate, needs little maintenance, and the 'blades' can be re-sharpened on site. They also have the added advantage of providing larger rock cuttings, so facilitating identification of the lithologies by the geologist logging the borehole. Blade bits are often changed for roller bits when drilling harder strata such as limestone.

Air-flush drilling rigs have higher penetration rates for noncore rotary drilling, where compressed air is used in place of water to cool the drill bit and flush the cuttings out of the hole. Air-flush also brings up the cuttings much faster, enabling the position of lithological changes to be located more accurately. Air-flush drilling is impeded by high rates of groundwater influx; small compressors then find difficulty in maintaining enough pressure to lift the rock cuttings to the surface. However, above the water table, air-flush drilling is particularly useful, although there is the drawback of producing large amounts of dust which may be environmentally unacceptable in certain areas. Again, air-flush drilling

Figure 6.13 Dando dual air–mud flush rig mounted on a bulldozer for use in difficult tropical terrain. Photograph by M.C. Coultas

may be a better option when drilling in zones of burnt coal or broken strata where there is a likelihood of loss of drilling fluids.

From the point of view of the geologists, water and mud circulation methods are messy, and do not allow rapid changes of lithology to be noted easily. The cuttings are slower to reach the surface, which allows a certain amount of mixing resulting in the less accurate positioning of lithological boundaries. The use of polymer foam has enabled this problem to be overcome and is now more widely practised in the industry.

There are numerous makes and types of rotary rigs; examples of those widely used are the Dando (Figure 6.13) and Edeco (Figure 6.14(a)) dual-air and fluid-flush rigs, and the Mayhew 1000 (Figure 6.14(b)), a fluid-flush rig.

6.3.2 Core drilling

Cored boreholes are drilled to obtain fresh coal samples and a detailed record of the complete lithological sequence associated with the coals.

Core drilling can be accomplished by using rotary rigs such as the Dando Mintec range of drills, or by diamond drill rigs such as the Boart Longyear and Edeco Stratadrill ranges. These rigs use a tungsten carbide or diamond bit attached to a series of metal rods, the lowest of which is designated a core barrel, and rotated under downward pressure; however, diamond drilling requires

the use of drilling fluids. A circle of rock is ground away and a cylindrical core remains in the hollow centre of the core barrel. The recovery of the rock core is facilitated by a nonrotary second metal tube within the core barrel; the core passes into this tube and is protected from damage. This is called a double-tube core barrel. Even better core recoveries can be obtained by placing a smooth metal tube, split longitudinally, inside the non-rotating inner segment of the double-tube core barrel; this type of equipment is called a triple-tube core barrel.

When the core is to be removed, the split inner tube is withdrawn with the core still inside; it can then be laid out horizontally and the upper half of the inner tube removed to expose the core for examination. It is then transferred to a segmented core box for logging at a later stage.

Removal of the core barrel can be time-consuming. Some equipment allows the central part of the core barrel to be drawn up the centre of the hollow drill rods on a steel cable; this technique is known as 'wireline' drilling. The rods themselves are only removed when the drill bit needs to be changed.

The different diameters of core barrels for rotary and diamond drill rigs are given in Table 6.1. As the diameter of the core decreases, there is a tendency for the core to break up and core losses to increase. It is general practice that the core recovery through a coal or coaly horizons should be not less than 95%. Boreholes drilled in soft sediments or unconsolidated deposits often have unstable sides, particularly in the top part of the hole.

(a)

(b)

Figure 6.14   (a) Edeco rig operating in UK; (b) Mayhew 1000 truck mounted rig, operating in tropical conditions, Indonesia. Photographs by M.C. Coultas

Table 6.1 Core sizes for wireline, conventional and air-flush drilling

1. Wireline core barrels

Q Series (Boart Longyear)	AQ	BQ	NQ	HQ	PQ	
Hole diameter	48.0	60.0	75.8	96.0	122.6	
Core diameter	27.0	36.5	47.6	63.5	85.0	
Diamand-Boart	ADBG.AQ	BDBG.BQ	NDBG.NQ	HDBG.HQ	PDBG.PQ	SDB.G
Hole diameter	47.6	59.6	75.3	95.6	122.2	145.3
Core diameter	27.0	36.4	47.6	63.5	85.0	108.2

2. Conventional core barrels

Double tube swivel type	HWF	PWF	SWF	WWF	ZWF
Hole diameter	98.8	120.0	145.4	174.5	199.6
Core diameter	76.2	92.1	112.8	139.8	165.2

3. Air-flush core barrels

	412F	HWAF
Hole diameter	105.2	99.4
Core diameter	75.0	70.9

Note: diameter sizes in mm.

Figure 6.15 Borehole core laid out in wooden boxes with depth markers, awaiting examination by the geologist. Photograph by LPT. Reproduced by permission of Dargo Associates Ltd

In order for the drill rods to rotate and drill correctly, casing is inserted into the hole to support the collapsing section of the hole. Casing may be metal or PVC, and is normally pulled out of the hole once logging has finished.

The core should be placed in a tube of polyethylene sheeting in the core box; this ensures that there is no loss of moisture and that no oxidation of the coal occurs prior to analysis. Ideally the core should be photographed before sealing in the polyethylene, using a measuring tape for scale. The core should be labelled with the borehole number and depth indicators.

Core boxes are usually made up of three or four one-metre compartments with lids, usually of wood, as shown in Figure 6.15, but metal ones are used in areas subject to fungal and termite damage.

It is essential that the core is placed in the core box in the correct stratigraphical order; occasionally drillers put core in boxes the wrong way up. A comparison with the geophysical log usually shows the error.

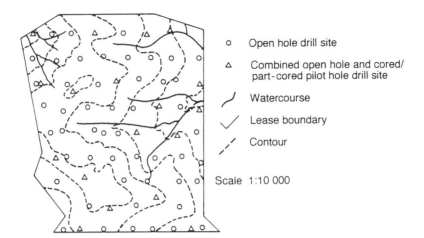

Figure 6.16 Exploration drilling grid showing distribution and position of open holes and part-cored pilot holes. Reproduced by permission of Dargo Associates Ltd

The polyethylene sheeting is opened up to log the core, and any core losses are calculated. The roof and floor of the seam are measured; the coal seam lithotypes are recorded in detail together with any partings or splits. In addition the degree of weathering, mineralisation and any structural features in the coal and associated strata are recorded.

Fully cored boreholes are rare in the exploration phase of a coal deposit; usually boreholes are only cored for the coals and the coal roof and floor, the depths of these being predetermined by previous openhole drilling. These part-cored boreholes are sited as 'pilot' holes next to completed openholes in order to accurately predict the depths and thicknesses of the coal seams. Figure 6.16. shows a typical exploration borehole grid with openholes and part-cored pilot holes.

Fully cored boreholes are usually drilled during the geotechnical investigation stage of the project to fully examine the strengths and structural character of the coals and the associated strata.

6.3.3 Portable drilling

Portable drilling, as the name suggests, involves the use of drilling equipment that can be dismantled into man-portable components; this is particularly useful in mountainous and jungle terrain where access with conventional drilling rigs is difficult. They normally operate by using a small motor (lawn mower or power saw type motor) which drives an axially rotating set of drill rods with a small blade or roller bit at the bottom; the holes are circulated with water or drilling fluid. If the rig motor is not very powerful, in order to obtain greater

penetration, the downward pressure may be increased by adding increasing numbers of personnel to sit/stand on the top of the rig as shown in Figure 6.17.

These small portable rigs are capable of drilling to depths of 60 m and are used to prove coal outcrops, to complete gaps in stream or road sections, and to delineate limits of underground burning zones in coal seams. There are several commercially produced rigs, the Minuteman and the Voyager 1000, 2000 and 2400 Series, as well as small drills made up by companies for specific use on their own projects.

6.3.4 Core and openhole logging

6.3.4.1 Core logging

Because core drilling is an expensive part of the exploration and development programme, cores that are obtained should be logged in as much detail as possible, particularly the coal seams and their roofs and floors.

Core logging is usually carried out in a core shed, where benching is provided on which are placed those core boxes currently being logged; Figure 6.18 shows typical conditions under which core logging takes place. The core shed also provides storage space for the cores already logged, as well as for new core awaiting examination. Core sheds are important for protecting the cores (and the geologist) from the elements; it is worth remembering that the quality of the core log can vary with the conditions under which the geologist has had to work.

The core box will be marked with the depths that the core run commenced and ended; great care should be taken in relating these markers to the depths and

Figure 6.17 Portable drilling rig using manpower to exert downward pressure on the drill bit. In use in East Kalimantan, Indonesia. Photograph by LPT. Reproduced by permission of Dargo Associates Ltd

thicknesses of the lithologies actually cored. Core losses may occur and these should be clearly marked; thickness and depth figures will have been reconciled to the geophysical log which will have been run after coring has been completed (see Section 7.5).

Where possible, a photograph of the cored material with way up and depths clearly marked should be taken for the record as shown in Figure 6.15, and any special feature of the coal or its contact with the beds above and

below should be photographed. Figure 6.19 shows core exhibiting an erosive sandstone roof to a coal seam. The use of digital photography allows photographs of the core to be reviewed on the computer and then printed as required.

The state and condition of the core should be described. Complete solid core is a solid core attached to the roof and floor; a fragmented solid core is a broken seam but all the fragments join up, so that there is no

Figure 6.18 Coal geologist logging borehole core in a core shed on site. Photograph by LPT. Reproduced by permission of Dargo Associates Ltd

doubt of the core recovery. Part core indicates that only part of a solid length of core has been recovered, but that all lithotypes are present. Fragments indicate that no cores fit together, and that it is not possible to state accurately what length of the core the fragments represent.

Once the measurements of seam roof and floor boundaries, major partings, core loss and any other significant features have been reconciled, and the individual ply sections have been identified, the core can then be split for detailed lithological logging. Splitting is achieved by using a wide chisel or bolster and hammer to split the core lengthwise, making sure that the split is a fresh surface and does not follow a joint.

Low rank coals such as lignite can be cut lengthwise using a saw; one half can then be examined for colour and lithological variations, whilst the other is left until the core has sufficiently dried out before the texture of the lignite can be described.

Core logging data can be recorded either as a written descriptive log on printed coding sheets designed for computer usage, or directly into a portable computer. Figure 6.20 shows a core logging sheet giving depths, lithological description, rock strength, weathering, bedding character, and sample numbers.

As a general guide, the following features should be recorded:

1. Lithology: dominant lithotype, colour and shade, grain size and sorting, distinctive mineralogy and

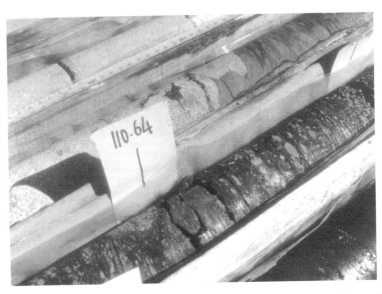

Figure 6.19 Borehole core photographed to show special features. In this instance, the erosive sandstone contact above the coal and the siderite nodules contained in the upper part of the coal seam are seen. Photograph by M.C. Coultas

Project code Type	Number	Suffix	Geophy. Co-ord. Datum	Easting	Northing	Elevation	Total Depth	Depth to Base of Weathering	Depth to Water Table	Core Size (mm)	Incl.	Azim.	Units
CCL	51					24.45	115.82	.	.	63.5			

Commenced	Completed	Geo.	Drill Contr.	Driller	Geoph. Contr.	Plugs	Casing Depth	Status	Other Records
12/1/87	15/1/87	JGC							

HOLE No. CCL 51
PAGE 2 OF 3
ENCODED BY JGC

Depth to Base of Unit	Recovered Thickness	Rock type	Shade	Colour	Lith. Adj.	Grain Size	Tex.	Weath	Mec. St.	Minor Rock Rel.	%	Bedding Sed. Str.	Base	Dip	Cond.	Disc. Spac. Dip	Qual. Adj. A Adj. Qual.	Form-ation	Seam	Split	Occur.	Sample Number	
.	.	CT		DEBDB				S	R2	IB													
60.98	0.61	LG		DB				S	R2			XLIL	DF	10				LNK				CCLS1HS2	
62.01	1.03	LG		DEB					S5		71	IL		10				LNK				CCLS1HS3	
.	.	CT		DB				S	R2	TM													
65.40	3.39	CL		LGMG	ST				S5		87							LNK					
.	.	IS		MGBF	SR	FM			R4	TN													
.	.			IRONSTONES AT: 62.84–62.92, 63.11–63.21, 63.50–63.57, 63.98–64.06, 64.84–64.96m																			
65.55	0.15	CL		MG	SF				S5									LNK				CCLS1LL4	
65.70	0.15	CL		MG	ST				S5				IN					LNC				CCLS1LL5	ROOF LOWER LIGNITE 65.70m
66.75	1.05	LG		DBBK				S	R2			XLIL		10			SCYLM	LNK	LLG	C		CCLS1LL6	
67.70	0.95	LG		DBRB				S	R2			XL				J3V	CWLLN	LNK	LLG	C		CCLS1LL7	
68.72	1.02	LG		DBRB				S	R2		68			5		F3L		LNK	LLG	C		CCLS1LL8	
.	.	CT		DBBK				S	R2	TU													
69.80	1.08	LG		DBRB				S	R2		81	XL		5		F3L		LNK	LLG	C		CCLS1LL9	
.	.	LG		DBBK	WD			S	R3	XN													
70.90	0.90	LG		DBRB				S	R2		69	XL		5		F4L		LNK	LLG	C		CCLS1LL10	
.	.	CT		DBMB				S	R2	BU													
73.19	2.29	LG		DBRB				S	R2		91	XLIL		5		F3W		LNK	LLG	C		CCLS1LL11	
.	.	CT		DEBMB				S	R2	XN													
74.52	1.33	LG		DBRB				S	R2		88	XL		5		F4W		LNK	LLG	C		CCLS1LL12	
.	.	CT		DB				S	R2	MU													
76.00	1.48	LG		DBRB				S	R2		89	XL				F3W		LNK	LLG	B		CCLS1LL13	
.	.	LG	D	OB	FC		d	S	R1	TN													
82.00	6.00	LG		DBRB				S	R2		95	MS				F3W		LNK	LLG	B		CCLS1LL14	
.	.	LG	D	OB	FC		d	S	R1	XN													
88.00	6.00	LG		DBRB				S	R2		94	XL		5		F3W		LNK	LLG	B		CCLS1LL15	
.	.	LG	D	OB	FC		d	S	R1	XN													
94.00	6.00	LG		DBRB				S	R2		89	XL		5		F3W		LNK	LLG	B		CCLS1LL16	LL2/LL1 BOUNDARY RE-ASSESSED AT 91.88m
.	.	LG	D	OB	FC		d	S	R1	TN													SAMPLE 16 THEREFORE INCLUDES SPLITS A AND B
95.18	1.18	LG		DB				S	R2		70	XL				F3L		LNK	LLG	A		CCLS1LL17	BOUNDARY IS TAKEN TO NEAREST SAMPLE i.e. 94.00m.

END OF RECORD

Figure 6.20 Example of a core logging sheet used by the coal geologist in the core shed. Key to core logging sheet: (CL) clay; (CT) clayey lignite; (IS) ironstone; (LG) lignite; (D) dark; (BK) black; (BF) buff; (DB) dark brown; (EB) grey–brown; (LG) light grey; (MB) medium brown; (MG) medium grey; (OB) orange–brown; (RB) red–brown; (FC) finely comminuted; (SR) sideritic; (SF) smooth; (ST) silty; (WD) woody; (FM) fine–medium; (S) slightly weathered; (R1) very weak rock; (R2) weak rock; (R3) moderately weak rock; (R4) moderately strong rock; (S5) very stiff soil; (IB) interbedded; (TN) thin interbeds; (BU) towards base of unit; (MU) towards middle of unit; (TM) towards middle and top of unit; (XN) very thin interbeds (20–60 mm); (IL) irregularly laminated; (MS) massive bedding; (XL) thinly laminated (<6 mm); (DF) diffuse base; (IN) inclined base; (F3) medium spaced (200–600 mm); (F4) closely spaced (60–200 mm); (L) low angled; (V) vertical; (W) low and medium angled; (C) common; (S) sparse; (CY) clayey; (LM) laminated; (LN) lignitic; (WL) woody lignite. Reproduced with permission of Antrim Coal Company Ltd

Figure 6.21   Coal geologist logging borehole chip samples in an on-site core shed. Photograph by M.C. Coultas

cementation, associated lithotypes and relative proportions.

2. Sedimentary characters: thickness and type of bedding and lamination, types and dimensions of cross-bedding, bioturbation, disturbed bedding, contacts with units above and below, fossil content.

3. Mechanical characters: degree of weathering, degree and types of fracturing, orientation and frequency, strength characteristics, mineralisation.

6.3.4.2 *Openhole logging*

The logging of open boreholes involves the identification of rock chippings collected for every metre drilled. The chippings are washed, laid out on polyethylene sheets on benches, and then examined by the geologist, as shown in Figure 6.21. The basic lithology is recorded, and the depth at which the predominant lithology in the chippings changes. The accuracy of the depths of the top and bottom of important lithologies, such as coals and thick sandstones, can be reconciled when compared with the geophysical log of the borehole. The lithologies are recorded on data sheets in exactly the same fashion as cored boreholes except that there will be less geological detail for the individual lithologies encountered.

Open borehole logs are important as they give the best indication of where to site cored boreholes, and to predict the depths at which coals can be expected to occur; this is particularly important for the siting of part cored boreholes in which only coal cores are required for analysis. The coal geologist can expect to spend the greater part of his/her time logging open boreholes in the exploration stages of a project.

6.4   GEOTECHNICAL PROPERTIES

The geotechnical logging of surface exposures in trial pits, dug sections and more particularly in fully cored boreholes is an integral part of the overall geological studies of a coal deposit prior to the engineering studies to determine mine design and the specification of the coal product to satisfy market requirements.

The geotechnical logging of surface excavations is similar to that of cored boreholes; those at the surface will tend to give greater information in a lateral sense, while borehole cores give better control in a vertical sense.

The fully cored boreholes drilled during the exploration and reserve-proving stages of mine development provide a large amount of information useful to the geotechnical engineer. Therefore it is particularly important to record as accurately and in as detailed a manner as possible all the relevant geological information. As well as the basic lithological data, the detailed recording of discontinuities (their type, attitude, spacing and density) will provide valuable data to ensure safe mine design and working methods.

The drilling history of each borehole may indicate where drilling difficulties, loss of circulation and core losses have occurred. These may take on a new significance as the data are evaluated. The potential for inducing fractures into the core is greatest during the extrusion and subsequent handling and transport of the core. The use of core barrel liners is recommended, as this not only enhances recovery but minimises damage to the core during extrusion and transport.

The general description of the rock materials can be as follows: strength, e.g. moderately weak; weathering, e.g. fresh; texture and structure, e.g. thinly cross laminated; colour, e.g. light grey; grain size, e.g. fine to medium; name, e.g. sandstone; other properties, e.g. slightly silty with mudstone laminae.

Definitions of the various levels of strength, weathering, texture, structure and grain size vary. Those given below are the definitions given in British Standard 5930. Description of the engineering properties of rocks is well documented in Hoek and Brown (1980) and Fookes (1997), and logging techniques are described by Deere (1964) and the Geological Society Engineering Group Working Party (1970).

6.4.1 Strength

There are several kinds of rock strength, which can only be accurately determined by testing in the laboratory. Testing is designed to measure both the stress needed to rupture a rock and the strain developed during the application of stress. It may be argued that in the absence of laboratory test results the estimates of material strength are subjective. However, they should always be made, using the guidelines provided, as it is possible to use the later laboratory results from a selected sample to 'calibrate' the logger's assessment of the material strength.

Field guidelines to the assessment of material strength by minimal inspection and handling are given in Table 6.2.

6.4.2 Weathering

The degree of weathering is an important element of the full description of the rock material and should always be included. Omission of any reference to weathering should not be taken as an implication of fresh material; this is particularly true of coals.

Weathering is important because it has a direct effect on the strength of the rock. It may indicate the movement and chemical action of groundwater, either through the rock fabric or along open discontinuities. The presence of weathering in the rock profile will indicate the likelihood of oxidation in any coal seam

Table 6.2 Terms used to assess material strength in the field

Description	Characteristic of rock
R7 Extremely strong	Great difficulty in breaking with hammer, hammer rings
R6 Very strong	Requires several hammer blows to break
R5 Strong	Requires one hammer blow on hand held sample to break
R4 Moderately strong	Hammer pick indents *c.* 5 mm. Cannot be cut with a knife
R3 Moderately weak	Hammer pick indents deeply, difficult to cut with a knife
R2 Weak rock	Rock crumbles under hammer blows, cuts easily with a knife
R1 Very weak	Broken by hand with difficulty
Cohesive soil and clay	
C4 Stiff	Can be indented by thumbnail, cannot be moulded with fingers
C3 Firm	Can be moulded by strong finger pressure
C2 Soft	Easily moulded with fingers
C1 Very soft	Exudes between fingers when squeezed
Noncohesive soils	
S4 Weakly cemented	Lumps can be abraded with the thumb
S3 Compact	Not cemented but would require pick for excavation
S2 Loose	Could be excavated with a spade
S1 Very loose	Hand could penetrate 'running sand'

within this sequence. In old mine workings, the degree of weathering will help to indicate the state of the rock mass around existing or closed voids.

In the examination of rock types the terms given in Table 6.3 can be used to describe the degree of weathering.

6.4.3 Texture and structure

Reference to the bedding spacing will be sufficient to describe the texture and structure of the rock mass (Table 6.4).

In rocks which are heavily fractured, sheared or faulted, mention of this could draw attention to a particularly weak mass strength, for example, strong

Table 6.3 Terms used to describe degree of weathering in the field

Description	Characteristics of rock
W6 Fresh	No discolouration and maximum strength
W5 Slightly weathered	Discolouration along major discontinuity surfaces, may be some discolouration of rock material
W4 Moderately weathered	Discoloured, discontinuities may be open with coloured surfaces, rock material is not friable, but is noticeably weaker than fresh material
W3 Highly weathered	More than half the rock material is decomposed, discolouration penetrates deeply and the original fabric is only present as a discontinuous framework, corestones present
W2 Completely weathered	Discoloured, decomposed and in a friable condition, but the original mass structure is visible
W1 Residual soil	Totally changed, original fabric destroyed

Table 6.4 Terms used to describe the bedding spacing in the field

Description	Spacing
Very thick	>2.0 m
Thick	0.6–2.0 m
Medium	0.2–0.6 m
Thin	0.06–0.2 m
Very thin	0.02–0.06 m
Thickly laminated	0.006–0.02 m
Thinly laminated	<0.006 m

(intact strength), weak (mass strength), slightly weathered, indistinctly thinly bedded, heavily sheared, light grey, fine sandstone with some soft clay associated with shears. Note that the term 'heavily sheared' is not a standard term but serves to draw attention to the weak rock mass.

6.4.4 Colour

Colour is the most subjective of any observation made during the logging of lithotypes, and uniformity between geologists is difficult to achieve. There are several published rock colour charts, e.g. the *Rock Color Chart* published by the Geological Society of America, and these can be used together with supplementary terms such as light, dark, mottled etc. and secondary descriptors such as reddish and greenish. Colour is important to note as it can be an aid to correlation and, during the course of an investigation, it may become apparent that a distinctive coloured horizon is of particular significance.

6.4.5 Grain size

Grain size is important in the description of rock and soil material; its omission can only be justified when describing mudstones, claystones, shales and siltstones in hand specimen. In conglomerates and breccias the sizes of clasts should be included.

Typical grain sizes are as follows: conglomerates, larger clasts in a finer grained matrix 2.0–>20 mm, sandstones 0.06–2.0 mm, siltstone 0.002–0.06 mm, and mudstone/claystone <0.002 mm.

Grain shapes include descriptions of angularity, e.g. angular, subangular, subrounded, and rounded, and of form, e.g. equidimensional, flat, elongated, flat and elongated, and irregular.

6.4.6 Total core recovery

This is the length of core recovered, both solid stick and broken, expressed as a percentage of the full core run. It is a simple percentage figure entered on to the log. When core is lost this will be less than 100%, but can be greater than 100% if dropped core is overdrilled and recovered. The percentage core recovery is important, as in general, recoveries less than 95% are not accepted, and a redrill may be required.

6.4.7 Solid core recovery

This is the total length of pieces of core recovered which have a full diameter, expressed as a percentage of the full core run. Like total core recovery, this can be less or more than 100%.

6.4.8 Rock quality designation (RQD)

Deere (1964) proposed a quantitative index of rock mass quality based on core recovery by diamond drilling. As a result, RQD has come to be very widely used and has been shown to be particularly useful in classifying rock mass for the selection of tunnel support systems (Hoek and Brown 1980).

The calculation of RQD is made by taking the total length of core recovered for a length of 100 mm or longer, and at least 50 mm in diameter, expressed as a percentage of the full core run, with only core lengths terminated by natural fractures being considered.

$$RQD\ (\%)\ =\frac{100\ \times\ \text{length of core in pieces} > 100\,\text{mm}}{\text{Length of borehole}}$$

As with total core and solid core recovery, percentages can be more than 100%, but RQD cannot be greater than solid core recovery. For core with one large fracture along the entire length, 0% should be recorded. RQD is sometimes expressed for lithological units rather than core runs. RQD descriptions are: 0–25% very poor; 25–50% poor; 50–75% fair; 75–90% good; and 90–100% excellent.

6.4.9 Fracture spacing index

Fracture spacing index is a measure of the number of fractures per metre of core; it is defined for lithological units and is independent of core run or recovery. If in one lithological unit there is a marked change in the fracture spacing, then the index for sub-units should be given. An upper limit to the fracture spacing index should be defined, above which the index is not calculated, but is recorded as being greater than the defined limit, e.g. >25. Also nonintact material should be recorded separately. This applies to sections of lost core, core badly broken and disorientated during drilling, and noncohesive broken core from fault zones or old mine workings.

6.4.10 Fracture logging

The clearest way of presenting fracture details is to draw a graphic log alongside the lithological description. The log shows the exact position of the discontinuity, which is numbered and described on the separate fracture log. It is common practice to describe each discontinuity as in Table 6.5.

In Table 6.5, the number is the reference number given on the lithology log, and type is the type of

Table 6.5 Terms used to describe discontinuities in the field

No.	Depth	Type	Dip	Azimuth	Description	Aperture	Infill
1	24.82	B	20	90	Planar, smooth	T	–
2	25.31	Fr	30	90	Irregular	2.0	Clay
3	26.18	Fr	30	320	Stepped, rough	T-0(4)	Broken rock

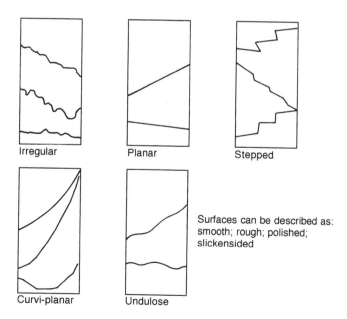

Figure 6.22 Types of discontinuities, including bedding and joints which have no displacement, and faults and shears which have a measurable or unknown displacement. Reproduced by permission of Dargo Associates Ltd

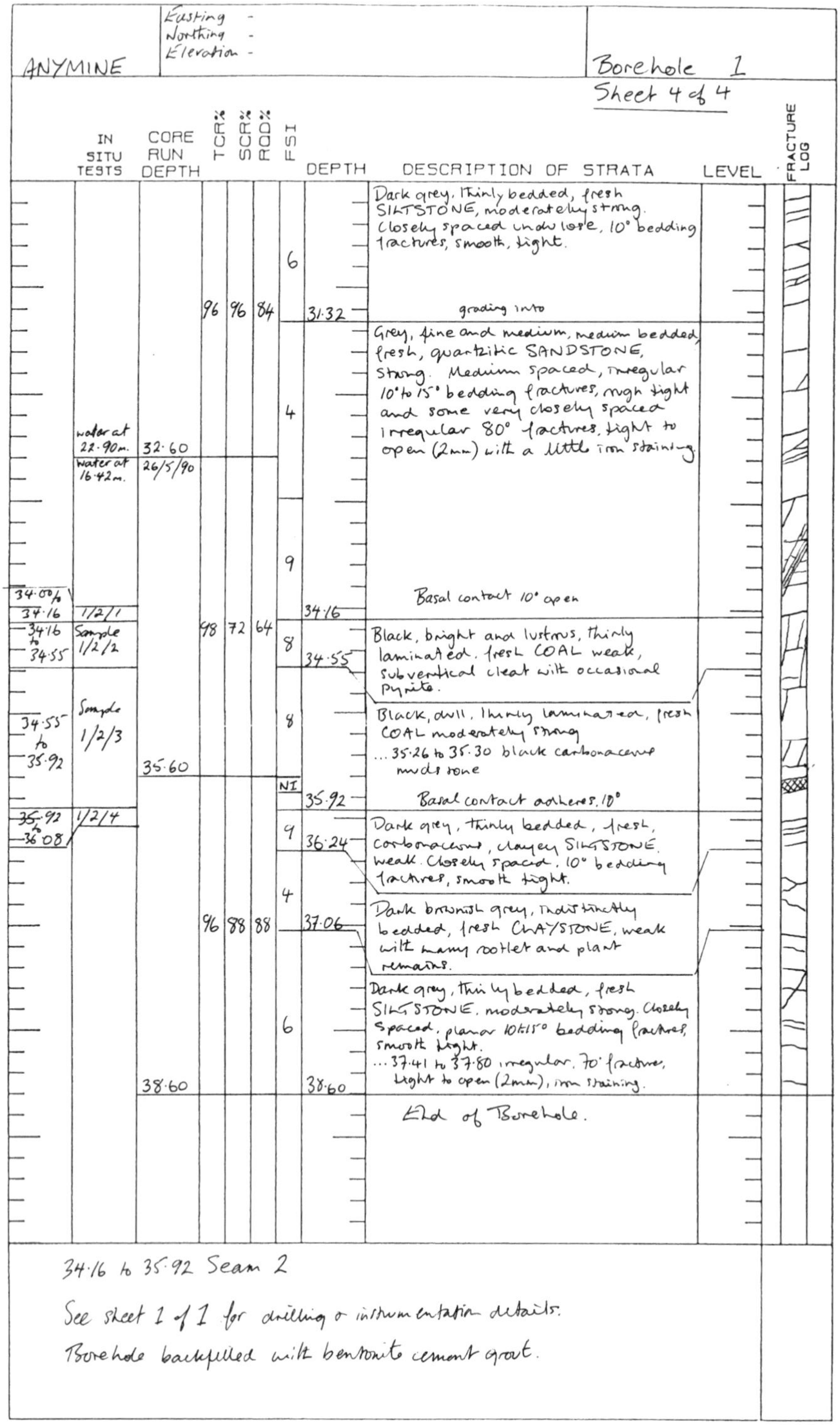

Figure 6.23 Example of a geotechnical logging sheet. Reproduced by courtesy of M.C. Coultas

discontinuity, e.g. B = bedding, J = joint, F = fault, S = shear, Fr = fracture, FrZ = fracture zone, SZ = shear zone, and FZ = fault zone.

The dip is the angle between the discontinuity and the plane perpendicular to the core axis, and the azimuth is the angle between the bedding and the dip of the discontinuity measured clockwise, looking from the top of the borehole. Discontinuities are commonly described as irregular, planar, stepped, undulose or curvi-planar, and if seen, the surface can be described as smooth, rough, polished or slickensided (Figure 6.22). The aperture is usually described according to its width, i.e. closed = <0.1 mm, tight = 0.1–1.0 mm, open = 1–5 mm, and wide = >5 mm (West 1991).

The infill is a description of any material filling the aperture, usually materials like clay, calcite, and broken rock.

An example of a combined geotechnical logging sheet is shown in Figure 6.23.

6.4.11 Rock mass rating (RMR)

No single method is adequate as an indicator of the complex behaviour of rock mass surrounding an underground excavation. Two classifications are generally used, one proposed by Bieniawski of the South African Council for Scientific and Industrial Research (CSIR) (1976), and one by Barton *et al* of the Norwegian Geotechnical Institute (NGI) (1974). The CSIR classification uses five basic parameters: strength of intact rock material, RQD designation, spacing of joints, condition of joints and groundwater conditions. A series of importance ratings are applied to the parameters, and the number of points or rating for each parameter are added together to give an overall rating with adjustments for joint orientation (Bieniawski 1976). The NGI proposed

an index for the determination of the tunnelling quality of a rock mass (index Q), whereby

$$Q = \frac{(RQD) \times (J_r) \times (J_w)}{(J_n)(J_a)(SRF)}$$

where RQD is Deere's rock quality designation, J_n is the joint set number, J_r is the joint roughness number, J_a is the joint alteration number, J_w is the joint water reduction number and SRF is a stress reduction factor (Barton *et al* 1974). Whilst RMR values are usually applied to civil engineering excavations, an understanding of the various parameters used is important when planning and implementing underground excavations in underground coal mines.

6.5 OTHER DATA COLLECTION

In addition to the field mapping, drilling, logging and geotechnical data collection, additional data are collected as an integral part of the exploration programme.

6.5.1 Hydrogeological data

As outlined in Chapter 9, the exploration drill holes may have piezometers installed in them and monitoring of groundwater levels at selected horizons will be carried out. Any surface water flow will also be measured and monitored. It is essential to understand the groundwater regime in any potential mine area.

6.5.2 Geophysical data

As a follow-up to the field mapping and drilling, downhole geophysical logging will be carried out and the drill logs reconciled with the geologs. Magnetometer surveys may be deemed necessary to determine zones of coal outcrop burning (see Chapter 8).

7

Coal Resources and Reserves

7.1 INTRODUCTION

The investigation of any coal deposit is carried out to ascertain whether coal can be mined economically, and that a coal product can be obtained that will be marketable. An essential requirement of any coal investigation is that an assessment is made of the coal resources within the area of interest. Such an assessment will influence the decision of whether to develop the deposit, to extend existing mine operations or conversely to curtail mining activity or even to cease development or operations altogether.

In the case of the sale of a lease or mine prospect, the coal resource assessment will play an important part in determining the success or failure of the transaction.

Resources can be divided on the basis of two points of view, namely according to:

1. their degree of geologic assurance;
2. their degree of economic feasibility.

There is a third subdivision which distinguishes between the coal in place and the amounts that can be technically recovered. Unavoidable losses during exploitation represent the difference between the two quantities: those in place and those recoverable.

The reliability of a coal tonnage estimate is based on the definition and expression of geological assurance and the methods of its estimation.

Geological uncertainties pertaining to coal arise from topographical and tectonic variations in the environment at the time when peat was being deposited, and from post-depositional erosion and structural alteration. As described in Chapter 2, the geometry and morphology of coals varies according to the depositional setting in which they were formed. For example, lenticular coals with great variations in thickness will need more data points than relatively undisturbed areally extensive coals of constant thickness. Data point spacing criteria should take into account such differences in the depositional settings and geological features specific to each coal deposit.

Coal resources categories range from the general evaluation of a coal basin to the calculation of specific reserves located within mine workings. The final result of geological investigation of a coal deposit will be to calculate all categories of coal resources, using the codes of practice adopted by the project management for the lease area under consideration.

It should be borne in mind that for providing information related to coal supply in the short term, reserve estimates have limitations; here one is more concerned with capacity and deliverability, whereas resource analysis is valid for longer term assessment, i.e. for ten years hence and longer.

7.2 COAL RESOURCES AND RESERVES CLASSIFICATION

There is no one internationally recognised and uniform method for the recording, categorisation and designation of coal reserves. A number of countries and organisations have developed sets of definitions and methods used to calculate coal resources and reserves. These include the principal coal-producing countries which have devised codes for the assessment of coal resources to meet their particular requirements; these vary in complexity and degrees of scale.

7.2.1 United States of America

In the USA, a coal resource classification system was published by the United States Geological Survey (Wood et al 1983). The system is based on the concept by which coal is classified into resource/reserve base/reserve categories on the basis of the geological certainty of the existence of those categories and on the economic feasibility of their recovery. Categories are also provided that take into account legal, environmental and technological constraints.

Geological certainty is related to the distance from points where coal is measured or sampled, thickness of coal and overburden, knowledge of rank, quality, depositional history, areal extent, correlations of coal seams and associated strata, and structural history.

The economic feasibility of coal recovery is affected not only by geological factors but also by economic variables such as the price of coal against mining costs, coal preparation costs, transport costs and taxes, environmental constraints and changes in the demand for coal.

The term resource is defined as naturally occurring deposits of coal in the earth's crust in such forms and amounts that economic extraction is currently or potentially feasible.

The hierarchy of coal resources/reserves categories outlined by the United States Geological Survey is given in Figure 7.1, and the application of the reliability categories based on distance from points of measurement, i.e. coal outcrops and boreholes, by the United States Geological Survey is demonstrated in Figure 7.2.

- *Original resources* is the amount of coal in place before production, the total of original resources is the sum of the identified and undiscovered resources plus the coal produced and coal lost in mining.
- *Remaining resources* include all coal after coal produced and coal lost in mining is deducted.
- *Identified resources* are those resources whose locations, rank, quality and quantity are known or estimated from specific geological evidence. The levels of control or reliability can be subdivided into inferred, indicated and measured resources.

  These subdivisions are determined by projecting the thickness of coal, rank and quality data from points of measurement and sampling on the basis of geological knowledge.
- *Inferred resources* are assigned to individual points of measurement which are bounded by measured and indicated coal for 1.2 km, succeeded by 4.8 km of inferred coal. Inferred resources include anthracite and bituminous coal 0.35 m or more in thickness and subbituminous coal and lignite 0.75 m or more in thickness to depths of not more than 1800 m. Coal resources outside these limits are deemed hypothetical in nature.
- *Indicated resources* are assigned to individual points of measurement bounded by measured coal for 0.4 km succeeded by 0.8 km of indicated coal. Indicated resources have the same thickness and depth limits as inferred resources.
- *Measured resources* are determined by the projection of the thickness of coal, rank and quality data for a radius of 0.4 km from a point of measurement.

Measured resources also have the same thickness and depth limits as indicated and inferred resources.

- The *reserve base* is identified coal defined only by physical and chemical criteria as determined by the geologist. The concept of the reserve base is to define a quantity of in-place or *in situ* coal, any part of which is, or may become, economic. This will depend upon the method of mining and the economic assumptions that will be used.

  The reserve base includes coal categories based on the same distance parameters given for coal resources but further defining the coal thickness and depth criteria, i.e. anthracite and bituminous coal to be 0.7 m or more, subbituminous coal to be 1.5 m or more, all to occur at depths not more than 300 m, lignite to be 1.5 m or more at depths not greater than 150 m.
- *Inferred reserves* include all coal conforming to the thickness and depth limits defined in the reserve base, and bounded by the same distance limits as given for inferred resources.
- *Indicated reserves* include all coal conforming to the thickness and depth limits defined in the reserve base, and bounded by the same distance limits as given for indicated resources.
- *Measured reserves* include all coal conforming to the thickness and depth limits defined in the reserve base, and bounded by the same distance limits as given for measured resources.
- *Marginal reserves* are those reserves that border on being economic, i.e. they have potential if there is a favourable change in circumstances, mining restrictions are lifted, quality requirements are changed, lease areas become available, or there is a newly created demand for the type of coal held in this reserve category.
- *Subeconomic resources* are those in which the coal has been lost in mining, is too deeply buried, or seam thickness becomes too thin, and/or the coal quality deteriorates to unacceptable limits.

7.2.2 Australia

The Australian code for calculating and reporting coal reserves is more simplistic and is easier for the geologist to apply. Figure 7.3 outlines the framework on which the classification is based.

Reserves should be estimated and allotted to one of the following categories, which are determined by an objective appraisal of the information available for the coal under consideration.

- *Inferred resources* are those for which the points of measurement are widely spaced, so that only an

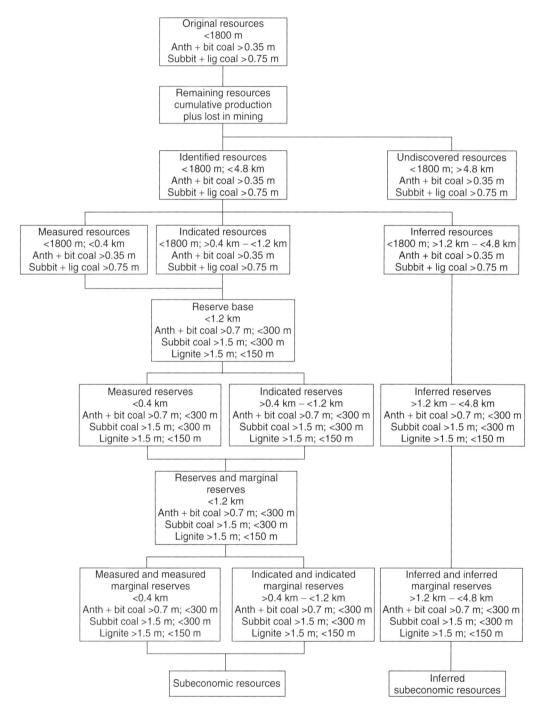

Figure 7.1   Criteria for distinguishing coal resource categories, adapted from United States Geological Survey (USGS) hierarchy of coal resources (Wood *et al* 1983). Reproduced by courtesy of the USGS. Anth = anthracite; bit = bituminous; subbit = subbituminous; lig = lignite

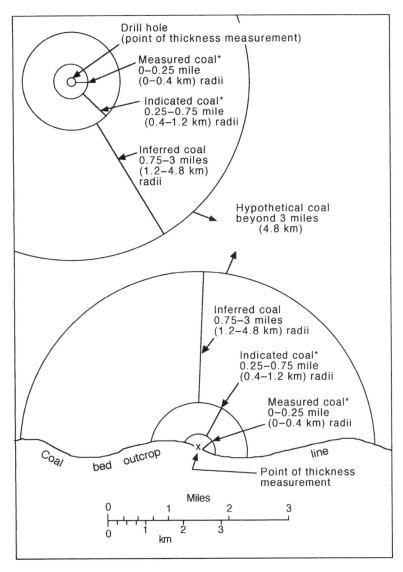

Figure 7.2 Reliability categories based solely on distance from points of measurement. From (Wood *et al* 1983). Reproduced by courtesy of the USGS. *Measured and indicated coal can be summed to demonstrated coal

uncertain estimate of the resource can be made. Points of measurement generally should not be more than 4.0 km apart, extrapolation of trends should extend not more than 2.0 km from points of measurement. Inferred resources can be subdivided into:

- *Inferred resources Class 1* are those resources for which the points of measurement allow an estimate of the coal thickness and general coal quality to be made, and for which the geological conditions indicate continuity of seams between the points of measurement.

- *Inferred resources Class 2* are those resources for which there is limited information and as a result, the assessment of this type of resource may be unreliable. Provided the coal thickness can be determined, the order of magnitude of inferred resources Class 2 may be expressed in the following ranges: 1–10 Mt, 10–100 Mt, 100–500 Mt, 500–1000 Mt, >1000 Mt.

- *Indicated resources* are those for which the density and quality of the points of measurement are sufficient to allow a realistic estimate of the coal thickness,

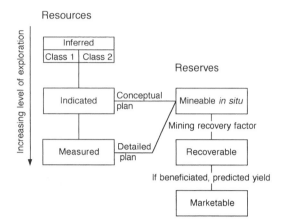

Figure 7.3 Australian code for reporting coal resources/reserves. From Galligan and Mengel (1986). Reproduced by permission of The State of Queensland. Department of Natural Resources and Mines

quality, depth and *in situ* tonnage, and for which there is a reasonable expectation that the estimate of resources will not change significantly with more detailed exploration. Points of measurement should not be more than 2.0 km apart. Where geological conditions are favourable, it may be possible to extend trends up to 1.0 km from the points of measurement.

- *Measured resources* are those for which the density and quality of points of measurement are sufficient to allow a reliable estimate of the coal thickness, quality, depth and *in situ* tonnage. Points of measurement should not be more than 1.0 km apart. Where geological conditions are favourable, it may be possible to extrapolate trends up to a maximum distance of 0.5 km from the points of measurement. The points of measurement should provide a level of confidence sufficient to allow detailed planning, costing of extraction and specification of a marketable product.

- *Mineable in situ reserves* are the tonnages of *in situ* coal contained in seams or sections of seams for which sufficient information is available to enable detailed or conceptual mine planning. Mineable *in situ* reserves exclude coal which is prohibited from mining. Mineable *in situ* reserves are calculated only from measured and indicated resources; measured resources are required for detailed mine planning, and are the preferred basis for mineable *in situ* reserves. Indicated resources may be used for conceptual mine planning. In general, further exploration will be required before the start of mining operations.

Mineable *in situ* reserves should be quoted separately for opencast and underground mines, together with an outline of the proposed mining method.

- *Recoverable reserves* (also termed extractable reserves) are the proportion of mineable *in situ* reserves that are expected to be recovered, i.e. the amount of coal which will be extracted. If dilution is added to the recoverable reserves tonnage, the total equates to the run-of-mine tonnages. In calculating recoverable reserves, a mining recovery factor must be applied to the mineable *in situ* reserves. This factor depends upon the mining method to be used. As a guide, a mining factor of 50% for underground reserves and 90% for opencast reserves may be applied.

- *Marketable reserves* (also termed saleable reserves) are the tonnages of coal that will be available for sale. If the coal is to be marketed raw, the marketable reserves will be the same as recoverable reserves (i.e. the run-of-mine tonnage). If the coal is beneficiated, e.g. by washing, the marketable reserves are calculated by applying the predicted yield to the recoverable reserves. The basis on which the predicted yield is calculated should be given.

7.2.3 United Kingdom

In the UK, the code used for classifying reserves was developed by the once publicly owned sector of the coal industry. It applies to black coal mined underground, and is orientated towards mining and the reliability of achieving planned outputs. Only those reserves are included that are judged to be economically viable, and likely to allow production within the maximum acceptable degree of risk (Elliott 1973). Classified reserves must be of suitable quality and thickness, and the effect of any mining hazards must be minimised.

Reserves are defined as Classified reserves, Classes 1, 2, and 3 and Unclassified reserves with decreasing levels of geological assurance. Figure 7.4 shows the classification of the UK black coal reserves for 1993 (from Cook and Harris 1998).

- *Class 1 reserves* are workable reserves of suitable quality which are sufficiently well proved in regard to thickness, depth, and tectonic and stratigraphic setting to warrant economic planning in accordance with a mine's future objectives. The effects and location of mining hazards can be predicted with sufficient accuracy that any significant annual production change can be avoided.

- *Class 2 reserves* are those reserves to which hazards are sufficiently great as to preclude an assurance that a forecast production can be achieved when they are developed. Constant revision will be required to review hazards inherent in this class of reserves.

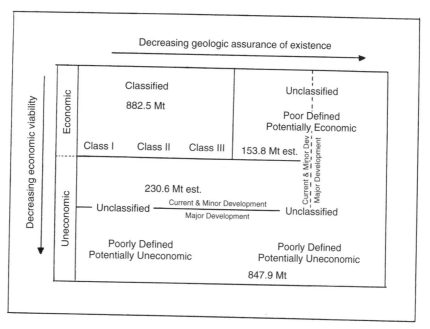

Figure 7.4 British Coal's reserve classification, based on Cook and Harris (1998). Crown Copyright material is reproduced under Class Licence Number Co1W000048 with the permission of the Controller of HMSO and the Queen's Printer for Scotland

- *Class 3 reserves* lack the degree of knowledge required to detect hazards that will stop or seriously hinder mining. Precipitant hazards are unpredictable in this class of reserves, and the mine will require spare capacity to insure against them. The mine's standby production potential is liable to be brought into operation because of the risks attached to Class 3 reserves. These classes of reserves are roughly comparable to recoverable reserves (class 1), to mineable *in situ* reserves (class 2), and to measured reserves (class 3).
- *Unclassified reserves* are divided into those that are poorly defined but are potentially economic, and those that are poorly defined and considered to be potentially uneconomic.

In opencast mining in the UK, British Coal uses the terms proved and unproved reserves within the mining lease area, dependent on the degree of geological certainty.

- *Proved reserves* are those areas of coal that have points of measurement with a spacing of 50 m or less.
- *Unproved reserves* are those areas of coal that have points of measurement with a spacing of greater than 50 m. In practice, these reserves will regularly be at least of mineable *in situ* reserve status.

Unlike the resource/reserve classifications of the USA and Australia, the system in the UK is based on resources/reserves pertaining to lease areas; this is very localised and is not applicable to those large areas which may contain coal but are not mined. The UK system is centred on a geographically concentrated mining industry with a long history, a somewhat different scenario from the developing countries.

In practice, the geologist will work to the codes of reserve assessment used in the country where exploration is taking place. The US Geological Survey system is comprehensive in that it covers a wide range of coal types and classes of reserves. The Australian system is a simplified version which is easy for the geologist to implement during exploration and development programmes, and still retains the advantage in that it allows coal reserve assessments to be made for all levels of coal deposit. The UK system, whilst excellent for mining operations, makes little allowance for general exploration categories of coal reserves.

7.2.4 Germany

The definition of geological assurance in terms of the measured–indicated–inferred system is considered a sound one. In order to further identify the geological assurance limits Germany has attempted to define all its geological assurance categories in terms of statistical confidence criteria, as shown in Table 7.1, rather than relying just on point of measurement spacing criteria. For each category, both a confidence level and a confidence interval are specified. It is considered that this defining of the assurance categories in terms of statistical confidence criteria gives a clearer expression of the geological uncertainties.

7.2.5 United Nations

The classification system recommended by the United Nations Economic and Social Council (ECOSOC) in 1979 is shown in Figure 7.5, after Fettweis (1979, 1985). As can be seen, the quantities in place (R) and the recoverable amounts (r) correspond to one another. The assessment starts with the resources in place and is divided into three categories 1, 2 and 3 according to the level of geological assurance that can be assigned to each category. New terms have been chosen to allow agreement with different national systems. Each of these categories can be further subdivided according to the viewpoint of economics.

E represents those resources thought to be exploitable in a particular country or region under the prevailing socioeconomic conditions with currently available technology.

S represents the balance of resources that was not considered to be of current interest but might become of interest as a result of foreseeable economic or technological changes.

In practice, the resources group r–1–E is of prime interest as it represents the reliable estimated economically exploitable and recoverable tonnages. They are frequently referred to as reserves following the recommendations of the United States Geological Survey given in Section 7.2.1.

Table 7.1 Degrees of geological assurance of coal reserves according to the guidelines of Germany. Adapted from Steenblik (1986), with permission from IEA Coal Research – The Clean Coal Centre

Assurance category	Confidence level (%)	Confidence interval (%)
In sight	>90	+10
Probable	70–90	+20
Indicated	50–70	+30
Inferred	30–50	+30
Prognostic	>10–30	No limit

The World Energy Conference Survey refers to this resources group as 'Proved Recoverable Reserves' and the total estimated resources in place is put together in a single category called 'Additional Resources'.

The United Nations classification introduces a main group termed Occurrences beside that of Resources. Occurrences cover any additional material with a lower economic potential, estimates of which would fall outside the boundaries of Resources and should be reported separately with some clarification as to the derivation and meaning of the estimates.

7.3 REPORTING OF RESOURCES/RESERVES

All factors used to limit resources and reserves that are necessary to verify the calculations must be stated explicitly. These will include the points of measurement, e.g. open hole, cored boreholes, and outcrops. The relative density value that is selected for the calculation of the coal tonnage should be stated, together with the reasons for its selection.

7.3.1 Coal resources and reserves

To report resources and reserves, the required data will be based on the following:

1. Details on each coal seam within the lease area.
2. On a depth basis, in regular depth increments if sufficient information is available.
3. On a seam thickness basis, the minimum coal thickness used and the maximum thickness of included noncoal bands should also be stated. Normally where a seam contains a noncoal band thicker than 0.25 m the two coal splits can be regarded as separate seams, and tonnages should be reported for each. The limits for noncoal bands in brown coal sequences may be greater, e.g. 1.0 m.
4. On a quality basis, maximum raw coal ash should be stated, and for marketable reserves, only that coal that can be used or beneficiated at an acceptable yield (which should be stated) should be included in the estimate. Other raw coal parameters, particularly those which affect utilisation should be given, e.g. total sulphur and calorific value. Subdivisions of the resources may be made for areas of oxidised coal and heat-affected coal.

A summary of all the factors relevant in determining the categories of coal resources/reserves assessment is shown in Figure 7.6. In the diagram, coal product value

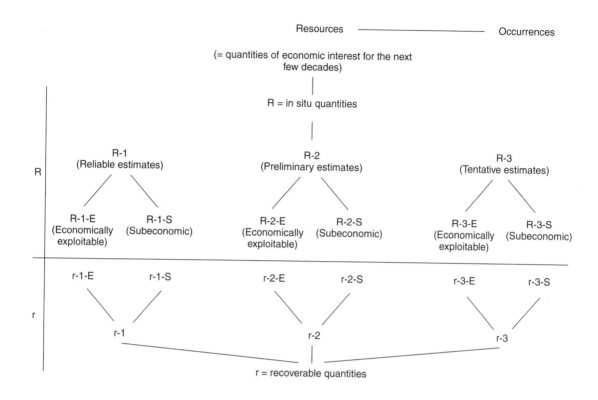

1 = reliable ~ demonstrated; reasonable assured; established; explored
2 = preliminary ~ inferred; estimated additional; possible
3 = tentative ~ undiscovered (hypothetical and speculative); potential; prognostic

Figure 7.5 International classification system of mineral resources. E: Exploitable resources in a country or region under prevailing socioeconomic conditions with available technology. S: Balance of resources not of current interest due to foreseen economic or technological changes (United Nations 1979), from Fettweis (1985). Reproduced with permission of the author

and yield are plotted against ease of mining, coal recovery and mining costs and against geological certainty.

7.3.2 Coal resources and reserves maps

Any report of resources/reserves should be accompanied by maps and plans at appropriate scales showing all the relevant data. Such maps and plans should show those areas assigned to each category of resources/reserves, seam depth contours, seam isopachs, quality contours for each seam, and all areas not to be mined.

The geological information required for resources/reserves assessments are based on the points of measurement. Both for quantity and quality assessments it is important that the recorded information at the points of measurement is correctly compiled and is

therefore reliable. If this is not the case, then making resources/reserves assessments is a useless exercise. If any point of measurement has a doubt against its reliability, this should be taken into account when the assessment is made.

Extrapolation from points of measurement up to the distance limits imposed by the resources/reserves category, is based on the judgement and knowledge of the local geology. Geological hazards that need to be taken into account are faulting, seam thickness variations, particularly rapid thinning and splitting of seams, washouts, sharp changes in dip, and the presence of igneous intrusions.

The distances between points of measurement used for the different resources/reserves categories is theoretically the same for underground and opencast coal.

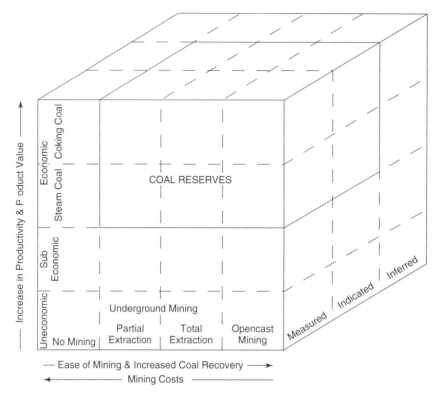

Figure 7.6 Variable factors in coal reserves assessment. Adapted from Ward (1984), by permission of Blackwell Scientific Publications

If the geology is similar between two points the same distance apart at depth or near the surface, then the confidence level must be similar. However, shallow drilling is relatively cheaper and this allows for more holes to be drilled, and therefore the confidence level will be greater. Also in coal occurring near the surface, there are the additional problems of oxidation and, in some cases, zones of burning that have to be delineated.

Deep boreholes are costly so that at the exploration stage a lower level of confidence may be achieved; however, geophysical surveys are used to supplement the drilling to ascertain any major structural disturbances and changes in thicknesses in the coal-bearing sequence.

7.3.3 Calculation of coal resources

7.3.3.1 In situ *tonnage calculations*

The basic formula to calculate coal resources is:

$$\text{Coal thickness(m)} \times \text{Area(m}^2) \times \text{R.D.}$$

$$= \text{Total tonnes (for each defined}$$

$$\text{resources/reserves block)}$$

Coal thickness is determined at each point of measurement.

The area of each resources/reserves block is measured on the map or plan either:

1. In the traditional way by using a planimeter. The boundaries of each block area are traced and the area is calculated automatically and given as a reading; Figure 7.7 shows such a planimeter being used in this fashion.
2. The coordinates of each block area are entered into a computer program specified for the purpose, and an areal calculation obtained (see Chapter 11).

The relative density is normally taken from a total seam section, i.e. that section of the seam to be mined, not a density of the cleanest portion of the seam. In opencast mines this will usually be a whole seam section, whilst underground this is not always so; coal quality constraints or mining difficulties may mean mining only part of the seam.

If no density determination is available, an estimated density can be adopted dependent upon the known

Figure 7.7 Digital planimeter being used to calculate coal reserve areas within a working mine. Photograph by M.C. Coultas

average ash content of the seam, this would be in the order of 1.3–1.4 in coals with reasonably low ash contents.

This means that an area of one square kilometre (one million square metres) underlain by a coal with a thickness of 1.0 m and with a density of 1.4 m will contain 1.4 million tonnes of coal. Resource tonnages are usually quoted in million tonnes (Mt) and are usually rounded off to the nearest hundred thousand, ten thousand or one thousand tonnes, depending on the degree of accuracy required.

A common method of calculation of coal resources has been the polygon or area of influence technique. This method assigns an area to a point of measurement which is a function of the distance to the immediate neighbouring points of measurement. A polygon is formed by joining the mid points between the point of measurement and those surrounding it, resulting in the original point of measurement forming the centre of the polygon. Figure 7.8 shows the construction of polygons for a series of points of measurement together with the calculation of reserve tonnages for selected polygons for a hypothetical example.

In constructing polygons of influence for the estimation of coal reserves, mistakes can occur; for example, in the construction of blocks of influence for a triangular grid, Figure 7.9(a) shows the correct selection of areas of interest, whereas Figure 7.9(b) illustrates how such areas have been selected incorrectly (Tasker 1985).

The weaknesses of using the polygon method are that if the drill holes in a deposit are widely and irregularly spaced, it is possible that some points of measurement will give undue emphasis in the calculation of reserves; also widely spaced drill holes leave uncertainty as to the continuity of coal seams. Polygons based on widely spaced data points give no indication of the accuracy of the results; it is possible that the actual and calculated reserves may differ by significant amounts.

A practical development of the polygon method is to outline reserve zones containing several points of measurement, calculate the area of the zone and use an average or weighted mean seam thickness for each zone.

Many areas are insufficiently well delineated, either by outcrops or drill holes, so that some boundary of arbitrary width needs to be placed around the drilled portion of the deposit. If the deposit is large, a subjective selection of the boundary zone width is satisfactory, because a moderate change in width will not substantially alter the total area of the deposit. However, if the deposit is small, a significant proportion of the reserves may lie within the boundary zone. The calculated reserves therefore will be unreliable because of their dependence on the subjective selection of the boundary zone.

One method of improving the reliability of the boundary width for a deposit is to estimate the average range of influence of the points of measurement by means of a variogram. The variance in thickness is calculated for

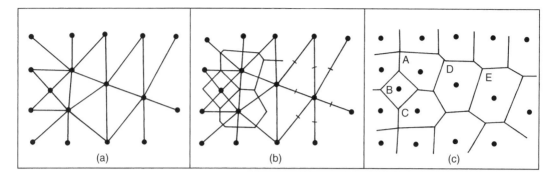

Calculation of reserves for polygons A – E:

	Area (m²)	Coal thickness (m)	Relative density	*In situ* tonnage
A	211 500	2.20	1.4	651 400
B	108 000	2.16	1.4	326 600
C	225 000	1.86	1.4	585 900
D	289 800	2.40	1.4	973 700
E	351 900	2.46	1.4	1 211 900

Figure 7.8 Polygon method for calculation of *in situ* coal resources/reserves. (**a**) Link all points of measurement; (**b**) At the midpoint between points of measurement, draw lines at right angles and join to form polygonal areas around each point of measurement; (**c**) Complete all polygons, e.g. polygons A–E can now be measured to calculate reserves using the central point of measurement as control for each polygon. Total *in situ* tonnage is 3 749 500 tonnes (3.75 Mt). Reproduced by permission of Dargo Associates Ltd

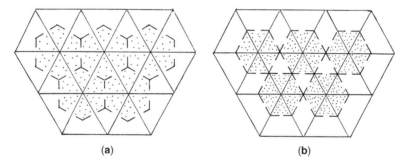

Figure 7.9 Construction of polygons (blocks) of influence for an equiangular triangular grid. (**a**) Correct construction of elementary blocks; (**b**) Incorrect construction of elementary blocks. From Tasker (1985)

pairs of points of measurement, where the pairs have a common distance of separation (as on a grid pattern). In general, as the distance of separation increases so does the variance, until a maximum variance is reached, as shown in Figure 7.10. The distance corresponding to this maximum may be taken as a measure of the range of influence for points of measurement within the deposit. Thus the areal extent of a deposit may be defined by assuming the deposit to extend (geological continuity permitting) a distance equal to the range of

influence in all directions beyond the points of measurement. The boundary zone width is therefore dependent on the data available; if these are insufficient, or the points of measurement are irregularly sited, this method is not applicable.

In an area which has been drilled extensively, detailed contouring of the coal seams can enable reserve estimates to be made as a further definition of the estimated reserve figure; this method is also useful to highlight areas of geological hazard such as washouts or faulting.

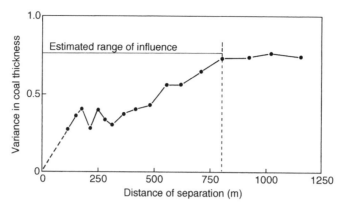

Figure 7.10 Variogram of coal thickness. From Ventner (1976). Reprinted with permission of the Canadian Institute of Mining, Metallurgy and Petroleum

An extension of the previous reserve calculation technique is to input geological data into a geologically significant computer model of the deposit (see also Section 13.2). In this way coal deposits can be subdivided into major block units, each equating to the proposed working district of a coal mine. Within these units, smaller structurally delineated blocks provide the basic geological data which are transformed into computer data. Within each of these smaller units the coal seams provide the basic data for the whole assessment, i.e. depth, thickness, dip and size of the area. For example, the minimum coal thickness to be evaluated could be taken as 0.30 m, whilst the minimum fault displacement to delimit a block could be 10.0 m. Such blocks consist of a series of vertically neighbouring coal seams and the projected stratigraphic column represents the series of coal seams within the block (Figure 7.11).

From these data, the spatial position, form and area of all intermediate coal seams are calculated using a mathematical model. The top and bottom of a block are digitised to include regularly distributed depth values. On all the intermediate planes, sectional areas are defined by closed polygons. Data input of these sectional areas is done by digitising seam projections at 1:10 000 scale. Straight lines joining corresponding points on the top and bottom planes define the lateral delineation of the block on all intermediate planes and also all in-between sectional areas, as shown in Figure 7.11.

In areas previously worked, the percentage of the worked areas is estimated for each sectional area, and has to be subtracted in the process of calculation.

In similar fashion, other coal seam data can be built into the computer model, such as seam thickness and coal quality data.

In Germany, where this method has been developed, experience has shown that for a reliable experimental variogram, it is necessary to have at least 100 data points per coal seam in a major block unit, and at least 25 data points per coal seam for the estimation of one sectional area. Such data points need to be less than 1.0 km apart to ensure a level of reliability acceptable to implement the computed reserve calculation.

It can be seen that computers now provide a means whereby statistical confidence criteria can be provided on a regular basis. The main limitation to these methods is the amount and quality of data required; most methods require data points relatively close together and evenly spaced, and such conditions are only likely to occur in fairly well-defined coal deposits.

7.3.3.2 *Opencast coal mining*

Stripping ratio calculation

In opencast mining operations, overburden to coal ratios are often quoted on a volumetric basis, i.e. bank (*in situ*) cubic metres (bcm) of overburden per tonne of coal *in situ*. The calculation of the stripping ratio (SR) is:

$$\text{Stripping ratio} = \frac{\text{overburden cubic metres (bcm)}}{\text{coal cubic metres} \times \text{coal R.D.}}$$

For coal deposits where the relative density of the coal is essentially constant, the stripping ratio is expressed simply as the thickness of overburden to that of the total workable coal section; however, the basis of the ratio must always be stated clearly. The most realistic results are achieved when the overburden thickness is calculated from the difference between the topographic surface and the structure contours of a seam at the

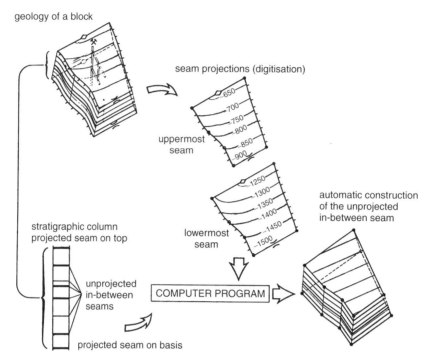

geology of a block

seam projections (digitisation)

uppermost
seam

stratigraphic column
projected seam on top

lowermost
seam

automatic construction
of the unprojected
in-between seam

unprojected
in-between
seams

COMPUTER PROGRAM

projected seam on basis

Figure 7.11   Schematic diagram of the calculation of sectional areas using the block model for coal resource/reserve calculations. From Juch *et al* (1983), reproduced with permission

selected data points within the area of interest. Where numerous data points exist, the stripping ratios are most conveniently calculated by computer and these data can later be plotted as stripping ratio contour plans.

In the UK overburden and stripping ratios are calculated somewhat differently, and can be considered as either:

overburden ratio (without batters)

= *in situ* vertical ratio

actual stripping ratio (including batters)

= working ratio

Batters are the amount of overburden which must remain in the pit as part of the angled walls to ensure pit stability.

The overburden ratio is calculated as follows:

overburden (*in situ* vertical) ratio

= *in situ* overburden thickness

(incl. coal thickness)

− *in situ* coal thickness

× *in situ* coal thickness

The overburden ratio can also be calculated on a volumetric basis:

overburden (*in situ* vertical) ratio

= *in situ* overburden volume (incl. coal volume)

− *in situ* coal volume × *in situ* coal volume

For mining a multiple coal seam sequence, the interburden, i.e. the thickness of noncoal between coal seams is treated as overburden for the coal seams below the highest coal seam. Figure 7.12 shows the effects of topography and geological structure on the overburden or stripping ratio. The stripping ratio selected for any coal deposit is based on the economics of removing and rehandling overburden, coal quality and geotechnical limitations.

Depth of planned opencast mining

In the UK the following equation is used for the rapid estimation of the highwall depth:

highwall depth = 2[(overburden ratio × *in situ*

thickness) + *in situ* thickness]

− low wall depth

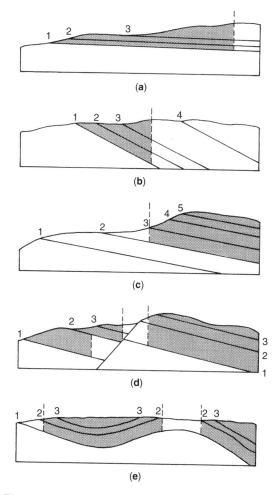

Figure 7.12 Effects of geological structure and topography on stripping ratios. For this example, the SR cut-off value is taken at 10:1 bcm; only shaded area(s) will be considered for mining. Coal seams are numbered in ascending order. (**a**) Effects of shallow dips; (**b**) Effects of steep dips; (**c**) Effects of topography; (**d**) Effects of faulting; (**e**) Effects of folding. Reproduced with permission of Dargo Associates Ltd

and for estimating the average depth of the coal excavation area

coal excavation area average depth

= highwall depth + low wall depth × 2

7.3.3.3 Geological losses

The calculation of recoverable or extractable reserves requires the identification of those geological constraints that are likely to inhibit mining. Such constraints include the identification and positioning of fault zones, changes in dip, washouts, seam splitting and thinning, losses in quality and igneous intrusions. In opencast workings, reserves will be affected by deleterious changes in the stripping ratio. All these factors contribute to a reduction in the mineable *in situ* reserve figure and are known collectively as geological losses.

In underground workings, the method of mining will influence those geological losses deducted from the mineable *in situ* reserve figure. If the method to be used is longwall mining, a larger geological loss will occur, due to the fact that longwall operations need hazard-free runs in a designated panel of coal. All faulted areas will need to be discounted if their amounts of throw displace the coal seam to be mined out of line with the preset coal shearer. If the coal panels between faults are too small, then the whole block may be discounted. If the bord and pillar (= room and pillar) method is adopted, small faults can often be worked through, and in some cases igneous intrusions can be worked around, as is the case in some South African coal mines. This method also allows for small blocks of coal still to be taken. All methods of mining are affected when coal is lost through washouts, and changes in seam thickness or seam quality.

Geological losses will vary considerably from mine to mine, but in general, in opencast operations a 10% geological loss can be expected, whereas in underground operations, geological losses of 25–50% may occur.

In addition, losses other than geological may need to be accounted for at this stage. Areas close to lease boundaries may be discounted, as well as reserve areas which run beneath railways, motorways and critical buildings and installations. Those reserves deemed recoverable may in certain circumstances have a minimum depth limit imposed. This is often the case where reserves are accessed by drivages in from the base of the highwall in opencast workings. Such limits will be determined according to the nature of the particular surface area, and are intended to reduce the effects of subsidence, particularly close to areas of urban population.

In opencast mines, the effects of depth can produce an increase in the SR, which will become uneconomic at some point, as well as imposing geotechnical constraints on deeper excavation.

7.3.3.4 Reserves reporting

Using the above calculations, the objective is to report, first, mineable *in situ* reserves. To do this, the following information is required:

1. An outline of the proposed mining method, together with a conceptual mine plan.
2. In underground mines, the physical criteria limiting mining such as maximum and minimum working section thickness, minimum separation of seams, the maximum dip at which the coal can be mined by the stated method, geological structure.
3. Overburden or stripping ratios in opencast operations.
4. Quality restrictions, maximum and minimum levels for ash, sulphur, volatile matter etc. In the case of coals that have quality problems and need to be beneficiated, the predicted yield needs to be given.
5. Depth limits, imposed by either physical or economic constraints or both.

Added to these are mining losses such as:

1. Areas where coal may not be mined, e.g. beneath motorways.
2. Stress fields in the coal seams, which may require the reorientation of reserve panels, and therefore loss of reserve.
3. Roof stability, affecting the thickness of pillars to be left in the mine, again producing loss of reserve.

These and other factors, once deducted from the measured *in situ* reserve, enable the measured recoverable (or extractable) reserves to be calculated. These are the reserves required by investors when considering any mine's potential.

7.3.3.5 Reserve economics

Where the mine configuration is known, and the production costs and sales figures are also known, it is possible to apply computer methods to determine pit areas of equal value, e.g. in an opencast pit whose walls earn a constant value on the investments made. Such analysis has been applied to metalliferous mines but is equally applicable to modern opencast operations. Such an analysis does require considerable details of the coals mined, and the condition of the mine itself, e.g. whether it is a series of discrete open pits, or separate parts of a mine producing from different seams.

From this it is possible to understand the following relationships:

1. physical (coal quantity and quality);
2. rate dependent relationships (quantity of coal A per quantity of coal B);
3. economic relationships and potential (cash flow against pit size).

Additional information, such as changes in mine access, and need for pit backfilling and restoration, all have an influence, as will primary drivers such as market price, market sustainability and environmental constraints.

7.4 WORLD COAL RESERVES AND PRODUCTION

7.4.1 World coal reserves

Estimates of proven coal reserves for black coal, i.e. for anthracite and bituminous coals, and for brown coal, i.e. subbituminous coal and lignite, are given in Table 7.2. These figures are taken from the World Energy Council's *Survey of Energy Resources* (2001) and from the BP *Statistical Review of World Energy* (2001). These proven reserves are those which can be regarded with reasonable certainty to be recoverable from known deposits under present-day conditions, and do not include those large resources of coal which are currently being, or will in the future, be fully evaluated.

The table gives an overall total of 985 828 Mt for those known deposits in the world today. Such a figure in itself is meaningless, but the regional totals do give an indication as to the geographical distribution of the bulk of the world's black coal (520 442 Mt) and brown coal (465 386 Mt) resources. Those countries with one million tonnes or less are omitted from the table.

In the regions containing Gondwana coal deposits, black coal reserves are greater than those for brown coal. This is true for Africa, Central and South America and the Indian Subcontinent. Australia is an exception to this in that its black and brown coal reserves are evenly divided.

In the Far East, total black coal reserves far exceed those for brown coal; however, if the figures for the People's Republic of China are excluded, then the reverse is true.

In Europe, the USA and the CIS brown coal reserves are larger than those for black coal.

7.4.2 World coal production

7.4.2.1 Coal production statistics

Production figures for black and brown coals are shown in Table 7.3 for the eight world regions. Total world coal production, based on figures for 2001, is 4 342 018 000 tonnes. Countries with very minor production levels are excluded.

It can be seen that the total production figure for black coal (3 010 627 000 tonnes) is more than double that of

Table 7.2   World coal reserves (million tonnes), based on WEC (2001) and BP *Statistical Review of World Energy* (2001)

Region	Bituminous + anthracite (Mt)	Subbituminous (Mt)	Lignite (Mt)	Total
Algeria	40			40
Botswana	4 300			4 300
Congo	88			88
Egypt		22		22
Malawi		2		2
Mozambique	212			212
Niger	70			70
Nigeria	21	169		190
South Africa	50 000			50 000
Swaziland	208			208
Tanzania	200			200
Zambia	10			10
Zimbabwe	502			502
Total Africa	**55 651**	**193**		**55 844**
Canada	3 471	871	2 236	6 578
Greenland		183		183
Mexico	860	300	51	1 211
United States of America	115 891	101 021	33 082	249 994
Total North America	**120 222**	**102 375**	**35 369**	**257 966**
Argentina		430		430
Brazil		11 929		11 929
Chile	31	1 150		1 181
Colombia	6 267	381		6 648
Ecuador			24	24
Peru	960		100	1 060
Venezuela	479			479
Total South America	**7 737**	**13 890**	**124**	**21 751**
Afghanistan	66			66
China	62 200	33 700	18 600	114 500
India	82 396		2 000	84 396
Indonesia	790	1 430	3 150	5 370
Japan	773			773
Korea (Dem. People's Rep.)	300	300		600
Korea (Republic)	78			78
Pakistan		2 265		2 265
Philippines		232	100	332
Thailand			1 268	1 268
Turkey	278	761	2 650	3 689
Vietnam	150			150
Total Asia	**147 031**	**38 688**	**27 768**	**213 487**
Austria			25	25
Bulgaria	13	233	2 465	2 711
Croatia	6		33	39
Czech Republic	2 114	3 414	150	5 678
France	22		14	36
Germany	23 000		43 000	66 000
Greece			2 874	2 874
Hungary		80	1 017	1 097

(continued overleaf)

Table 7.2   (*continued*)

Region	Bituminous + anthracite (Mt)	Subbituminous (Mt)	Lignite (Mt)	Total
Ireland	14			14
Italy		27	7	34
Netherlands	497			497
Poland	20 300		1 860	22 160
Portugal	3		33	36
Romania	1	35	1 421	1 457
Serbia (incl. Bosnia) and Montenegro	64	1 460	14 732	16 256
Slovakia			172	172
Slovenia		40	235	275
Spain	200	400	60	660
United Kingdom	1 000		500	1 500
Total Europe	**47 234**	**5 689**	**68 598**	**121 521**
Kazakhstan	31 000		3 000	34 000
Kyrgyzstan			812	812
Russian Federation	50 000	97 472	10 450	157 922
Ukraine	16 274	15 946	1 933	34 153
Uzbekistan	1 000		3 000	4 000
Total CIS( ) Former Soviet Union	**98 274**	**1 13 418**	**19 195**	**230 887**
Iran	1 710			1 710
Total Middle East	**1 710**			**1 710**
Australia	42 550	1 840	37 700	82 090
New Zealand	33	206	333	572
Total Oceania	**42 583**	**2 046**	**38 033**	**82 662**
TOTAL WORLD	**520 442**	**276 299**	**189 087**	**985 828**

Table 7.3   World coal production (thousand tonnes), based on WEC (2001) and BP *Statistical Review of World Energy* (2001)

Region	Bituminous	Subbituminous	Lignite	Total
Algeria	25			25
Botswana	945			945
Congo	50			50
Egypt	200			200
Malawi		44		44
Morocco	129			129
Mozambique	18			18
Niger	168			168
Nigeria		20		20
South Africa	223 510			223 510
Swaziland	426			426
Tanzania	5			5
Zambia	128			128
Zimbabwe	4 977			4 977
Total Africa	**230 581**	**64**		**230 645**
Canada	36 538	24 300	11 659	72 497
Mexico	2 366	7 678		10 044

Table 7.3   (*continued*)

Region	Bituminous	Subbituminous	Lignite	Total
United States of America	568 260	352 260	76 570	997 090
Total North America	**607 164**	**384 238**	**88 229**	**1 079 631**
Argentina	337			337
Brazil	5 602			5 602
Chile	170	470		640
Colombia	32 754			32 754
Peru	20			20
Venezuela	6 500			6 500
Total South America	**45 383**	**470**		**45 853**
China	985 000		45 000	1 030 000
India	292 203		22 212	314 415
Indonesia	70 703			70 703
Japan	3 906			3 906
Korea (Dem. People's Rep.)	60 000	21 500		81 500
Korea (Republic)	4 197			4 197
Pakistan		3 307		3 307
Philippines		1 028		1 028
Thailand			18 270	18 270
Turkey	1 990		65 050	67 040
Vietnam	8 830			8 830
Total Asia	**1 422 632**	**30 032**	**150 532**	**1 603 196**
Albania			33	33
Austria			1 137	1 137
Bosnia-Herzegovina			1 850	1 850
Bulgaria	90		25 940	26 030
Croatia	15			15
Czech Republic	14 419	44 278	512	59 209
FYR Macedonia			8 400	8 400
France	4 533		558	5 091
Germany	40 500		161 282	201 782
Greece			61 900	61 900
Hungary	700	6 500	7 700	14 900
Italy			19	19
Poland	110 200		60 800	171 000
Romania		2 751	20 131	22 882
Serbia, Montenegro	49		30 451	30 500
Slovakia			3 748	3 748
Slovenia		758	3 804	4 562
Spain	13 200	3 700	8 500	25 400
United Kingdom	37 077			37 077
Total Europe	**220 783**	**57 987**	**396 765**	**675 535**
Kazakhstan	56 436		1 763	58 199
Kyrgyzstan	135		280	415
Mongolia	1 423		3 529	4 952
Russian Federation	166 000		83 400	249 400
Ukraine	34 871	46 176	1 182	82 229
Uzbekistan	89		2 864	2 953
Total CIS( ) Former Soviet Union	**258 954**	**46 176**	**93 018**	**398 148**
Iran	1 500			1 500
Total Middle East	**1 500**			**1 500**

(*continued overleaf*)

Table 7.3   (*continued*)

Region	Bituminous	Subbituminous	Lignite	Total
Australia	222 000	16 200	65 800	304 000
New Zealand	1 630	1 670	210	3 510
Total Oceania	**223 630**	**17 870**	**66 010**	**307 510**
TOTAL WORLD	**3 010 627**	**536 837**	**794 554**	**4 342 018**

Table 7.4   World coal consumption (thousand tonnes), based on WEC (2001) and BP *Statistical Review of World Energy* (2001)

Region	Bituminous	Subbituminous	Lignite	Total
Algeria	490			490
Botswana	945			945
Congo	100			100
Egypt	2 000			2 000
Kenya	100			100
Malagasy Rep.	14			14
Malawi	17			17
Mauritius	75			75
Morocco	3 200			3 200
Niger	168			168
Nigeria		20		20
South Africa	153 460			153 460
Swaziland	180			180
Tanzania	5			5
Zambia	121			121
Zimbabwe	4 750			4 750
Total Africa	**165 625**	**20**		**165 645**
Canada	23 700	26 600	10 200	60 500
Cuba	20			20
Dominican Rep.	160			160
Jamaica	25			25
Mexico	2 716	9 469		12 185
Panama	65			65
Puerto Rico	185			185
United States of America	520 800	350 000	76 600	947 400
US Virgin Islands	260			260
Total North America	**547 931**	**386 069**	**86 800**	**1 020 800**
Argentina	1 300			1 300
Brazil	12 286	6 690		18 976
Chile	4 130	870		5 000
Colombia	4 200			4 200
Peru	500			500
Venezuela	164			164
Total South America	**22 580**	**7 560**		**30 140**
Bangladesh	300			300
Bhutan	75			75
China	1 035 000		45 000	1 080 000
Cyprus	20			20

Table 7.4 *(continued)*

Region	Bituminous	Subbituminous	Lignite	Total
Hong Kong (China)	6 393			6 393
India	308 160		22 200	330 360
Indonesia	17 000			17 000
Japan	137 000			137 000
Korea (Dem. People's Rep.)	61 680	21 500		83 180
Korea (Republic)	54 137	4 992		59 129
Malaysia	1 150	1 500		2 650
Myanmar (Burma)	16		27	43
Nepal	300			300
Pakistan		4 370		4 370
Philippines		6 416		6 416
Taiwan	40 023			40 023
Thailand	3 230		18 840	22 070
Turkey	11 200		64 080	75 280
Vietnam	5 500			5 500
Total Asia	**1 681 184**	**38 778**	**150 147**	**1 870 109**
Albania			33	33
Austria	3 440		1 640	5 080
Belgium	9 710	310		10 020
Bosnia-Herzegovina			1 850	1 850
Bulgaria	3 400		25 940	29 340
Croatia	284		81	365
Czech Republic	10 402	41 454	512	52 368
Denmark	7 804			7 804
Estonia	80			80
Finland	5 368			5 368
FYR Macedonia	250		8 400	8 650
France	22 416		612	23 028
Germany	64 500		163 335	227 835
Greece	1 382		61 000	62 382
Hungary	1 400	6 500	7 700	15 600
Iceland	100			100
Ireland	1 839		40	1 879
Italy	17 100			17 100
Latvia	126			126
Lithuania	200			200
Luxembourg	151			151
Moldova	500			500
Netherlands	11 800			11 800
Norway		2 100		2 100
Poland	89 000		60 800	149 800
Portugal	5 000			5 000
Romania	2 411	2 752	20 131	25 294
Serbia, Montenegro	100		30 400	30 500
Slovakia	12 282		5 042	17 324
Slovenia	80	1 237	3 770	5 087
Spain	28 200	17 400	8 500	54 100
Sweden	3 000			3 000
Switzerland	140			140
United Kingdom	55 529			55 529
Total Europe	**357 994**	**71 753**	**399 786**	**829 533**

(continued overleaf)

Table 7.4   (*continued*)

Region	Bituminous	Subbituminous	Lignite	Total
Georgia	25			25
Kazakhstan	41 650		1 600	43 250
Kyrgyzstan	350		350	700
Mongolia	1 500		3 200	4 700
Russian Federation	154 000		83 000	237 000
Tajikistan	100	19		119
Ukraine	61 785		1 240	63 025
Uzbekistan	1 150		2 850	4 000
Total CIS( ) Former Soviet Union	**260 560**	**19**	**92 240**	**352 819**
Iran	1 900			1 900
Israel	9 200			9 200
Lebanon	200			200
Total Middle East	**11 300**			**11 300**
Australia	44 900	16 200	65 800	126 900
Fiji	24			24
New Caledonia	170			170
New Zealand	230	1 660	260	2 150
Total Oceania	**45 324**	**17 860**	**66 060**	**129 244**
TOTAL WORLD	**3 092 498**	**522 059**	**795 033**	**4 409 590**

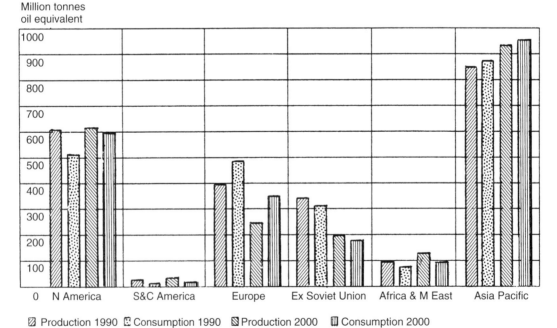

Area Coal Production and Consumption 1990 and 2000

Million tonnes
oil equivalent

◪ Production 1990   ▤ Consumption 1990   ◩ Production 2000   ▥ Consumption 2000

Figure 7.13   Coal production and consumption for 1990 and 2000 (BP *Statistical Review of World Energy* 2001). Reproduced by permission of BP PLC

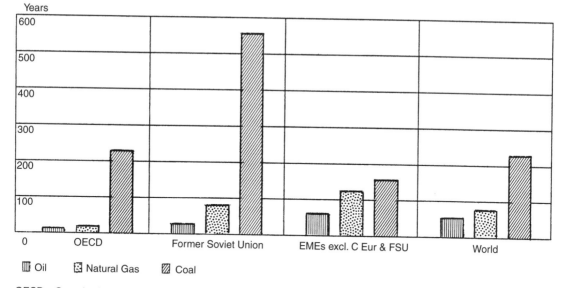

Fossil fuel R/P ratios at end 2000

OECD – Organisation of Economic Cooperation & Development
EME – Emerging Market Economies
C Eur – Central Europe
FSU – Former Soviet Union

Figure 7.14   Fossil fuel R/P ratios at end 2000 (BP *Statistical Review of World Energy* 2001). Reproduced by permission of BP PLC

brown coal (1 331 391 000 tonnes). This is in part due to the greater usefulness of black coals as both steam and coking coals, and its widespread development in those countries that are more economically advanced. Black coals also make up all internationally traded coal in the world.

The People's Republic of China is the largest producer of black coal and Germany produces the greatest tonnage of brown coal.

The development of the coal industry in countries such as Indonesia and Venezuela has increased production in their regions, whilst the political and economic reorganisation in Eastern Europe may influence future production figures in Europe.

7.4.2.2 Regional production and consumption

Table 7.4 shows the principal areas of coal consumption, some 4 409 590 000 tonnes in total. This can be compared with Table 7.3 to illustrate that the bulk of all coal mined is actually consumed in the area in which it is produced. Figure 7.13 shows a comparison of coal consumed against coal produced worldwide for the period 1990–2000. Those countries in which coal

production is greater than consumption and where the coal quality is marketable are large coal exporters, e.g. the USA, Australia and South Africa. Smaller producers such as Colombia export a large percentage of their total production, whilst others such as the People's Republic of China and the Democratic People's Republic of Korea produce large tonnages but use almost all domestically. Those developing producers such as Venezuela and Indonesia will produce more coal than they will consume; this production is intended for the export market.

7.4.2.3 Reserves/production ratio (R/P)

The reserves/production ratio is based on those reserves remaining at the end of any year divided by the production in that year. The result is the length of time that those remaining reserves would last if production were to continue at the present level.

Figure 7.14 illustrates R/P figures projected over 600 years, and it can clearly be seen that at today's production rates, the R/P ratio for coal is nearly six times that for oil and four times that for natural gas. Coal's dominance in R/P terms is most pronounced in the OECD countries and the CIS or former Soviet Union.

8

Geophysics of Coal

8.1 INTRODUCTION

The use of geophysics in the exploration for coal basins and in the delineation of coal seams and geological structures, in particular coal deposits, is now well established.

Since the oil crisis in 1973, there has been an enormous increase in the use of geophysical methods to identify coal deposits, and to further determine their economic potential. Several of the techniques have been used initially in the exploration for oil and gas and adapted where applicable to coal exploration.

Large scale studies make use of regional gravity, deep seismic and aeromagnetic surveys to determine the sedimentary and structural framework of the area under consideration. Smaller scale, more detailed examination of the coal deposits utilises shallow seismic, ground magnetic, electrical resistivity and microgravity coupled with the geophysical logging of all boreholes, which in turn involves the use of density, electrical, electromagnetic and radiometric techniques.

In established coal mining areas, the combination of geophysical logging with high resolution seismic and ground magnetic surveys contributes significantly in the delineation of economic mining areas, both for opencast and underground operations. In underground operations, the use of in-seam seismic is now used as a tactical tool in planning the orientation of mining areas, in particular the siting of longwall panels.

This combination is used on both large and small scale investigations; the drawback is that it can be expensive to carry out such investigations. When planning an exploration programme, any use of geophysics, whether as a field survey or in borehole logging, will be a high cost item on the exploration budget. The benefits of using such techniques, such as whether the amount of drilling required will be reduced, will be set against such costs.

The background principles of physics governing the various geophysical techniques employed in coal exploration and development are not covered in this book as they are available in standard geophysical texts (Milsom 1989, Reynolds 1997). Rather, a simple outline is given of the basic physical properties of coal-bearing sequences together with an outline of the field methods used to locate and quantify coal deposits.

8.2 PHYSICAL PROPERTIES OF COAL-BEARING SEQUENCES

Coal as a lithology responds well to most geophysical methods in that its physical properties contrast with those of other lithologies commonly found in coal-bearing sequences. Coal has in general a lower density, a lower seismic velocity, a lower magnetic susceptibility, a higher electrical resistivity and low radioactivity compared with surrounding rocks in typical coal-bearing sequences.

8.2.1 Density

Density measurements of rocks are not usually measured *in situ*, but in the laboratory on some small outcrop or drill core samples. Such results rarely give the true bulk density because samples may be weathered or dehydrated; consequently density is not often well known in specific field situations. Table 8.1 gives saturated density ranges and averages for coals, sediments in coal-bearing sequences, and igneous and metamorphic rocks which may be associated with coal basins either as underlying basement or as intrusives into the coal-bearing strata.

Low and high rank coals ($1.1-1.8\,\text{gm/cm}^3$) are less dense than the surrounding sediments ($1.6-2.9\,\text{gm/cm}^3$), which in turn are less dense than igneous and metamorphic rocks ($2.1-3.1\,\text{gm/cm}^3$). In sedimentary rocks, the wide range of density is due to variations in porosity, nature of pore fluids, age and depth of burial as well as mineralogical composition. Some igneous rocks, such as volcanics, have high porosities and therefore lower density; for example, pumice can have a density less than $1.0\,\text{gm/cm}^3$. Density also increases with the degree of

Table 8.1   Table of Physical Properties of Coals and Associated Sedimentary and Igneous Rocks. (Based on Telford *et al* 1990); Applied Geophysics; with permission of Cambridge University Press

Lithology	Density(wet) (g/cm^3)		Seismic Velocity (km/s)	Magnetic Susceptibility ($\times 10$ SI Units)		Electrical Resistivity (Ω m)
	Range	Average		Range	Average	Range
Sandstone	1.61–2.76	2.35	3.6	0–20	0.4	1–6.4 $\times 10^8$
Shale	1.77–3.20	2.40	2.8	0.01–15	0.6	20–2 $\times 10^3$
Limestone	1.93–2.90	2.55	5.5	0–3	0.3	50–1 $\times 10^7$
Lignite	1.10–1.25	1.19				9–200
Bituminous Coal	1.20–1.80	1.32 ⎫	1.8–2.8	–	0.02	0.6–1 $\times 10^5$
Anthracite	1.34–1.80	1.50 ⎭				1 $\times 10^{-3}$–2 $\times 10^5$
Acid Igneous rock	2.30–3.11	2.61	4.0–5.5	0–80	8.0	4.5 $\times 10^3$ (wet granite) 1.3 $\times 10^6$ (dry granite)
Basic Igneous rock	2.09–3.17	2.79	4.0–7.0	0.5–100	25.0	20–5 $\times 10^7$ (dolerite)
Metamorphic rock	2.40–3.10	2.74	5.0–7.0	0–70	4.2	20–1 $\times 10^4$ (schist)

metamorphism, as recrystallisation reduces pore space to form a denser rock as well as converting some minerals to more dense forms.

8.2.2 Seismic velocity

The seismic velocity of a rock is the velocity at which a wave motion propagates through the rock media. As shown in Table 8.1, the seismic velocity of coal is in the range 1.8–2.8 km/s; mudrocks such as shales have similar values. Sandstones have a higher value which increases with increased quartz content, while dense limestones, igneous and metamorphic rocks have much higher velocities of 4.0–7.0 km/s.

8.2.3 Seismic reflection coefficients

The seismic reflection coefficient determines whether an interface gives a reflection and depends upon the density as well as the seismic velocity. Coal seams with a low density and low seismic velocity often have high reflection coefficients and can be picked up well on seismic sections.

8.2.4 Magnetic susceptibility

The magnetic susceptibility of a rock depends primarily on its magnetite content. Weathering generally reduces susceptibility because of the oxidation of magnetite to hematite. As in the case of rock density, measurements of magnetic susceptibility in the field do not necessarily give a bulk susceptibility of the formation;

however, outcrop magnetic susceptibility measurement by portable instruments has led to improved bulk susceptibility measurements. Although there is great variation in magnetic susceptibility, even for a particular lithology, and wide overlap between different types, sedimentary rocks generally have the lowest average susceptibilities, with coals having among the lowest susceptibility within the sedimentary suite (see Table 8.1). Basic igneous rocks have high susceptibility values. In every case, the susceptibility depends on the amount of ferromagnetic minerals present, mainly magnetite, titano-magnetite or pyrrhotite. It is worth noting that the sulphide minerals such as pyrite, which is a common mineral in coals and associated sediments, have a low susceptibility value; like many of the sulphide minerals, pyrite is almost nonmagnetic. Table 8.1 gives the range and average values in rationalised SI units for those rocks associated with coal.

8.2.5 Electrical conductivity

Electrical prospecting involves the detection of surface effects produced by electric current flow in the ground. It is the enormous variation in electrical conductivity found in different rocks and minerals that requires a greater variety of techniques to be used than in the other prospecting methods.

Several electrical properties of rocks and minerals are significant in electrical prospecting, of these, by far the most important in coal prospecting is electrical conductivity or the inverse electrical resistivity, which is expressed in ohm-metres (Ω m), the others being of less significance. This has been shown by Verma and

Bandyopadhyay (1983), who employed the resistivity method for the geological mapping of coal in India.

As most rocks are poor conductors, their resistivities would be extremely large were it not for the fact that they are usually porous, and the pores are filled with fluids, mainly water. The conductivity of a porous rock varies with the volume and arrangement of the pores, and the conductivity and amount of contained water. Water conductivity varies considerably depending on the amount and conductivity of dissolved chlorides, sulphates and other minerals present, but the principal influence is usually the sodium chloride or salt content.

8.2.6 Radiometric properties

Trace quantities of radioactive material are found in all rocks. Small amounts of cosmic radiation passing through the atmosphere produce a continuous background reading which may vary from place to place. In general, the radioactivity in sedimentary rocks and metamorphosed sediments is higher than that in igneous and other metamorphic types, with the exception of potassium-rich granites.

In coal-bearing sequences, the contrasts in natural radioactivity in coals and surrounding sediments have led to the development of the use of nuclear well logging instruments for measuring radioactivity of formations encountered in boreholes.

Coals have very low radioactivity, as do clean sandstones, sandstones with high contents of rock fragments and clay matrices; siltstones and nonmarine shales have low to intermediate values, whereas marine shale and bentonite (tonstein) have high radioactivity due to the presence of uranium/thorium minerals in the shale and potassium in the bentonite.

8.3 SURFACE GEOPHYSICAL METHODS

The petroleum industry has used various seismic geophysical methods for a number of years as an aid in the exploration for geological structures suitable for hydrocarbon entrapment. In order to locate sedimentary basins, electrical, electromagnetic, gravity and magnetic surveys, together with reflection and refraction seismic surveys are used; these are usually large scale operations involving a great deal of equipment, manpower and finance. Although of use in broad regional investigations, they are little used in the examination of coal-bearing sequences for small selected areas.

In the investigation of mine lease areas high resolution seismic reflection surveys are the most effectively employed. Other methods used are cross borehole seismic techniques and seismic refraction, which are particularly useful in opencast mine development.

8.3.1 Seismic surveys

Exploration for mineable coal is generally concerned with the top 1.5 km of strata; because of this shallow nature of the investigations, high resolution seismic profiling is required to detect relatively thin coal seams. The recording system is designed to retain as much of the high frequency reflections as possible.

The efficiency of surface seismic reflection surveying has steadily improved over the last twenty years, and it is now applied to coal mining with increased accuracy and confidence.

8.3.1.1 Seismic reflection surveys

The principle of seismic reflection is that an acoustic signal or seismic wave produced by an explosion or other impulse source is introduced into the ground at selected points, and this signal radiates through the ground. The velocity at which the signal travels depends upon the rock type encountered. Typical velocities are 3.6 km/s in hard, dense limestone and ranges between 1.8 and 2.8 km/s for most types of hard coals. These and other velocities of typical lithologies encountered in coal-bearing successions are given in Table 8.1. The velocity of the seismic wave is a function of the lithology through which it passes; when the wave reaches a boundary marking a lithological change, a reflected and a refracted ray result. When the change in velocities and density at a boundary is large, there is a large reflection coefficient and a strong reflection is generated; this reflection is detected by receivers or geophones, which produce an electrical signal which is recorded, as shown in Figure 8.1(a) and (b).

The instant the signal is generated it is also recorded, and by recording the time it took for the signal to reach the reflection point and return, referred to as two-way travel time (TWT), the depth of the reflection point can be determined, providing the velocities of the traversed lithologies are known.

The physical property of coal and its surrounding strata which makes seismic surveys feasible is its acoustic impedance, which is defined as the product of its density and seismic velocity. Coal has a much lower density and velocity relative to other sedimentary rocks normally encountered in coal-bearing sequences; the contrast in acoustic impedance between coal and other sediments may be between 35% and 50%. This produces large reflection coefficients (Hughes and Kennett 1983).

In seismic surveys of coal deposits, reflections from most features of interest return to the surface during the first second after the seismic wave has been generated, i.e. the TWT. This would give a maximum reflection depth of 1.5 km if the overlying rocks have an average seismic velocity of 3.0 km/s.

The advance in the identification of smaller and smaller disturbances in coal seams is mainly due to the increased power of computers, which has enabled interactive data processing, in particular the opportunity to use combinations of selected parameters, with the capacity to refine the data along each seismic section and between sections. This technique has been successfully carried out in the UK, where surface seismic exploration is combined with in-seam seismic methods (see Section 8.4.1). The latter can map smaller faults which are below the resolution of surface seismic (Carpenter and Robson, 2000).

The identification of faults, folds, washouts, seam splits and thickness changes using seismic reflection techniques is an effective method whereby potential geological hazards can be pinpointed. This can then be built into mine planning and design.

The source commonly used in seismic reflection surveys for coal is either the detonation of an explosive charge such as dynamite, but this produces environmental problems in populated areas, or a lower energy impulse produced by an earth compactor known as the mini-SOSIE (Pinchin *et al* 1982). Another energy source used for shallow reflection surveys is a 'gun' firing blank ammunition into the ground, but this is only suitable for investigation to around 150 m in rocks with good transmission properties.

One of the main problems with any shallow seismic survey is the effect of the total travel time when the waves have to pass through a low velocity weathered zone near the ground surface. Variations in the thickness and velocity characteristics of the weathered zone produce variations in the arrival times of the wave reflections. This can be partly overcome by placing the shot point and, if possible, the geophone in holes drilled to below the low velocity zone, although this is often not feasible due to the extent and depth of weathering. This can be further complicated by the presence of superficial deposits masking the rockhead. Weathering is a particular problem in subtropical and tropical countries such

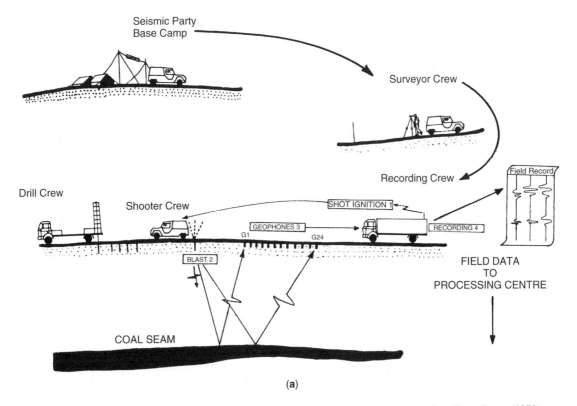

Figure 8.1 Seismic reflection survey. (**a**) Field data acquisition; (**b**) Seismic data processing; From Peace (1979)

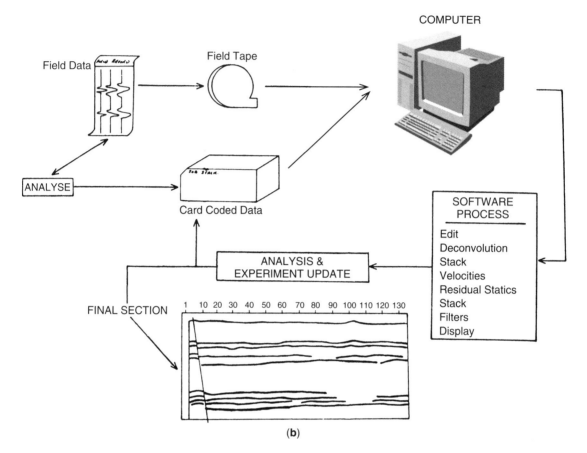

(b)

Figure 8.1   (*continued*)

as Africa and Australia, whilst superficial deposits are a common problem in Europe and North America.

The ease of processing and interpreting the data from surface seismic surveys is naturally influenced by the local geology. Where coal deposits are geologically uncomplicated, i.e. thick seams with low dips, little faulting and close to the surface with few weathering effects, interpretation will be relatively easy. Contrast this with other geological scenarios where coal seams exhibit complexities in splitting and variation in thickness, have a high incidence of faulting or the presence of washouts in the coal seams, and perhaps lie at depths up to 1.0 km; then the interpretation of seismic data from surveys in such areas is much more difficult.

Figure 8.2 illustrates a seismic reflection profile across a coalfield in the USA. The section shows the coal seam reflection to be robust and continuous from SP20 to SP100, indicating uniform coal seam thickness across the entire section.

Figure 8.3(a) shows a seismic reflection profile across a lignite deposit in Northern Ireland, UK; this is a product of a shallow reflection seismic survey using a 'gun' energy source. In this survey, in order to provide depth resolution, the higher frequencies of seismic waves were used. The sequence consisted predominantly of saturated Tertiary clays which have a very low attenuation for seismic energy; this suggests an expectation of good high frequency transmission. This is also dependent upon the effects of the superficial deposits in which the energy source is coupled. The geology of the area consists of superficial heterogeneous glacial deposits, overlying Tertiary clays and lignite; these in turn lie on a zone of weathered basalt with fresh basalt at depth. In this instance there is little seismic velocity contrast between the clays and lignites, but because there is a large density contrast, the acoustic impedance is sufficiently large to produce a large reflection coefficient and a detectable reflection.

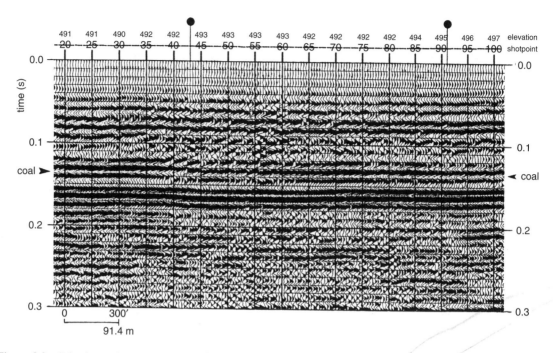

Figure 8.2 Seismic section showing a robust and continuous coal seam reflection. This indicates uniform coal seam thickness with no detectable geological disturbance. From Gochioco (1991)

The interpretation of Figure 8.3(a) is shown in Figure 8.3(b). The good reflectors at about 0.07 s and 0.1 s (TWT) on the SE margin mark the top and bottom of the lignite. The weathered/fresh basalt interface is seen on the SE margin at about 0.13 s (TWT) and can be seen clearly to shallow towards the NW. In addition two faults with downthrows to the SE can be detected. The irregularly shaped body at X on the section is considered to be a raft of lignite that has become detached from the main lignite due to frost action during the Quaternary glacial period.

Surface reflection seismic has been carried out in the Kladno Coalfield in Bohemia. Oplustil *et al* (1997) describe the results of seismic measurements across the Kladno Basin, and Figure 8.4 shows the depth seismic profiles and the interpreted faulting. Up to 75% of seismic indications correspond with observations in underground mines. The number of normal faults in the seismic profiles exceeds the number of observed faults, and in one case, a large fault with a downthrow of tens of metres has been interpreted with no equivalent in any mine. There is also some suggestion of faults being the result of synsedimentary movements in the deposit.

In Wyoming, USA, high resolution reflection profiles were recorded over a prospective underground coal gasification test site. The target seam was the Wyodak

coal, some 180 m below the surface with a thickness of 30 m. Seismic reflection profiling was considered the most effective on technical and financial grounds. This survey used a dynamite energy source, and one profile line is shown in Figure 8.5. The top and bottom of the Wyodak coal give strong reflections, as does the overlying Badger coal (3 m thick). All the sections show a gentle anticlinal structure in the sequence, particularly in the top of the seam; this may be due to differential compaction of the coal seam or movement within the coal seam as a response to gravity in a down dip direction before coalification occurred. No faulting was observed in the profile. The base of the Wyodak coal has a zone of anomalous amplitude (Figure 8.5), interpreted as a washout in the basal section of the seam. This was later proved by drilling to be the case.

Figure 8.6 shows a distinct washout structure in the roof of a lignite in the Texas Gulf Coast region. The edges of the strata surrounding the washout are quite distinct; from this the dimensions of the channel can be measured and located within any proposed mine development. In this example, the washout is approximately 35 m (115 ft) thick, extending from 141 m (465 ft) to 176 m (580 ft) on the section.

The dynamite shot hole method has been widely used, and although it gives higher frequency content

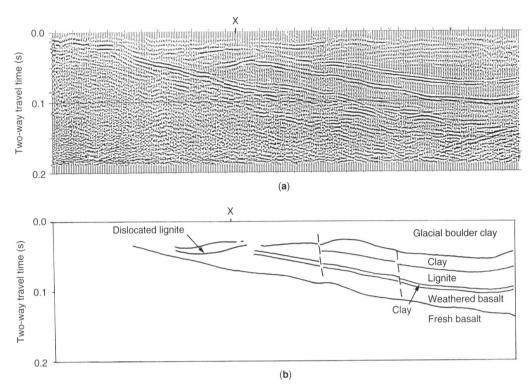

Figure 8.3 (**a**) Shallow seismic reflection survey, Northern Ireland, UK; (**b**) Interpretation of seismic section A, clearly showing interbedded lignite and underlying basalt. Unpublished figure produced by permission of Antrim Coal Company Ltd

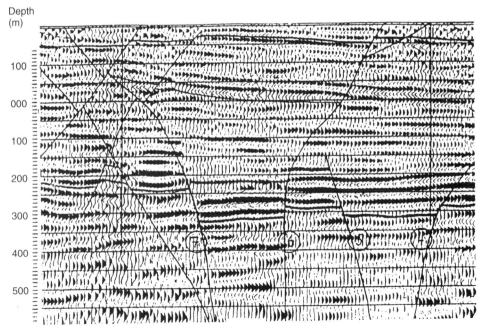

Figure 8.4 Depth seismic profile with interpreted faults. From Oplustil *et al* (1997), with permission of the authors

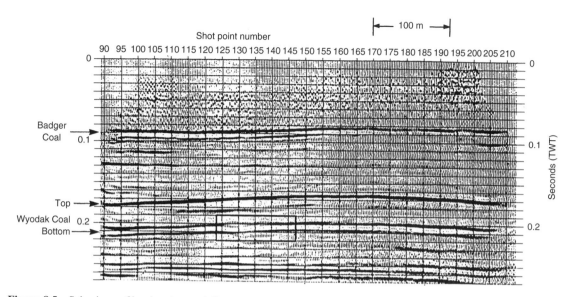

Figure 8.5 Seismic profile showing anticline structure of Wyodak Coal and the good reflection of Badger Coal. Seismic anomaly (boxed area) is interpreted as a channel cut into the base of the Wyodak Coal, Wyoming USA. From Greaves (1985)

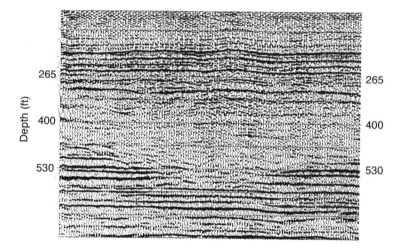

Figure 8.6 Seismic reflection profile showing a buried channel at a depth of 152.4 m (500 ft), Texas Gulf Coast, USA. From Peace (1979)

and better resolution, such surveys are expensive and environmentally problematic. The cheaper mini-SOSIE technique has been applied and can provide data of equal quality in the right circumstances. The method uses a similar recording spread as for dynamite, and the mini-SOSIE energy source can put 800 to 1500 impulses into the ground for each record.

Mini-SOSIE has been widely used, particularly with success in Australia; difficulties have, however, been encountered in using the technique in dissected hilly terrain combined with deep weathering profiles.

In the UK, seismic reflection techniques using an energy source situated at depth in a borehole have been developed for use in investigating opencast coal operations.

The hole-to-surface reflection data are processed with standard vertical seismic profiling (VSP) techniques. Small down-hole shots are fired at 2.0 m spacing below the water table in the borehole, and geophones with a spacing around 4.0 m are deployed at the surface along a line intersecting the top of the borehole. The travel time along a seismic ray path is independent of the direction of travel. Using this method, the rays are traversed in opposite directions from conventional reflection surveys and can be processed accordingly. A seismic depth section obtained from such a survey is shown in Figure 8.7; here is shown the seismic depth section, coal seam stratigraphy, and the velocity profile used for migration. The coal seams at 30 m, 54 m, 58 m, 70 m and 128 m all give a good reflection, but the thin seam at 80 m is hardly seen at all. Other reflections

at 100 m and 110 m may be weaker reflections from differing lithotypes in the sequence.

This technique has been used to correlate between boreholes; Figure 8.8 shows the stratigraphy and the combined seismic section from three hole-to-surface surveys with the migration interval velocities used in seismic data processing. In Figure 8.8, the coal seam at 20 m has a good reflection, and the worked-out seam at 50 m has a very good reflection because the air-filled void in the old workings produces a very large reflection coefficient. This worked-out seam also shows two small faults with downthrows 1–2 m to the right. The weaker reflections below 50 m are interpreted as a

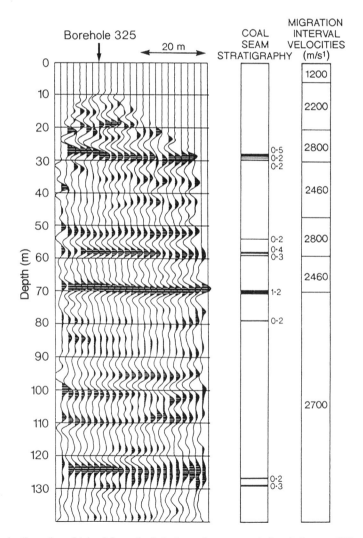

Figure 8.7   Seismic depth section obtained from the hole-to-surface survey in boreholes at a UK opencast site, with the coal seam stratigraphy and velocity profile used for migration. From Kragh *et al* (1991), with permission of EAGE

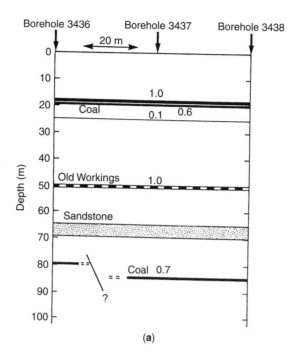

(a)

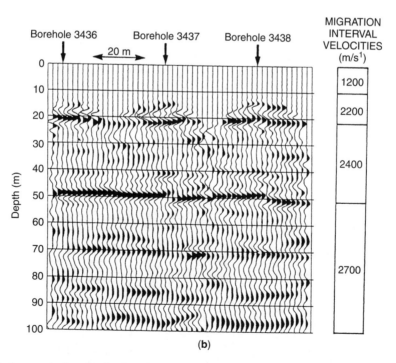

(b)

Figure 8.8 (a) Coal seam stratigraphy for three boreholes at an opencast site, UK; (b) Combined seismic section from the three hole-to-surface surveys conducted in these boreholes with the velocity profile used for migration. The section is true scale, zero phase, normal polarity, and an equal energy trace normalisation has been applied. From Kragh *et al* (1991), with permission of EAGE

sandstone/mudstone interface at 70 m and a coal seam at 85 m; the fault in this seam is inferred from the borehole data.

Cross-hole seismic reflection surveys are possible in certain opencast exploration sites which have closely spaced boreholes (Goulty *et al* 1990, Kragh *et al* 1991). This method involves down-hole seismic sources and detectors sited below the water table (hydrophones); this provides better resolution than surface seismic surveys and even vertical seismic profiling, but requires the availability of numerous boreholes.

These seismic methods are particularly useful in identifying old mine workings and worked-out coal seams as well as illustrating the geology and structure of the area. In areas of the UK which have a long mining history, the position and extent of shallow underground workings have not always been recorded. In planning opencast operations, hole-to-surface seismic reflection surveys and cross-hole seismic reflection methods can help to identify such potential problems before detailed mine planning begins.

In underground mines, the planning of the mining of new reserve blocks of coal will lead to a reassessment of the extent and quality of any existing seismic data. This will then result in either the requirement for additional high resolution seismic data over the planned mine area, and/or the reprocessing of existing seismic data using modern processing techniques. The reprocessing can yield an improvement of the seismic sections and often identifies structure previously undetected. This is particularly important where the geology and surface conditions vary considerably over a short distance. Figure 8.9 shows part of a seismic section in which the original processing is compared to reprocessing. The coal seam reflector is seen to be more continuous and identifiable. The static is removed and a medium sized fault appears on the right hand side of the section which was not detected on the original section (Carpenter and Robson 2000).

Mining activity itself can induce seismic movements, especially along fault zones, and in areas with a long mining history. In areas where this is an established

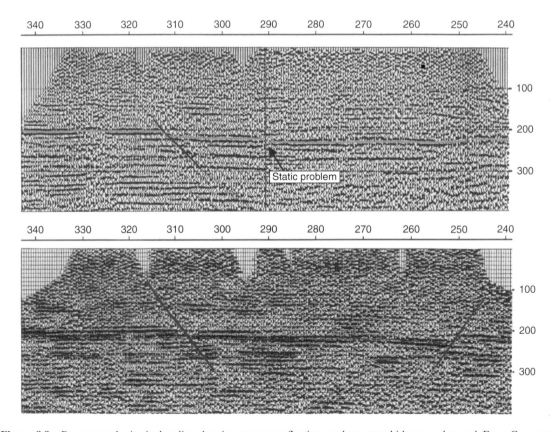

Figure 8.9　Reprocessed seismic data line showing stronger reflections and structure hitherto undetected. From Carpenter and Robson (2000), with permission of EAGE

phenomenon, seismic monitoring may be carried out. Mining-induced seismic events may be intense rock-bursts or seismic activity induced by mining equipment and/or blasting. Kalab (1997) and Holub (1997) describe such seismic activity in the Ostrava–Karvina coalfield in the Czech Republic. Redmayne *et al* (1998) describe similar seismic effects in the Midlothian Coalfield, UK.

8.3.1.2 Seismic refraction surveys

The energy input to the ground must be stronger for refraction shooting; consequently explosives continue to be the dominant energy source, although other sources are also used, such as falling weight for shallow studies. Refracted waves differ from reflected waves in that the principal portion of the refracted wave path is along an approximately horizontal interface between two rock layers before refraction to the detectors at the surface.

Refraction methods have been used in opencast coal exploration to locate previous workings (Goulty and Brabham 1984), to determine the variation in thickness of glacial deposits overlying coal-bearing sequences and to locate faults at shallow depths. Refraction methods have limitations when applied to the location of sub-surface positions of coal seams, but it is successful in locating previous workings because of the contrast in seismic velocity between backfilled mine workings and unworked areas.

8.3.2 Gravity surveys

The distribution of rock masses of different densities in the earth's crust gives rise to local and regional variations in the earth's gravitational field.

Gravity measurements are made using a gravimeter, taking readings at stations whose spacing may vary from 1 m to 20 km. The station interval is usually selected on the basis of assumed depth and size of the anomalous bodies sought.

Areas with an anomalously high Bouguer gravity value (a positive anomaly) can indicate relatively dense rock such as crystalline basement. A low Bouguer gravity value (a negative anomaly) is associated with the presence of less dense material, such as a thick succession of sedimentary strata. The magnitude and form of anomaly is related to the shape, orientation and depth of the feature, together with the contrast in density between the different rock types involved.

Gravity surveys are important, particularly on a regional scale, and are used in coalfield exploration both to detect the presence of sedimentary basins and to provide information on the overall structure of individual sedimentary basins. The results of a gravity survey are usually supported by additional geological data such as density determinations on rock samples and field mapping results. Typical densities of coal and coal-bearing sediments together with igneous and metamorphic rocks are given in Table 8.1.

The main use of gravity surveys has been in the location of sedimentary basins that could be coal bearing and which may be concealed by younger strata.

Those areas containing Gondwana coalfields, i.e. Australia, South Africa, India and Brazil, are especially suited to the use of gravity surveys. Most Gondwana sediments are preserved in basins lying directly on crystalline basement. These produce negative Bouguer anomalies which contrast with the surrounding positive values over areas of crystalline rocks.

In Western Australia, the Collie coal measures occur in a basin of Gondwana age, extending for about 180 km^2. The basin is a remnant of a once exten-sive deposit preserved by downfaulting or folding in the Pre-Cambrian basement. The surrounding granites and the coal measures themselves are almost wholly covered by laterite and Pleistocene or Recent lacus-trine deposits. Figure 8.10 shows the negative Bouguer gravity anomaly map of the Collie coal basin. The principal feature is the low value of the Bouguer gravity anomaly which represents the less dense sedimentary coal-bearing sequence. The boundary of the coal sedi-ments is indicated by a large Bouguer gravity anomaly gradient where anomaly values increase as they pass across from the lighter sediments to the denser granite basement. The gravity survey indicates that the Col-lie Basin is divided into two main troughs, separated by a basement ridge extending from the southeast end, through the Bouguer gravity anomaly high to the north-ern boundary (Figure 8.10). A drilling programme has confirmed these results and discovered a new coal-bearing sequence covering approximately 25 km^2 in the eastern trough, containing a coal seam 10.0 m in thick-ness, which is now being mined.

In India and Bangladesh, similar coalfield basins have been identified. In northwest Bangladesh, Gond-wana coal-bearing sediments are present in a series of small basins, now preserved as graben structures in the underlying crystalline basement, with the whole area concealed beneath Tertiary sediments. Gravity (and magnetic) surveying has identified these areas and sub-sequent drilling has confirmed the presence of thick Gondwana coals down to depths of 300 m.

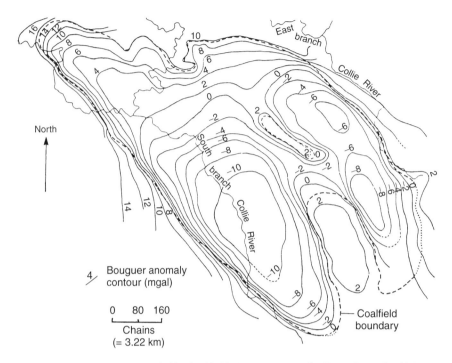

Figure 8.10 Bouguer anomaly map of the Collie Coalfield, Western Australia. From Parasnis (1986), with permission of Kluwer Academic Publishers

8.3.3 Magnetic surveys

The magnetic properties of rocks may differ by several orders of magnitude rather than by a few tens of percent. Typical values for coal and coal-bearing sediments together with igneous and metamorphic rocks are given in Table 8.1.

Coal-bearing sequences have relatively low magnetic susceptibilities in contrast to the higher magnetic properties of basement igneous and metamorphic rocks. Magnetic surveys are used to delineate the broad structural framework of a coal-bearing area. Such surveys do not detect coal, but help in locating sedimentary sequences likely to contain coals at accessible depths (Evans and Greenwood 1988).

In northwest Bangladesh, aeromagnetic surveys have indicated in some areas that the depth to crystalline basement is less than 250 m (Busby and Evans 1988). The aeromagnetic survey together with subsequent drilling has delineated coal-bearing sediments preserved in graben structure in the basement. This has enabled those areas identified as accessible by mining to be targeted, and further drilling has identified sediments of Gondwana age, containing a number of thick bituminous coals.

Such regional aeromagnetic survey results are often combined with regional gravity data to confirm the presence of sedimentary basins.

Distinct from large scale aeromagnetic surveys, detailed ground magnetic surveys are used to locate the presence of basic igneous (dolerite) dykes in mine areas, and also to detect the limits of burnt coal seams.

To locate dolerite dykes, a series of profiles is surveyed and plotted approximately perpendicular to the strike of each dyke and magnetic readings are taken every few metres. An example of a lenticular shaped magnetic anomaly from Causey Park, northeast UK, shows the configuration of a dolerite dyke as shown in Figure 8.11 (Goulty *et al* 1984). The anomaly contours are expressed in 100 nanoTeslas (nT) intervals, a nanoTesla being a unit of magnetic field strength. The location of dykes of significant size is important, particularly in opencast mining; coals which are in contact or close proximity to dykes will have undergone devolatilisation due to baking and therefore many have deteriorated (or in some instances improved) in rank and quality. This phenomenon is particularly important in South Africa, where dyke and sill swarms are intruded into the coal-bearing Karroo sediments. Undetected dykes cause problems in underground mining by making

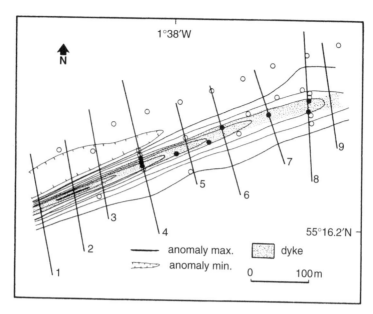

Figure 8.11   Magnetic anomaly map for part of the Causey Park dyke, UK. Contours are at 100 nT intervals. Boreholes which encountered dolerite at rockhead are shown as solid circles. From Goulty *et al* (1984), with permission of EAGE

tunnelling through them difficult and expensive and by affecting coal quality.

In the delineation of burnt zones in coal seams, the magnetic susceptibility of unbaked sedimentary rocks is quite low, whilst the magnetic susceptibility of the baked rocks is variable (Hooper 1987). Most of the maghaematite and magnetite in the baked rocks is derived from the thermal alteration of sedimentary minerals; some baked areas have undergone iron enrichment; iron is mobile during thermal metamorphism and can be redeposited in the baked rocks. On heating, shales and siltstones undergo significant reductions in volume; however, shales tend to separate into small pieces, so exposing a greater surface area available for iron enrichment. The sediments around the edges of a burnt seam may contain appreciably more magnetite if the coal fire is extinguished due to the lack of oxygen, which reduces more iron oxides and hydroxides to magnetite. In these cases larger magnetic anomalies may be expected along the margins of the baked zones. Figure 8.12 shows a magnetic profile over a burnt coal zone in East Kalimantan, Indonesia; the zone of burnt coal is some 160 m wide extending in from the outcrop. The magnetic profile shows a distinct magnetic anomaly of over 1000 nT amplitude when passing across the burnt mudstones.

Similar anomalies produced by baked sediments over burnt coal seams in the Southland District of New Zealand are shown in Figure 8.13.

By carrying out a series of traverses perpendicular to the strike of the coal, a zone of burnt coal can be accurately determined and the loss of coal within the area of investigation can be calculated.

8.3.4 Electrical methods

8.3.4.1 Electrical resistivity methods

Resistivity values for coal and coal-bearing sediments are given in Table 8.1.

Electrical resistivity mapping is primarily used for detecting local, relatively shallow inhomogeneities, and is employed typically in delineating geologic boundaries, fractures and cavities.

In the Raniganj Coalfield, India, a combination of resistivity depth sounding and surface electrical mapping has been used to identify lithological contacts and the presence of faults and dykes under tropical weathering conditions.

Electrical resistivity to locate coal seams has also been applied to the Pennsylvanian anthracite field and in Wyoming USA, where resistivity has detected the splitting of a coal seam into two, separated by a sand body; it has also been used in Australia. In the UK, electrical resistivity has been used to locate old concealed mine shafts and other mining cavities.

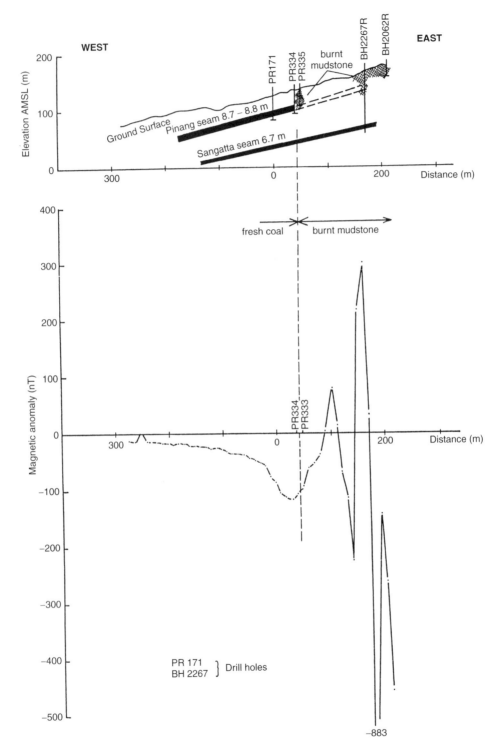

Figure 8.12   Magnetic profile over burnt coal and geological cross section, East Kalimantan, Indonesia, reproduced with permission of P.T. Kaltim Prima Coal

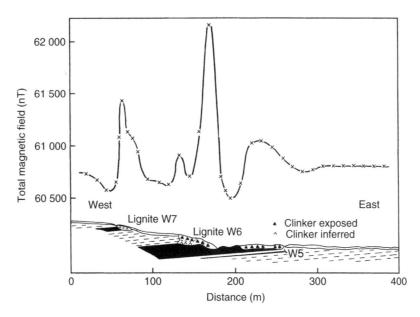

Figure 8.13 Total magnetic field profile and geological cross-section, southern end of New Vale Coal Co., eastern Southland, New Zealand. From Lindqvist *et al* (1985)

8.3.4.2 *Pulse radar methods*

Pulse radar systems have been used from the surface to locate and calculate the thickness of coal pillars in abandoned mines in the USA, usually to depths of less than 20 m. The method is also being applied to the detection of cavities and abandoned workings beneath opencast operations.

8.3.5 Radioactive methods

In coalfields, radioactive surveys can be used to trace marine shales high in radioactivity; this can be useful as an indirect method of mapping coal seams when the position of the radioactive shale is known to occur in the vertical sequence and also its position in relation to known coal seams.

Topographic irregularities, dispersion of radioactive materials due to weathering and 'background radiation' affect instrument readings.

8.4 UNDERGROUND GEOPHYSICAL METHODS

The use of geophysical techniques underground encounters difficulties. Space restrictions and safety requirements, particularly in the use of some electrical equipment, limit the use of certain geophysical methods

underground. Nevertheless, in-seam seismic and pulse radar methods have been used in these locations.

8.4.1 In-seam seismic surveys (ISS)

These surveys involve the use of channel waves propagating in the coal seam to detect discontinuities in advance of mining. ISS uses seismic waves which travel parallel to the bedding planes of sedimentary strata. They are restricted to travelling along beds within which the seismic velocities are lower than in the stratigraphic units above and below, and since they travel with the bulk of their energy confined to the low seismic velocity layer, they are referred to as 'channel' waves. Coal seams are excellent mediums for channel waves as they invariably have lower seismic velocities than the sediments which surround them. Table 8.1 shows the seismic velocity of coal to be about half that of sandstone and limestone. The targets for ISS are reflections from obstructions within the wave channel, namely faults, dykes and washouts. Such discontinuities are of vital importance to the economics of longwall mining.

An explosive source is normally used for seismic surveys, but for underground operations alternative sources are required, so mechanical devices such as Vibroseis, mini-SOSIE and Land Air Gun techniques may be used. In the UK, a pneumatic piston impactor has been tested for this purpose in mines where the use of roof and rib bolting as roadway support acts as anchor points for the geophones. Additional tests have used surface seismic sources to indirectly produce channel waves propagation in underground coal seams, in order to identify faults. These tests have shown some success in locating such seam discontinuities.

As the seismic recorder is not flameproof, it is located either on the surface or in an intake roadway in which methane concentration is less than 0.25%. In Germany a flameproof digital recording unit has been developed which will help to overcome this problem.

The in-seam seismic (ISS) method can be used as shown in Figure 8.14 (Jackson 1981), which illustrates the behaviour of channel waves reflected from or transmitted through a variety of discontinuities. Depending upon the target structure, shot holes and receivers can be located in the same or different roadways. The recorded seismic data are then processed in order to construct a map of the distribution of faults or washouts in the coal seam.

The ability to detect small scale faulting by ISS is dependent on the particular attenuation characteristics of the coal seam concerned. ISS can detect faults which are below the resolution of surface seismic, and can improve the positioning of faults interpreted from the surface seismic, both of which have economic significance, and faults that disrupt the coal seam can be detected within 200–300 m. Enhancement of the ISS technique is particularly useful where longwall mining techniques are employed. Where advancing longwall faces are used, it is essential to have prior knowledge of the nature of the coal seam through which the advance will be made. The high financial investment in establishing longwall faces is at risk due to loss of production if structurally affected ground is encountered. The prediction of faults in a panel of coal is important in maintaining the lifetime of existing longwall faces. Surveys from the face carried out at regular intervals, dictated by the rate of advance, can give early warning of impending dangers. For example, a longwall advance

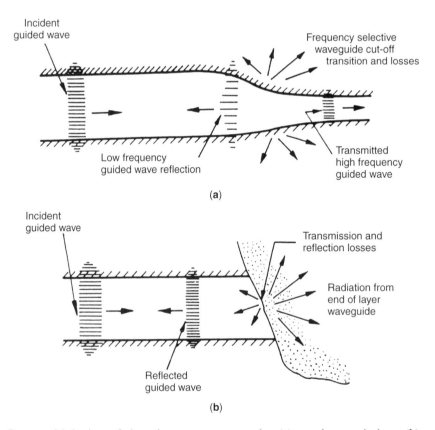

(a)

(b)

Figure 8.14   Conceptual behaviour of channel waves on encountering (**a**) a coal seam pinchout; (**b**) a channel sand cutout; (**c**) a fault with a throw less than the seam thickness; (**d**) a fault with a throw greater than the seam thickness. From Jackson (1981), reproduced by permission of IEA Coal Research – The Clean Coal Centre

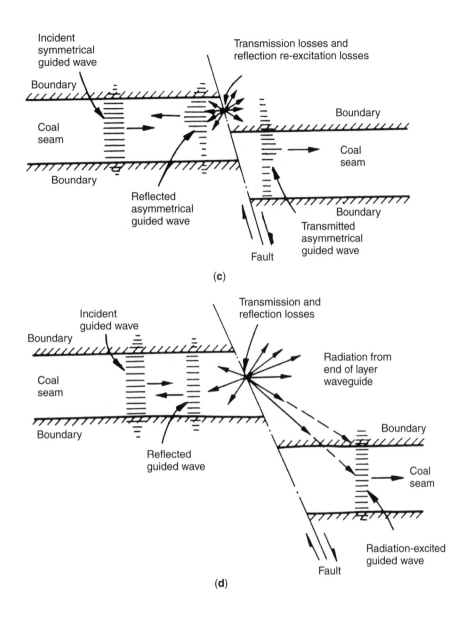

Figure 8.14 (*continued*)

face was expected to terminate due to the constraint of a 3.5 m fault encountered in a neighbouring worked-out panel (Figure 8.15(a)). An ISS reflection survey was undertaken to locate the exact position of the fault before the face came into contact with it. The ISS survey located the 3.5 m fault successfully, but also imaged a closer fault which was hitherto undetected (Figure 8.15(b),(c)). On this evidence, the face was stopped short to avoid damage to working equipment.

8.4.2 Pulse radar techniques

Pulse radar work in underground mine roadways is used to detect hazards in advance of mining, and to measure the thickness of the coal layer remaining in the roof of mine roadways, and to detect geological discontinuities in the coal panel to be worked. This is important in longwall face operations in controlling the cutting position of the continuous coal cutting machines,

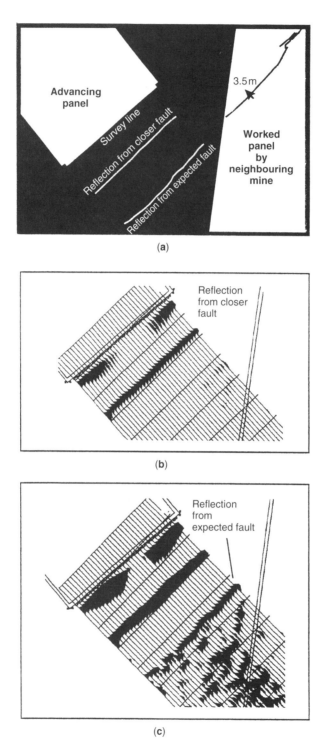

Figure 8.15   In-seam seismic reflection survey used to detect faulting in advance of longwall mining. By permission of IMC Geophysics Ltd

and for safety reasons. Pulse radar methods have been employed in underground workings in the UK and USA to determine coal thickness. In the USA, pulse radar has been used to determine mine roof stratigraphy, enabling clay and shale layers within the coal to be identified.

8.5 GEOPHYSICAL BOREHOLE LOGGING

Geophysical logging (geolog) is the measurement of the variation with depth of particular physical properties of surrounding rocks with geophysical measuring tools (sondes) located in boreholes. Measurements are made by lowering a sonde attached to the end of a cable to the bottom of the borehole, and then raising the sonde back out of the borehole at a constant rate to record the geolog. It is easier to maintain a constant rate by raising rather than lowering the sonde, which is important for data quality.

The sonde is connected by electrical conductors within the cable to recording and control instrumentation situated on the surface. This instrumentation is referred to as the logging unit, which also contains the powered winch used to lower and raise the sonde. In coal exploration, such units are small and portable. The logging unit makes a permanent record of the log data on a paper chart and on magnetic tape or disk, the latter suitable for future computer analysis.

The objective of geophysical logging is to determine *in situ* the rock type, and other properties such as porosity, fluid content and ash content, which may characterise the sedimentary lithotypes and igneous rocks intersected in boreholes.

All exploration and development drilling programmes will have a logging unit ready to log cored and open-holes within the area of interest. The depth and thickness estimates on the drilling lithological logs will be reconciled with the corrected depths recorded on the geolog. All features of interest on the logs will be used to site additional exploratory boreholes, and to site additional boreholes alongside selected logged boreholes (pilot holes) in order to take coal cores for quality analysis.

In coal exploration and mining, it is necessary to measure and identify one or more of the physical properties of the coal-bearing sequence. The appropriate geologs are selected to obtain the required geological information. The geological information sought includes: the identification of coal; the identification of depths to coal seam roof and floor; and the coal seam thickness; the identification of partings within the coal seams, and quality variations within the seam; the determination of geological features such as faulting, jointing, washouts, and thick sandstones and igneous intrusions; and the

determination of hydrogeological and geotechnical characteristics.

Once this information has been obtained, it can be built into the geological database, so that assessment of coal quality and geotechnical properties can be made, together with coal resource/reserves calculations with geological losses (see Chapter 7). This information will then be incorporated into the mine planning programme.

Geolog interpretation is essentially a three-phase exercise. The first phase is log calibration, converting the measured log units into either standardised log units or recognised physical properties. The second phase is basic interpretation, locating and measuring the bed boundaries, the depths and thicknesses, and an average value of log units for the formation. Finally, there is log analysis, relating the standard log units or physical properties to formation characteristics.

The logs most useful and most used to identify coal and coal-bearing sequences are gamma ray, density, neutron, calliper, sonic and resistivity. Of these, the first two, gamma ray and density, are usually sufficient to identify coal horizons and other common lithotypes in coal-bearing sequences. Additional logs are used to identify the structural attitude of coal-bearing strata and their inherent stress field orientation. These are dipmeter and acoustic scanning and image processing techniques.

8.5.1 Radiation logs

Gamma ray, density and neutron logs measure nuclear radiation emitted from naturally occurring sources within geological formations, or emitted from sources carried on the logging tool. Unlike electric logs, radiation will work in the absence of a borehole fluid (air-filled boreholes) and through casing.

In coal exploration, where boreholes tend to be shallow (less than 350 m), narrow and often dry, with walls in poor condition, radiation logs are often the only tools available for coal identification.

8.5.1.1 Gamma ray log

The gamma ray log measures the naturally occurring radiation in geological formations. The principal source of radioactivity in rocks is usually the potassium isotope K40 associated with clay minerals, and therefore found more abundantly in mudstones and clay-rich siltstones.

Conversely, good quality coals and clean sandstones have a very low level of natural radiation. As the amount of included clay material increases, in the form of clay partings in coal and as clay clasts and clay matrix in sandstones, so the natural radiation increases. In the case of marine mudstones, higher levels of potassium, together with other radioactive isotopes in the form of uranium and thorium, may be preserved. This causes the natural radiation levels to be much higher than in the more typical nonmarine mudstones.

The occurrence of horizons exhibiting high levels of radiation is extremely useful for correlation purposes; this is also the case for very low radiation levels in clean coal. Figure 8.16 shows the relationship of the gamma ray to selected lithotypes found in coal-bearing sequences.

The use of natural radiation to determine the ash content of coals is an unreliable technique, as coals with the same ash content may emit differing amounts of natural radiation, due to the make-up of the mineral content of the ash fraction in the coal.

The gamma ray log is not wholly diagnostic, and in normal practice, it is used in conjunction with other geophysical logs in order to fully distinguish formations.

Gamma ray logs are calibrated according to the American Petroleum Institute (API) Standards and adjusted so that the log gives values expressed in API units.

Gamma ray logs have relatively poor vertical resolution, as the gamma ray tool 'senses or sees' a fairly large area (up to 40 cm vertically). As adsorption increases as density increases, the depth of investigation becomes lower in high density formations such as basic igneous rocks. As a guide, coal seam thickness can be interpreted by taking the point on the gamma ray curve one-third down from the base of a typical mudstone. Such interpretations are asymmetrical as gamma rays travel further in the less dense coal medium.

Gamma ray performance is not impaired by borehole caving or loss of borehole fluids, as air, water and mud are not high absorbers of gamma rays. In addition, gamma rays can be run through casing, as casing tends only to attenuate the radiation received, the shape of the log being preserved although the base level is altered. For accuracy, however, an adjustment to allow for

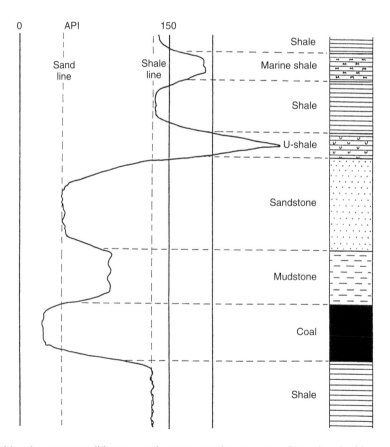

Figure 8.16   Coal-bearing sequence lithotypes and gamma ray log response. Reproduced with permission of Reeves Oilfield Services Ltd

casing has to be made.

As illustrated in Figure 8.16, the trace of the gamma ray log identifies a sharp geological boundary between two formations as a curve with a vertical height equal to the average vertical resolution of the gamma ray detector, and a horizontal length equal to the difference between the gamma ray count of the formations either side of the geological boundary.

8.5.1.2 Density logs

In coal exploration, the density log is used as a principal means of identifying coal, mostly because coal has a uniquely low density compared to the rest of the coal-bearing sequence (see Table 8.1). In certain circumstances, there is the additional benefit of an approximate linear relationship between ash content and density for a given coal seam in a given area.

In the density sonde, two detectors are used to measure gamma rays passed into the formation from the source and reflected to the detector by scattering. The long and short spacing logs are shielded from direct radiation from the source used, and measure the gamma rays which have been reflected, or back-scattered, from the rock.

Induced gamma rays in the energy range used for density logging are usually scattered in a forward direction. This means that density logs only respond to the formations between the source and the detector. The gamma rays can be considered to have 'diffused' through the material between the source and detector, and the density logs will equally be affected by all this material. The vertical resolution of the density logs is thus approximately equal to their spacing of the source and detector.

The density log is calibrated by measuring sonde output in homogeneous blocks of material of known density, and plotting a calibration curve, the results of which are applied automatically by a surface logging unit.

The density log will respond not only to the formation but also to the fluid in the borehole. As the borehole diameter increases, so the effect of the borehole fluid increases. This adverse effect is overcome by designing the sonde so that the measurement system is focused to give a narrow beam directed into the formation and forced against the borehole wall by a spring-loaded arm (caliper), so that it is always in contact with the formation. This removes the effects of borehole fluid except where irregularities in the borehole diameter occur during drilling due to material being washed out (caving), or deviations in borehole diameter. The problem of caving is significant for density logging in coal exploration,

as a short spaced density log can produce a response in a caved mudstone which resembles a response from a coal seam. Usually reference to the three-arm caliper log printout should highlight such anomalies.

A density log will show a sharp boundary between two formations as a curve with a vertical height approximately equal to the source to detector spacing (S) of the log and a horizontal length equal to the difference in densities between the formations either side of the boundary. Figure 8.17 shows the response of short and long spaced density logs. If the relationships between gamma ray intensity measured at the detector and the formation density is linear (short spaced density log), the boundary is taken as the halfway point along the curve. If the response is nonlinear (long spaced density log), the boundary point can be read off using the log calibration scale in gm/cc on the log or, if count rate only is available, it is assumed to be two-thirds along the curve from the high count rate value as shown in Figure 8.17. In coal seams with thicknesses less than S (source to detector spacing), the full log value of the thin bed is not recorded. Accurate thickness and log values for thin beds are more difficult to evaluate; this is particularly so for coal horizons with multiple thin partings.

With the development of a parallel sided drill rod, it has become possible to run geophysical logs through the rod. This has resulted in gamma ray and density logs being used in this way. This has minimised the likelihood of a radioactive source being lost down a borehole, with serious financial and environmental consequences. If there are such problems, the drill rod can be removed, bringing the tool up with it.

8.5.1.3 Neutron log

Neutron logs respond primarily to the hydrogen content of saturated rocks. The neutron log consists of a source which provides a continuous spectrum of high energy neutrons and a detector sensitive only to low energy (thermal) neutrons. Hydrogen is the most effective element in the slowing down or moderation of neutrons. Once slowed down, the neutrons diffuse away from the source, and are gradually captured. Therefore as one moves away from the source, the thermal neutron population first increases as more fast neutrons are moderated and then decreases as they are absorbed.

Hydrogen is found in the rock matrix itself, in water chemically bound to the rock molecules, and also in the fluid in the pore spaces of the rock. The last of these is a measure of the porosity of the rock and the amount of fluid it contains.

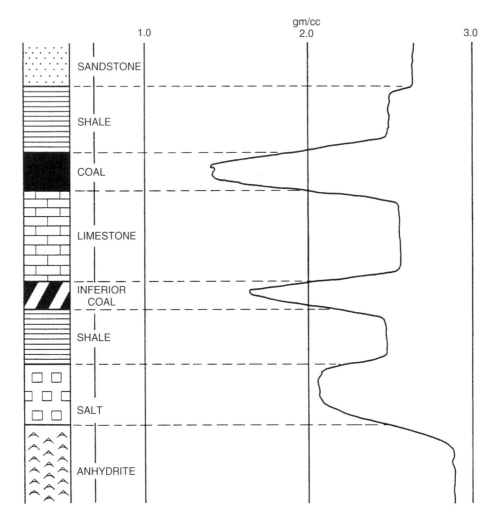

Figure 8.17 Long spaced density log response to coal. Reproduced with permission of Reeves Oilfield Services Ltd

In sandstone, the neutron response is logarithmic with porosity, such that at low porosities it is sensitive but at high porosities it is less so. Coal gives a response of around 60% effective porosity due to its structure of hydrogen and carbon; any change in count rate can reflect changes in calorific values, which on an ash-free basis can be considered as a coal rank parameter. Where moisture is relatively constant, the neutron log can give an approximate guide to the amount of volatile matter present.

The response of a neutron log over a formation boundary is not as simple as a density log response, and cannot be used for such accurate interpretation

of thickness. The same approach as used with density logs must be taken to arrive at average formation log values for use in quantitative work. From experience, bed boundary location can be approximately interpreted at about one-fifth from the high count rate value of the curve for a coal/sandstone interface as shown in Figure 8.18.

Neutron logs have been used to synthesise rock quality indices which can be directly related to mining problems. The hydrogen held in micro and macro fractures in the borehole and bound to rock molecules is measured (known as the Hydrogen Index), and empirical relationships are established between it and the observed

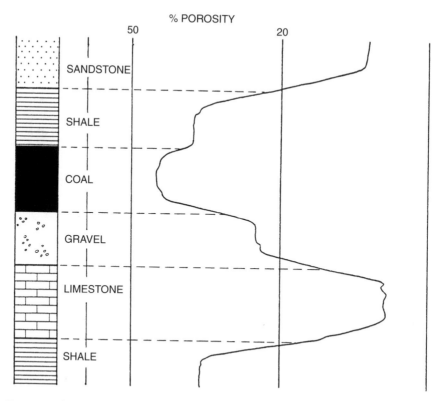

Figure 8.18 Response of neutron log over a coal seam. Reproduced with permission of Reeves Oilfield Services Ltd

fracture density for those rock types most commonly encountered in coal-bearing sequences. This technique has been applied particularly in Europe and North America for geotechnical studies.

8.5.1.4 Gamma spectrometry

Experiments to determine coal seam sulphur contents using well logging methods have been largely unsuccessful. However, Gregor and Tezky (1997) describe the measurement of sulphur content in brown coal by means of well logging equipment. It is based on spectral analysis of prompt gamma radiation generated by the capture of thermal neutrons using a spectrometric logging probe. The probe can also detect other elements such as Fe, Si, Al, Mn, Ti, Zn, K and Ca. This development is experimental but if successful and made available to operate with the established log suites, it would be a valuable tool in future coal exploration and mine planning.

8.5.2 Caliper logs

The caliper log measures the borehole diameter, and its main use is for correcting long and short spaced density readings. The caliper measuring system can either be part of the density logging tool, where it is a single arm used to force the logging tool against the side of the borehole, or as an individual tool with three arms at equal spacing. It is calibrated by measuring the log output at arm extensions fixed in place using a calibration plate marked out in borehole diameters.

The three-arm caliper gives an average of borehole diameter measured at three points. Difficulties arise when using the single arm tool where the hole size is enlarged due to caving in front of the density logging face, but not on the side of the borehole where the caliper is travelling.

The caliper log in association with the density log can be used to indicate rock strength; Figure 8.19 shows the response of caliper and density logs across a coal seam which has a good sound floor, but has a soft roof.

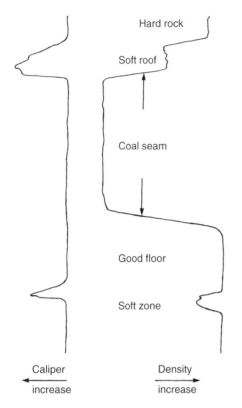

Figure 8.19   Use of the combination of caliper and density logs to determine coal seam roof and floor characteristics. Reproduced with permission of Reeves Oilfield Services Ltd

Such coal seam profiles are of importance to the mining engineer in estimating the mining conditions likely to be encountered in underground workings.

8.5.3 Electric logs

In the examination of coal-bearing sequences electric logs may be used to support radioactive logs, but are rarely used alone. This can be illustrated by the fact that coal possesses high resistivity, but in boreholes coal is difficult to distinguish from many rock types. Single point resistance log and self potential logs have been superseded by the gamma ray log due to the latter's reliability in a variety of borehole conditions. Ideally, shales give a low reading and all other lithologies give a high reading, but both coal and sandstone can read low and therefore cause serious errors.

Coals in the form of lignite and anthracite give very low resistivity readings, whilst subbituminous and bituminous coal can vary from low to high. A typical response of the resistivity log in a coal-bearing sequence is shown in Figure 8.20. In addition, resistivity logs are sensitive to the volume and salinity of groundwater, and also to the clay mineral content of lithologies encountered.

Resistivity logs can distinguish coal that is burnt close to an intrusion, or is oxidised due to weathering. Burning has the effect of reducing resistivity close to an intrusion; Figure 8.21 shows such a resistivity response across a burnt coal section.

8.5.4 Dipmeter logs

Dipmeters make high resolution micro-resistivity measurements around the borehole circumference which are correlated to produce apparent dip information. This is combined with tool orientation data to provide formation dips. Dipmeters are in two sections: a lower caliper arm holds the dipmeter pads against the borehole wall; these pads contain the micro-resistivity electrodes. The upper part contains the magnetometers and level cells needed to define the orientation of the tool in three dimensions. A minimum of three circumferential measurements is needed to define a plane, so that dipmeters have three arms 120° apart, and the intervals measured are overlapped by up to 50%. Dipmeters are used not only to calculate the structural dip of the strata in the borehole, but also dip patterns at the time of deposition, by subtracting the structural dip from the observed dip in the borehole. Figure 8.22 shows dipmeter tadpoles plots for a shallow dipping coal seam in the UK.

8.5.5 Sonic log

The sonic log has a similar response to the density log, as a result of the close relationship between compaction and density. In lithological interpretations it is not better than a density log, and is rarely run as a simple lithology log. The response of a sonic log in a typical coal-bearing sequence is shown in Figure 8.23.

The operational disadvantage of the sonic log is that it requires an open, fluid-filled hole; in addition it is adversely affected by caving. Nevertheless, the sonic log is a useful indicator of rock strength. The log interprets the velocity of sound waves in different lithotypes, which is of great value in the processing

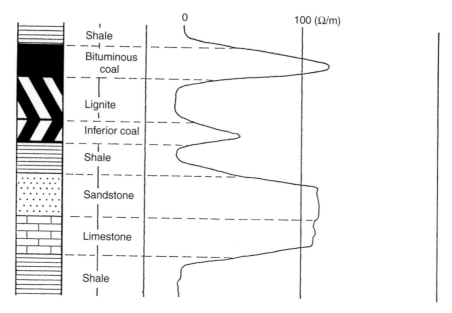

Figure 8.20 Response of resistivity log to coal-bearing lithotypes. Reproduced with permission of Reeves Oilfield Services Ltd

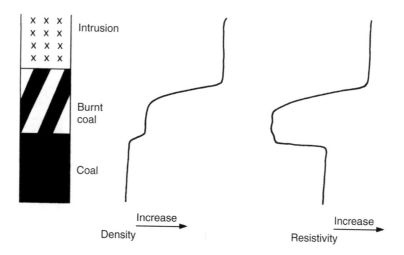

Figure 8.21 Response of density and resistivity logs over a burnt coal zone. Reproduced with permission of Reeves Oilfield Services Ltd

and interpretation of seismic data. As the velocity is related to the geomechanical properties of the rocks, the sonic log may also be used to predict the engineering characteristics of the strata for mine planning purposes.

8.5.6 Acoustic scanning tools

Acoustic scanning tools contain a rapidly rotating transducer which emits repeated short bursts of sound energy. Each burst produces a borehole wall reflection whose

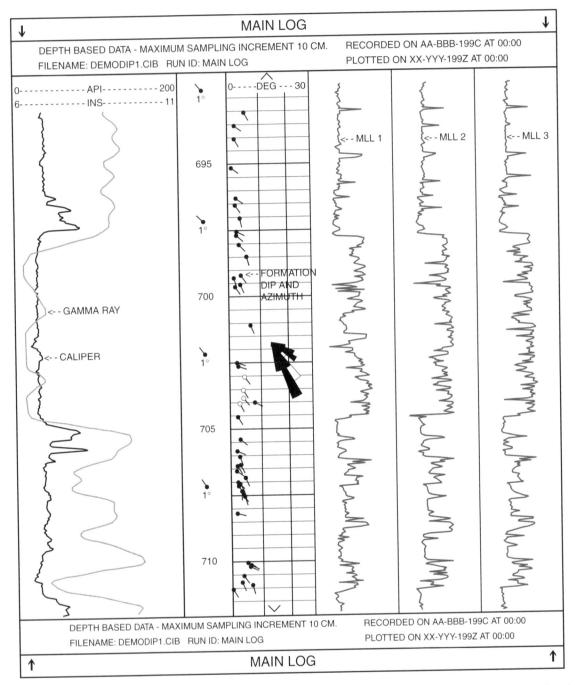

Figure 8.22 Dipmeter tadpole plot at the target seam depth, Fillongly Hall borehole, UK. From Firth (1999), reproduced with permission of Reeves Oilfield Services Ltd

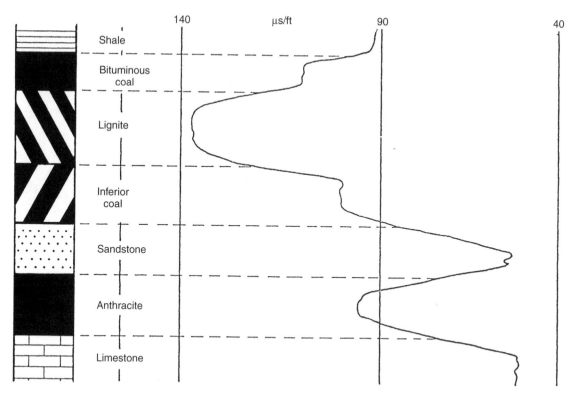

Figure 8.23 Response of sonic log to coal-bearing lithotypes. Reproduced with permission of Reeves Oilfield Services Ltd

amplitude and travel time characteristics are measured by the tool and recorded at the surface. As the tool traverses the borehole, a continuous helical scan is made. This is transferred into a series of circumferential scan lines which are then rotated into a common frame of reference to remove the effects of tool orientation and borehole trajectory (Firth 1999). Continuous false colour images are constructed by adding successive scan lines above one another on a plotter or display screen.

The maximum diameter of the tool is 57 mm, and it is important that the tool is properly centralised in the borehole. The tool is equipped with a pair of optical in-line centralisers, each centraliser comprising three articulating polished steel arms. The arms expand and contract as the borehole changes shape, thereby maintaining centralisation, even when tilted. The acoustic transducer is mounted in a rotating head assembly where it is exposed directly to the borehole fluid. A magnetometer adjacent to the transducer provides the azimuth information needed to orientate the image in a vertical

borehole. Two level cells allow the tool to be orientated in the case of an inclined borehole. The process also contains a natural gamma ray measurement which facilitates depth correlation to core data and other openhole logs.

Acoustic scanning tools are used in the identification of stress fields in coal-bearing strata by portraying breakouts as dark patches on the amplitude image. In Figure 8.24 (Firth 1999), the 360° caliper shows that the borehole has caved in a particular orientation. This corresponds to the direction of minimum horizontal stress. The plotting of stress directions in a series of boreholes will determine the final orientation of underground working areas by maximising roof and wall stability (see Section 10.2.2.3).

8.5.7 Temperature log

The standard bottom hole temperature is important in planning underground mining operations and, in

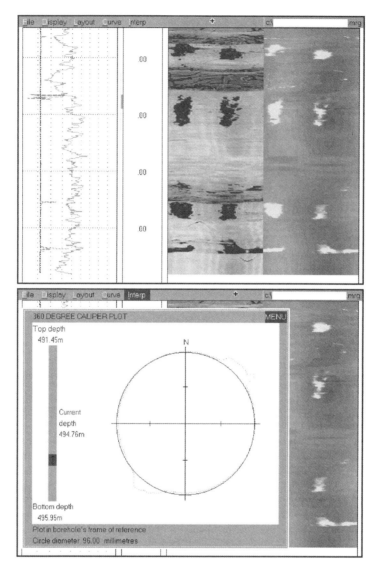

Figure 8.24   Breakout as it appears on amplitude (left) and transit time (right) images (upper picture), and its portrayal on a 360° caliper plot (lower picture). Firth (1999), reproduced with permission of Reeves Oilfield Services Ltd

particular, ventilation systems. Changes in the temperature log can also indicate the levels in the borehole at which significant groundwater inflow occurs.

8.5.8 Advanced interpretation

Further information on the lithology of the strata and the characteristics of the coal can be obtained by combining data from several different types of logs. For example, sonic and density logs can be combined for interpretation of coal rank as illustrated in Figure 8.25.

Attempts have been made to develop a simple method of measuring *in situ* moisture content in coal, but moisture levels in coal deposits are usually too high and render conventional neutron tools insensitive.

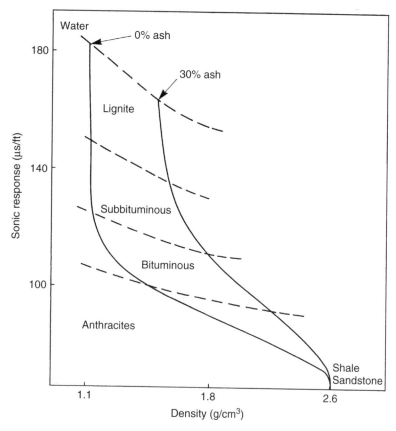

Figure 8.25 Interpretation of coal rank from sonic and density logs. Reproduced with permission of Reeves Oilfield Services Ltd

The combination of gamma ray and density logs with lithological logs from boreholes is used for correlation within coal deposits. Figure 8.26 shows a density analysis for a coal seam with bulk density, raw ash and raw calorific value calculations together with density and caliper logs. A series of such printouts for each seam in a lease area can be used for correlation and for estimations of likely raw coal quality across the area. Also, where horizons with high gamma readings are present they will readily show up on gamma ray logs, as will coal seams and clean sandstones which have low gamma radiation, all of which facilitates correlation. Figure 8.27 shows a combination of lithological, sonic and gamma ray logs for the coal-bearing Gidgealpa Group, Cooper Basin, Australia. Here the purpose is to identify not only the coal horizons but also the sandstones as possible hydrocarbon reservoir rocks.

This style of correlation is carried out during the exploration drilling phase of any project, together with ensuring the reconciliation of coal seam roof and floor depths on the geophysical logs with those on the lithological logs, the latter based on the open and cored borehole records.

The use of geophysical logs in coal exploration is now established practice, essential for any reserve assessment of a coal deposit. Using the various logs described, the required information can be summarised as follows: correlation of coal seams and other horizons across a deposit; accurate seam depths and thicknesses; coal seam structure details; control of drill core sampling; assessment of core recovery percentages and the indication of coal quality and quality variation across the deposit. A summary of log responses in a variety of lithologies is shown in Figure 8.28.

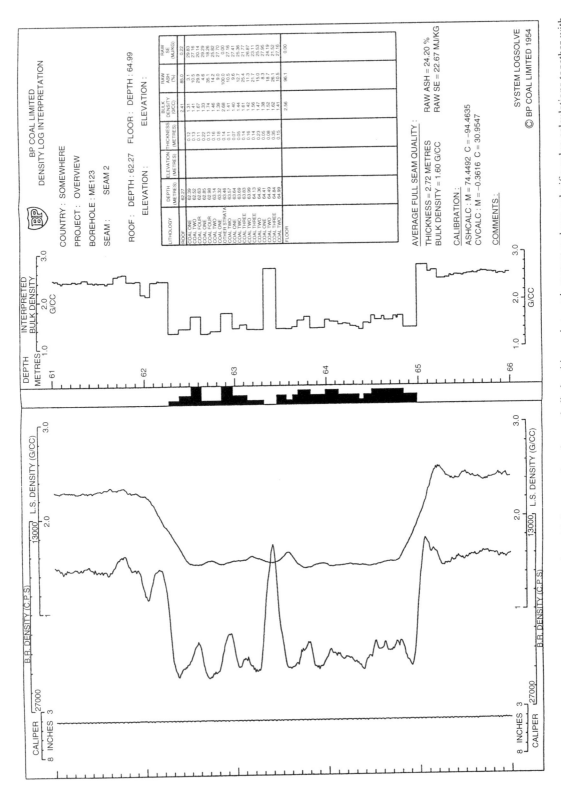

Figure 8.26 Density log interpretation together with interpreted coal seam bulk densities and raw ash content and raw calorific value calculations together with density and caliper logs. Reproduced with permission of BP Coal Ltd

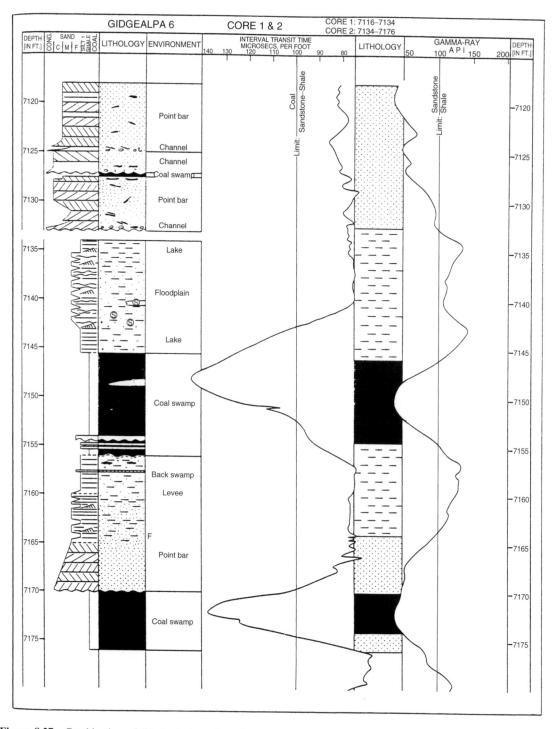

Figure 8.27 Combination of lithological, sonic and gamma logs to identify coal seams and sandstones as potential hydrocarbon reservoirs. From Thornton (1979), with permission of Primary Industries and Resources, South Australia

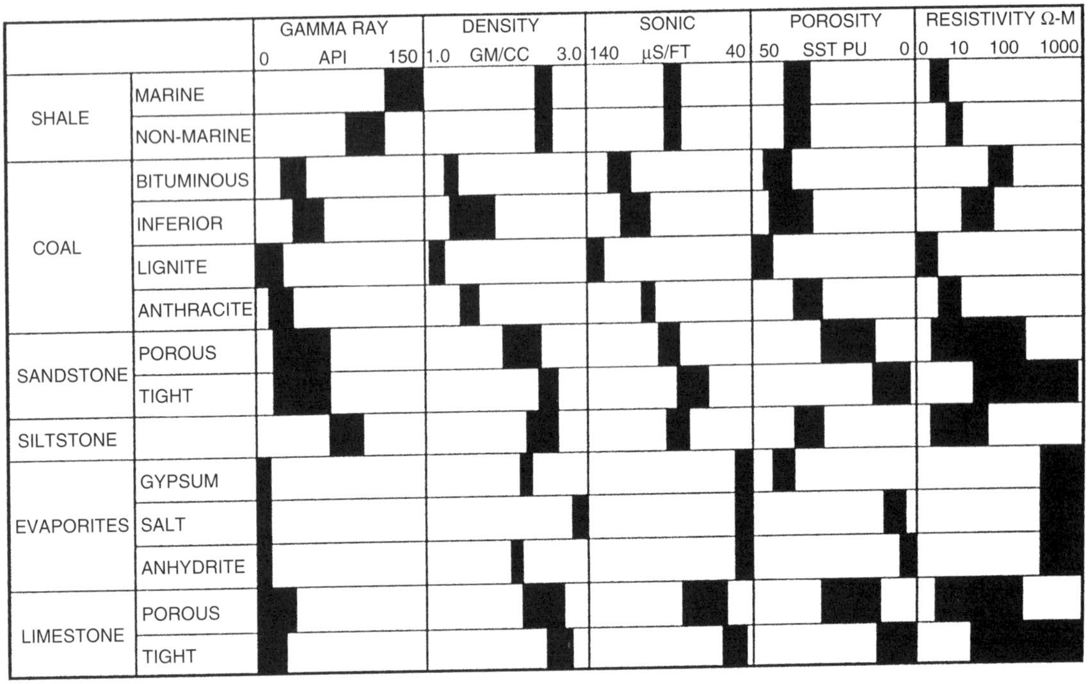

Figure 8.28 Summary of log responses in various lithologies. Reproduced with permission of Reeves Oilfield Services Ltd

More sophisticated geophysical logging techniques can provide the following additional information: lithology interpretation; assessment of geotechnical properties of formations, principally rock strength; the orientation (verticality and direction of deviation) of boreholes; and the measurement of stratigraphic dip of formations for use in the determination of geological structure.

9

Hydrogeology of Coal

9.1 INTRODUCTION

The influence that water plays in the mining of coal is significant, both in terms of surface water and groundwater movement. Water poses some of the major problems in mining in opencast and underground coal operations, and the presence of water, or sometimes the lack of it, in planning exploration and development programmes needs to be carefully assessed.

Field operations are strongly influenced in this way; for example, high and low water levels in stream and river sections will reduce or increase exposure. An abundant supply of water is required for water and mud-flush drilling, but not necessary for air-flush drilling, which is inhibited by a large head of water. Fundamental logistics such as transport are affected, dirt roads within the field area can become waterlogged and impassable, whilst very dry roads can create dust problems.

Of greater significance is how surface and ground-water will affect mine operations. There are numerous texts which cover this subject in detail, for example the National Coal Board (UK) publication, *Technical Management of Water in the Coal Mining Industry* (1982). A summary of the basic properties of groundwater and those aspects of hydrogeology directly related to coal-bearing sequences and coal mining is given in this chapter.

9.2 THE NATURE OF GROUNDWATER AND SURFACE FLOW

The hydrological cycle begins and ends with the oceans: water is evaporated from the ocean which then vaporises to form clouds. Water is precipitated from clouds, some of which falls on to the land surface and collects to form streams, rivers and lakes, and eventually flows overland back to the oceans. A portion of the rainfall passes through the soil to reach the water table and so becomes groundwater.

9.2.1 Groundwater

The upper part through which percolation occurs is known as the vadose zone or zone of aeration, and water movement is primarily under the influence of gravity. The phreatic zone or zone of saturation is below the water table in which pore spaces within the rock are filled with water. Water movement is primarily under the influence of hydrostatic and hydrodynamic pressures. These two zones are separated by the groundwater table, which will vary in position as changes in the groundwater level occur. These changes can be negative, resulting from groundwater movement and discharge, or positive, resulting from groundwater recharge by percolating water from the vadose zone.

Rocks which contain groundwater and allow it to flow through them in significant quantities are termed aquifers. Under normal circumstances water flows to a natural discharge point such as springs and seepages. This process can be interrupted if wells are sunk into the aquifer and water is abstracted.

This ability to allow water to flow through the aquifer is termed permeability, and is controlled largely by geological factors. When the properties of the fluid are considered, the permeability or the ease at which the water can move through the rock is referred to as the hydraulic conductivity, expressed in metres per day. The change in height or head that the water can attain naturally are known as the hydraulic gradient; the steeper the gradient, the faster the flow of water.

Groundwater may be contained in, and move through, pore spaces between individual grains in sedimentary rocks. Where rocks are fractured, this can significantly increase the hydraulic conductivity of the volume of the rock. The ratio of the volume of voids in the rock to the total volume of the rock is termed porosity. Some lithotypes do not allow the passage of fluids through them at significant rates, or may allow only small quantities to pass through. These are termed aquicludes and aquitards respectively, but are more commonly referred to as confining and impermeable horizons.

Groundwater usually flows under a hydraulic gradient, i.e. the water table. Where an aquifer is overlain by impermeable rocks, the pressure of groundwater may be such that the rest level of water would normally be well above the base of the impermeable layer. In such circumstances, the aquifer is said to be confined, and the surface is known as the potentiometric or piezometric surface.

The relationship of permeable and impermeable strata in confined and unconfined conditions in a coal-bearing sequence is illustrated in Figure 9.1.

In studying coal-bearing sequences, it is essential to identify the horizons which will act as aquifers, and those which will remain impermeable. In the case of aquifers, it will be necessary to calculate how much water passes through in a given time. To study this phenomenon, Darcy's law is used. This states that a fluid will flow through a porous medium at a rate (Q) which is proportional to the product of the cross-sectional area through which flow can occur (A), the hydraulic gradient (i) and hydraulic conductivity (K), i.e.

$$Q = KiA$$

An aquifer's effectiveness to transmit water, as calculated by Darcy's law, is known as its transmissivity, usually expressed in square metres per day.

In the case of an unconfined aquifer, the slope of the water table is a measure of the hydraulic gradient; in this case, the transmissivity is the product of the hydraulic conductivity and saturated bed thickness, the latter not being a constant feature. Because the water table is sloping, water flow is not purely horizontal, which means that the hydraulic gradient has a vertical as well as a horizontal component.

In the case of a confined aquifer, the aquifer remains fully saturated. When water is removed from the pore spaces, the water pressure is lowered, and downward pressure on the pore spaces within the aquifer causes slight compression. This reduction in pressure also causes a slight expansion of the water. The volume of water released from or taken into storage per unit surface area of the aquifer for each unit change of head is known as the storage coefficient.

In practical terms, changes in water levels within the designated area of interest must be monitored and recorded. This is accomplished by installing piezometers and observation and pumping wells (boreholes). Boreholes which are sealed throughout much of their depth, so that they can measure the head at a particular depth in the aquifer, are known as piezometers. The information they supply forms an essential part of the geotechnical investigations carried out prior to the mine feasibility study.

Boreholes which pass through the sequence to be mined act as monitoring holes from which the water table levels within the area can be measured on a regular basis.

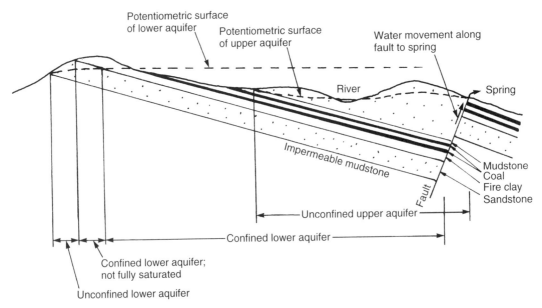

Figure 9.1 Groundwater conditions in a coal-bearing sequence, showing an upper unconfined aquifer and a lower confined aquifer with an intervening coal/mudstone sequence. Reproduced by permission of Dargo Associates Ltd

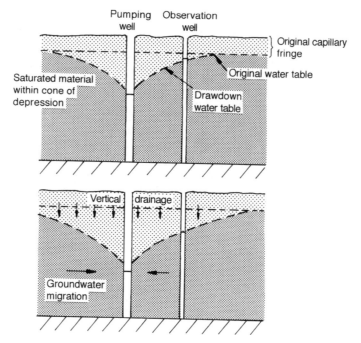

Figure 9.2 Drawdown of the water table due to pumping in an unconfined aquifer. Reproduced with the permission of Nelson Thornes Ltd from *Introducing Groundwater* (2nd Ed) Michael Price isbn 0748743715 (1996)

The action of pumping water from a borehole causes a reduction in pressure around the pump; this creates a head difference between water in the borehole and that in the aquifer. Water then flows from the aquifer towards the borehole to replace the water pumped out. Gradually water flows toward the borehole from further and further out in the aquifer; this has the effect of lowering the hydraulic gradient, so that around the borehole the hydraulic gradient becomes steeper, forming a characteristic lowering of the water table. This is known as the cone of depression; the reduction in head or lowering of the water table at the borehole itself is called the drawdown. These features are illustrated in Figure 9.2.

9.2.2 Surface water

The area of land that drains to a river is called the catchment area. These areas are separated by high ground called a watershed or divide. Streams that flow throughout the year are termed perennial streams, those that flow only occasionally are termed ephemeral streams, and those that flow only after the wet season are known as intermittent streams.

The discharge of a river or stream is the volume of water flowing past a given point in a unit of time, i.e. it equals the cross-sectional area of the flow section times the speed at which the water is flowing. Most useful is the record of flow plotted against time, which is shown on a discharge hydrograph.

Two components contribute to river flow: a baseflow component consisting of groundwater flow and slow interflow; and a quickflow component, derived from rapid interflow, surface runoff and any rain that falls directly on to the river or stream.

Surface water flow measurements are usually made during the geotechnical studies stage following the geological exploration stage, particularly where it is likely that watercourses will need to be rerouted during the construction and working life of the mine.

9.3 HYDROGEOLOGICAL CHARACTERISTICS OF COALS AND COAL-BEARING SEQUENCES

The majority of coal-bearing sequences contain sandstones, siltstones, mudstones, fireclays and coals. Of these, sandstones have the greatest potential for storing and transmitting groundwater; the other lithotypes have characteristically low permeabilities.

In older coal-bearing sequences, such as Carboniferous–Permian strata, the sediments are well indurated with low permeabilities; sandstones are often well cemented, so reducing their porosity and permeability. In younger formations, such as in the Tertiary, sandstones may still be only partially cemented or totally uncemented, and therefore have the potential to hold and transmit large amounts of groundwater by connected intergranular flow, i.e. they have primary permeability.

This is not to say that coals and coal-bearing strata do not allow the passage of groundwater. Most sequences are tectonically disturbed, and contain numerous discontinuities such as joints and faults, which, if open, will hold and allow groundwater flow, i.e. they have secondary permeability. In addition, it is a common feature for groundwater to flow along inclined bedding surfaces and appear in workings as a series of seepages, often staining the underlying strata with mineral precipitates.

The permeability of coal in general can be regarded as highly stress dependent, decreasing as the level of stress is increased. Coals react differently to stress due to their composition and rank. Coals with a high degree of elasticity and no apparent fractures are usually relatively unaffected by fluctuations in stresses exerted upon them. On the other hand, highly fissured and/or low mechanical strength friable coals tend to microfracture under stress. In the case of the latter, subsequent release of stress will leave the coal permanently microfractured and this then creates an increase in the overall permeability of the coal. There is also a relationship between the compressibility of coal and the volatile matter present. There seems to be an increase in compressibility with increasing volatile content up to around 36%, and then a decrease towards the lower rank coals.

Coals can hold significant amounts of water which, upon being breached by mine workings, is released and can cause mining difficulties; this mainly occurs in underground workings as opencast operations tend to be dewatered prior to mining.

Fireclays or seatearths and other clay-rich sediments have the ability to hold water, both in those clay minerals which have expanding lattice structures and as adsorbed water on the individual clay particles. This relatively high porosity does not, however, result in high permeability. The water is retained around the clay particles, held by surface tension; this phenomenon is known as specific retention. Although such sediments have low permeability, their water content is important as they have geotechnical significance as horizons of weakness when subjected to increased stress or sudden depressurisation.

Table 9.1 shows a list of indicative porosities and hydraulic conductivities for selected unconsolidated and consolidated sediments found in coal-bearing sequences

Table 9.1 Indicative porosities and hydraulic conductivities for unconsolidated and consolidated sediments which characterise coal-bearing sequences. From Brassington (1988), with permission of The Open University

Lithotype	Grain size (mm)	Porosity (%)	Hydraulic conductivity (K) (m/d)
Unconsolidated sediments			
Clay	0.0005–0.002	45–60	$<10^{-2}$
Silt	0.002–0.06	40–50	$10^{-2}–1$
Sand	0.06–2.0	30–40	$1–500$
Gravel	2.0–64.0	25–35	$500–10\,000$
Consolidated sediments			
Shale/ mudstone	<0.002	5–15	$5 \times 10^{-8}–5 \times 10^{-6}$
Siltstone	0.002–0.06	5–15	$5 \times 10^{-8}–5 \times 10^{-4}$
Sandstone	0.06–2.0	5–30	$10^{-4}–10^{b}$
Limestone	Variable	$0.1–30^{a}$	$10^{-5}–10^{b}$

^{a} Secondary porosity.
^{b} Secondary permeability.

(Brassington 1988). In the case of peats, brown and black coals, it is difficult to give such indicative values. Porosity values for peat will be high whereas brown and more particularly black coal will have low porosity values due to the increasing effects of compaction and coalification. Permeability values are difficult to quantify due to the fact that coals are dominated by discontinuities which may or may not allow the passage of water. High ash coals are known to have porosities of around 20%, with permeability values of less than one metre per day.

Studies of the hydraulic conductivity of peats have indicated that estimates showed time dependence, but that highly humified peat does not appear to transmit water strictly in accordance with Darcy's law; this may be due to air entrapment, whereas low humified peats tend to conform to Darcy's law.

In low rank coals all the water does not reside in pores alone; some must actually be included in the organic structure.

Experimentally, pore size distribution can be determined by forcing mercury into coals at increasing pressures and measuring the volume of mercury intrusion (mercury porosimetry). However, corrections have to be made for the compressibility of coals. Also, the high pressure may open/close pore space; this can be ascertained by measuring helium density of the coal samples before and after mercury intrusion.

The experimental work of Gan *et al* (1972) can be used to illustrate this. Twelve coals were tested by

Table 9.2 Gross open pore distributions in coals. See text for definitions of distributions. From Gan *et al* (1972), with permission of Elsevier Science

Sample	Rank[a]	C (% daf)	VT (% cm^3/g)	V1 (cm^3/g)	V2 (cm^3/g)	V3 (cm^3/g)	V3 (%)	V2 (%)	V1 (%)
PSOC-80	Anthracite	90.8	0.076	0.009	0.010	0.057	75.0	13.1	11.9
PSOC-127	lv	89.5	0.052	0.014	0.000	0.038	73.0	nil	27.0
PSOC-135	mv	88.3	0.042	0.016	0.000	0.026	61.9	nil	38.1
PSOC-4	hvA	83.8	0.033	0.017	0.000	0.016	48.5	nil	51.5
PSOC-105A	hvB	81.3	0.144	0.036	0.065	0.043	29.9	45.1	25.0
Rand	hvC	79.9	0.083	0.017	0.027	0.039	47.0	32.5	20.5
PSOC-26	hvC	77.2	0.158	0.031	0.061	0.066	41.8	38.6	19.6
PSOC-197	hvB	76.5	0.105	0.022	0.013	0.070	66.7	12.4	20.9
PSOC-190	hvC	75.5	0.232	0.040	0.122	0.070	30.2	52.6	17.2
PSOC-141	Lignite	71.7	0.114	0.088	0.004	0.022	19.3	3.5	77.2
PSOC-87	Lignite	71.2	0.105	0.062	0.000	0.043	40.9	nil	59.1
PSOC-89	Lignite	63.3	0.073	0.064	0.000	0.009	12.3	nil	87.7

[a] Bituminous coals: lv = low volatile, mv = medium volatile, hv = high volatile.

mercury porosimetry, and the results are shown in Table 9.2. The pore volume distributions are given for the following pore ranges;

1. Total pore volume VT accessible to helium as estimated from helium and mercury densities.
2. Pore volume V1 contained in pores >300 Å in diameter.
3. Pore volume V2 contained in pores 12 Å–300 Å in diameter.
4. Pore volume V3 contained in pores <12 Å, V3 = VT − (V1 + V2).

The proportion of V3 is significant for all coals; its value is a maximum for the anthracite sample (PSOC-80) and a minimum for the lignite sample (PSOC-89). From Table 9.2 it can be concluded that:

1. Porosity in coals with carbon content <75% is predominantly due to macropores.
2. Porosity in coals with carbon content 85–91% is predominantly due to micropores.
3. Porosity in coals with carbon content 75–84% is associated with significant proportions of macro- meso- and microporosity.

9.4 COLLECTION AND HANDLING OF HYDROGEOLOGICAL DATA

During any mining operation it is important to minimise any disturbance of the surface hydrology or groundwater regimes. These regimes comprise the dynamic equilibrium relationships between precipitation, runoff, evaporation and changes in the groundwater and surface water store. They can also be extended to include erosion, sedimentation and water quality variations. It is necessary therefore to know the pre-mining conditions that exist in the area of interest. The intensity of investigations will be influenced by the particular circumstances existing in the area of interest, e.g. the rainfall characteristics, drainage characteristics, presence or absence of aquifers, and the geological and structural character of the area.

The collection of data relating first to surface water flow and second to groundwater flow together with water quality analysis will enable a hydrogeological model to be constructed which will form an integral part of the development studies prior to mine design. The field techniques required to measure both surface and groundwater are outlined in detail in Brassington (1988).

9.4.1 Surface water

In the majority of countries in which coal is mined, the bulk of precipitation is in the form of rainfall. To measure rainfall over the designated area, a network of rain gauges is sited and monitored on a regular basis, in order to build up a detailed record of rainfall for the area; such rainfall is expressed as millilitres.

The flow of most springs is measured by filling a calibrated vessel in a given period of time. This can be a difficult operation in tropical terrain where the area around the spring may need to be cleared of vegetation before any measurement is taken.

The flow of rivers and streams is calculated by measuring the water velocity and the river cross-sectional area, or by installing a weir. The former is more suitable for rivers, the latter for streams. This flow is measured in litres per second.

9.4.2 Groundwater

In order to ascertain the potential groundwater problems that may be encountered during mining (particularly opencast operations), it will be necessary to determine the groundwater characteristics of the area; in particular, groundwater flow patterns, flow rates and depth to the water table.

Flow patterns will be affected by lithotypes, their disposition, relative permeabilities, and the presence of faults, joints and open bedding planes. For a full understanding of the groundwater for a proposed mining area a site specific hydrogeological model would need to be developed.

In order to achieve this, a system of monitoring boreholes or wells will need to be constructed. Their siting should be based on all the information gathered on the geology of the area plus all surface water occurrences. This information will be used to determine the expected direction and flow rate of groundwater, and enable the monitoring boreholes to be favourably sited.

These boreholes will provide information on the position of the water table, which in turn will define a baseline condition of any seasonal or climatic fluctuations against which the impact of mining can be assessed. In addition, samples taken from them will indicate general water quality. This monitoring programme is particularly important if aquifers are present in the designated mining area.

Boreholes used as monitoring points may be of different types. Those boreholes that are sealed throughout much of their depth in such a way that they measure the head at a particular depth in the horizon selected are known as piezometers. Piezometers are installed in areas which have opencast mining potential; they monitor the groundwater conditions in both shallow formations such as superficial deposits, and formations likely to influence surface mining operations. Piezometers can also be drilled at an angle to intercept vertical fractures in less permeable strata. Piezometers are usually sited where they can function throughout the life of the mine. Figure 9.3 shows piezometers set to measure levels in two formations in an opencast site.

Boreholes that are used as observation wells will include those that have already been drilled in the area

Figure 9.3 Piezometer group measuring water levels in two formations in an opencast working. Photograph by LPT. Reproduced by permission of Dargo Associates Ltd

for stratigraphical purposes and have been kept open for such a use, plus new site specific boreholes. Of these, at least one observation borehole should be placed intersecting each aquifer both upstream and downstream of the mine site; further boreholes should be placed around the periphery and within the mine site to ensure that a reliable estimation of the water level in each aquifer can be determined, and additional boreholes should be located at any geological discontinuities that may affect groundwater flow such as faults, folds, and abrupt changes in aquifer thickness.

Observation boreholes and piezometers are regularly placed to determine groundwater levels and the pressurisation field around the proposed position of a highwall in an opencast pit. Figure 9.4 shows the siting of such boreholes in a proposed opencast mine.

Water levels will be measured and recorded in all boreholes and piezometers on a regular basis. If any borehole is used to pump water, then the drawdown and

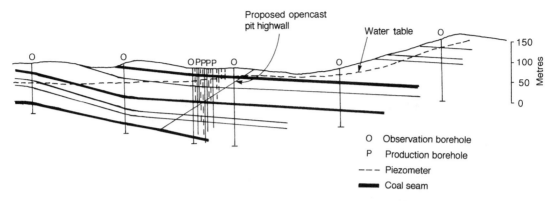

Figure 9.4 Section across proposed opencast pit showing position of observation boreholes, production boreholes and piezometers, placed to determine groundwater levels and pressurisation field around proposed position of highwall in opencast pit. Reproduced by permission of Dargo Associates Ltd

rest levels in the pumped borehole together with rest levels of water in surrounding observation boreholes will be recorded.

Once the network of piezometers and boreholes has been established in the designated area, regular monitoring and recording of field data will become an important routine operation.

Data obtained from piezometers and observation boreholes can be used to construct groundwater contour maps and flow nets. Groundwater contours are constructed from groundwater field levels related to a common datum plotted on a scale plan. Points of equal height are joined to form contours; flow lines are drawn at right angles from each contour. These give a plan view only, whereas in actual cross-section flow paths curve towards a discharge point such as a spring, stream or a pumping well.

The groundwater contour map may represent a water table surface or a potentiometric surface, which will be derived from the geological and well information. The spacing of contours gives an indication of aquifer permeability values: when they are close together, this indicates low permeabilities, as a steep hydraulic gradient is needed to impel the water through the aquifer; whereas widely spaced contours indicate a more permeable aquifer. Flow lines indicate the overall direction of groundwater flow and where such flow is concentrating.

Current investigations utilise this hydrogeological data to produce computer models of the groundwater movement patterns likely to exist in a proposed mining site. This has the advantage that the model can be modified as more geological and hydrogeological data can be input into the system.

9.5 GROUNDWATER INFLOWS IN MINES

When an aquifer is excavated during both open pit and underground mine operations, groundwater may enter the mine. Some aquifers have a finite flow; others are more persistent, particularly if the aquifer is being constantly recharged. Small aquifers may have insignificant water inflows, but in the case of open pit operations can seriously affect slope stability during dewatering and mining. Geological discontinuities such as faults, joints and bedding planes provide either pathways for groundwater flow or conversely act as hydrogeological barriers. The inflow of water can be controlled and prevented by dewatering the mine. This dewatering will also reduce the inflow of groundwater from prolific recharge zones such as rivers or lakes.

In open pit mines, water may enter through the pit floor, caused by the upward pressure in a confined aquifer fracturing overlying rocks which have become thin due to the deepening of the pit (Figure 9.5). In underground mines, a similar phenomenon can occur where substantial stress relief has taken place. Floor heave is a serious problem as it can result in the sudden flooding of the mine and cause disruption or even cessation of coal production. In old underground workings, flooding may have occurred; this can produce a large volume of water which may act unpredictably if the configuration of the workings is not fully known due to poor mining records.

9.5.1 Dewatering of open pit mines

Open pit mines where aquifers are known to be present either in the overburden, interburden or immediately below the target depth of the pit have to be dewatered

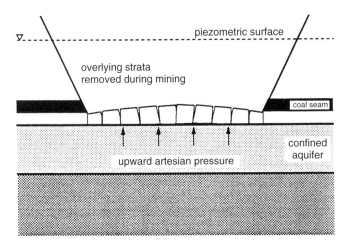

Figure 9.5   Floor heave resulting from upward artesian pressure. From Clarke (1995) with permission of IEA Coal Research

in order for the mine to operate successfully and safely. Dewatering is designed to stop water inflows into the pit, to maintain slope stability and to protect groundwater for abstraction in the area around the mine workings. The choice of dewatering method is dependent on the geology and hydrology of the mine site.

The standard situation is where the mine will intersect and excavate below the water table. This means that the groundwater level around the mine needs to be depressed to avoid flooding. This can be achieved by using pumps installed in a sump at pit bottom, but this does not allow for dewatering in advance of mining. Vertical wells are used extensively for dewatering, the exact pattern of wells depending on the site specific hydrogeological characteristics. Water can be removed from an aquifer by gravity or by pumping using submersible pumps in boreholes. Gravity wells drain water from an upper aquifer into a lower one below the level of the pit bottom. Pumping wells raise water from the aquifer to the ground surface to be disposed of. The water is pumped at a rate that maintains a steady cone of depression, the level of which is constantly monitored by piezometers. The cone of depression will extend beyond the mine boundary into the surrounding area. Figure 9.6 shows the dewatering of open pit mines using several methods of excavation (Clarke 1995); the wells are installed through the overburden, coals and footwall sequence, and also in the back-filled box cut area when up-dip mining. A series of interconnected well points may be used to lower the water table to a level below the proposed base of the excavations. In Figure 9.7 a two stage dewatering scheme is depicted

which is designed to lower the water table below the two levels of excavation in the site (Price 1996).

In Saskatchewan, Canada, dewatering is required in order to opencast mine brown coals overlain by 15–35 m of overburden. Here, the coal itself plus sands occurring both above and below the coal are the principal aquifers, with the coal acting as the major aquifer conducting water from the overburden sediments by means of joint and fissure flow. Dewatering tests included pumping from the coal and measuring the response in the overburden, and the measurement of the response of potentiometric levels in the overburden during excavation of test pits. The water levels in the coal were rapidly drawn down by pumping from structural lows in the coal seam; Figure 9.8 shows the migration of the 5.0 m drawdown contour from the pumping centres within the mine area. Pumping and test pit excavations caused relatively rapid reductions in potentiometric levels in the overburden. These tests showed that the overburden could be dewatered by directly pumping from the coal, the secondary permeability being sufficient to provide drainage from the overburden if enough lead time was provided prior to mining. A lowering of potentiometric pressure by further dewatering would be required in the sands below the coal to eliminate floor heave in the pit (Clifton 1987).

A large area may need to be dewatered for very large open pit mines to create a large cone of depression in the piezometric surface. The brown coal mines of Germany produce over 100 Mtpa and operate at depths up to 400 m. At Hambach, the pit area is up to $20\,km^2$ and the depth is up to 400 m. From this mine alone, 350 million m^3 of water has to be pumped each year to maintain the piezometric surface below a depth of

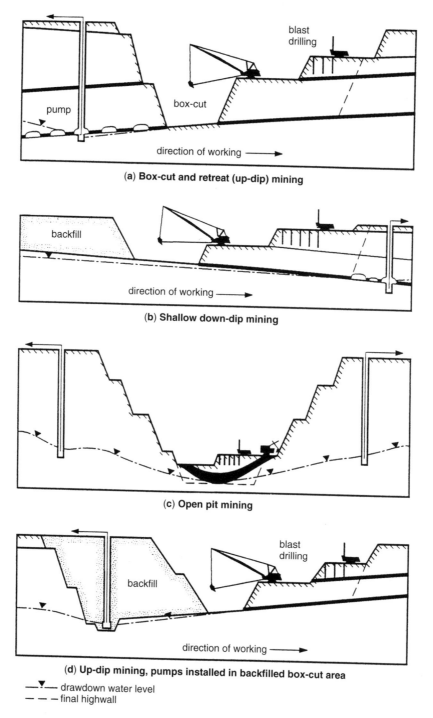

(a) Box-cut and retreat (up-dip) mining

(b) Shallow down-dip mining

(c) Open pit mining

(d) Up-dip mining, pumps installed in backfilled box-cut area

Figure 9.6 Types of opencast mining in which advance dewatering methods are used. From Clarke (1995) with permission of IEA Coal Research

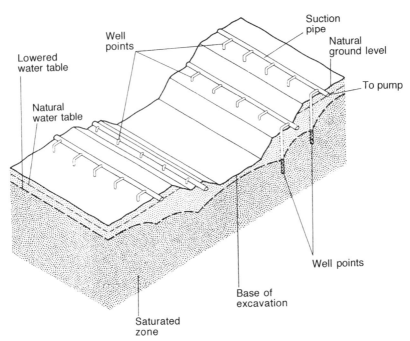

Figure 9.7   Site dewatering. A two-stage dewatering scheme using well points to lower the water table below the base of an excavation. Reproduced with the permission of Nelson Thornes Ltd from *Introducing Groundwater* (2nd Ed) Michael Price isbn 0748743715 (1996)

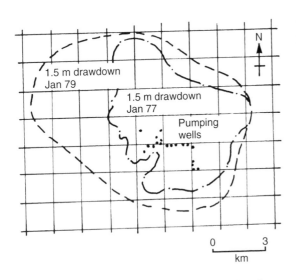

Figure 9.8   Migration of the cone of depression due to pumping. From Clifton (1987)

500 m. This has produced a cone of depression which extends for up to 30 km from the mine (Clarke 1995). The abstracted water is used for drinking water, cooling water in local power stations, industrial water supplies, irrigation, dust suppression within the mine and to protect the wetland areas close to the mine. Any remaining water is reintroduced into local streams and rivers. The abstraction of groundwater in this fashion can cause significant water management problems. The resultant cone of depression can result in reduced flow from springs and rivers, and lower levels in local wells, producing dry wells in some cases. The removal of water in the overburden can also produce settlement, which can affect buildings, underground conduits, and roads and railways in the area. Such changes can have strong influences on local populations, e.g. in India, where there are thousands of villages dependent upon shallow wells for water supply. Local open pit mining has to be conscious of this and provide alternative supplies and/or compensation. Figure 9.9 shows the predicted effect of dewatering an area proposed for opencast mining in India. Figure 9.9(a) shows the existing water table which slopes westwards towards a major watercourse. Figure 9.9(b) shows the probable effect of dewatering the proposed site after ten days, which has resulted in the cone of depression lowering the level of the water table by a maximum of 22 m. After two years of pumping, the likely effect is shown in Figure 9.9(c), where a

WATER TABLE

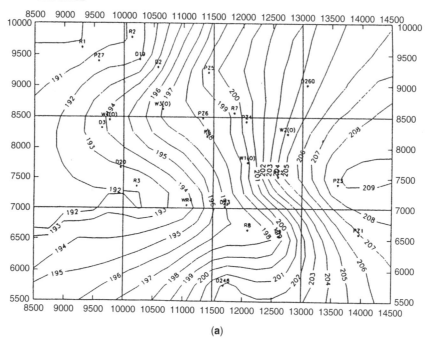

(a)

DECLINE IN WATER LEVEL AFTER 10 DAYS PUMPING.

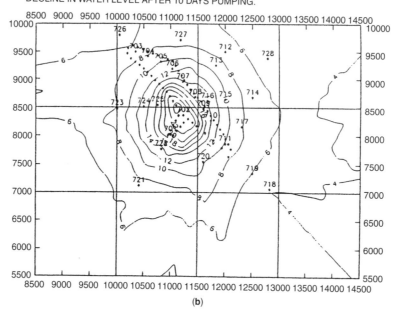

(b)

Figure 9.9 Theoretical study to predict the decline in water level after selected periods of pumping, for a proposed opencast mine in India. Drawdown levels are predicted to increase from 22 m after 10 days pumping to 64 m after two years' pumping

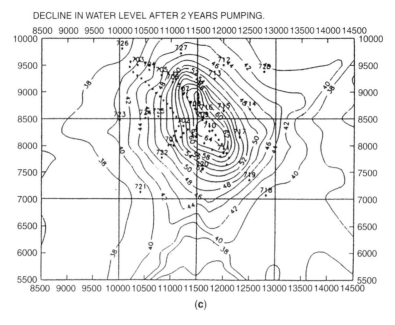

Figure 9.9   *(continued)*

maximum lowering of the water table is 64 m. The low-ering of the water table will affect an area $10-12\,km^2$ and will cause the local shallow wells to be dry. In this instance, the mining company will have to provide alternative sources of water for the local population.

In mines where strata are steeply dipping, horizontal or inclined drains are used for dewatering. In areas where the geology is more complex, wells may be concentrated in areas with potential water problems, e.g. in close proximity to a large fault. This will result in a local depression of the potentiometric surface, whereas the rest of the site may be served by a more general drawdown achieved by a grid of pump wells. Wells can also be used as depressurisation wells, to reduce hydraulic pressure in an aquifer, perhaps to improve slope stability or prevent floor heaving. A more specialised technique is the use of sealing walls; this involves the construction of an impermeable barrier between a prolific source of groundwater and the pit. They are designed to intercept groundwater and prevent it discharging into the pit or cone of depression. Sealing walls do not actively remove groundwater but can re-duce the pumping requirements needed to maintain the cone of depression.

Large scale dewatering projects can produce land subsidence. Unconsolidated or partially unconsolidated aquifers, or parts of aquifers that have been dewatered, are subject to lower pressures which can cause an increase in effective stress within the sediments; this per-mits greater consolidation and so produces subsidence, particularly in the floor of the opencast pit.

Conversely, dewatering will be required to depres-surise the pit floor, the highwall and end wall of an opencast pit. This is carried out to not only reduce inflows of groundwater, but to relieve pressure on the pit walls which may otherwise be subject to failure, partic-ularly when shear zones and other discontinuities have been identified.

9.5.2 Dewatering of underground mines

Numerous underground mining operations require the removal of water from the mine workings. Many mines, especially in old mining districts, are connected under-ground, so that pumping in one mine can control water levels in neighbouring mines. Water entering under-ground coal operations can be from several sources: from groundwater flow; from natural precipitation at the surface; and from abstractions from rivers, wells etc. The latter two occurrences are from infiltration of the work-ings via shafts and adits. Water may also enter from old abandoned workings situated in close proximity to the current mine.

In order to achieve effective mine drainage, detailed knowledge of the groundwater flow pattern within and around the mine is desirable. To achieve this can be a problem in that the changes in permeability of the coal-bearing sequence with depth, and the presence of unforeseen structural discontinuities, can make quantitative analysis difficult. In shallow mines (<100 m depth), rainfall, particularly tropical storms, can quickly affect water levels in the mine, as well as recharging groundwater which may enter mines at greater depths. The latter effect may occur after a time lag, depending on the permeability of the strata and rate of recharge.

The planning and design of effective mine drainage systems requires the best possible knowledge of underground water flow patterns and reliable forecasts of future yields. It is important that comprehensive records of mining be maintained and coordinated. All mine areas to be abandoned must be sealed off and clearly recorded on the mine plans, the locations of all mine interconnections, boreholes, shafts and adits must also be plotted. In addition to this information, the multiplicity of seams, faults and interseam connections must be recorded. Effective use of all of this data will allow long term planning of underground water control to be implemented.

The amount of water that may be removed from mining areas can be considerable. In the UK, mine waters pumped from all active underground mines and adjacent abandoned workings can total up to 1 million m^3 per day, and include the Nottinghamshire Coalfield totalling 14 million m^3 in 1991, and the Durham Coalfield totalling 70 million m^3 in 1994. Similarly in Germany, the Ruhr Coalfield pumped 114 million m^3 in 1994 (Clarke 1995). It is an ongoing operation and expense to drain water from underground workings both to facilitate working conditions but also to minimise potential hazards that water build-up can create.

9.5.3 Water quality

During the mining operation it is necessary to monitor the quantities and qualities of the waters flowing into the workings. Water pumped out of the mine should be utilised to the greatest practicable extent, but groundwater pumped from deep mine workings is often acidic or saline in nature and unsuitable for immediate discharge at the surface. Therefore it may be necessary to improve the quality by treatment (see Section 12.2.1).

In opencast workings, surface waters and shallow groundwaters do not have the concentration of elements found in deep groundwaters. This is not to say that water quality can be taken for granted; quality measurements are just as important an exercise as for deep mines; it is just that the problems of water quality are usually more acute in underground workings.

The groundwater that flows into mine workings is normally free from suspended particles, but always contains dissolved substances in concentrations related to the depth and hydrogeological conditions. However, such waters can become contaminated with fine-grained particles of coal and other lithotypes through contact with underground working operations.

It is the contaminated groundwater that creates the most problems, and its treatment is an additional cost to the mining operation. Methods used are given in Section 12.2.1.

Oxygenated groundwater when in contact with coal results in the oxidation of the organic and inorganic constituents of the coal. Water is adsorbed on to the cleat faces and if it is held there for a period of time it will oxidise that coal immediately in contact with it. This has the effect of reducing the coal quality (see Section 4.3.4).

9.6 GROUNDWATER REBOUND

In some traditional mining areas, regional dewatering of underground mines may have been in operation for over 100 years. Any change to this regime will have a profound effect on the water levels in the region. Figure 9.10 shows the effect of dewatering a mine and the resultant water table rebound effect once pumping has ceased. The original situation in Figure 9.10(a) shows an upper unconfined aquifer and a lower confined aquifer. Their respective piezometric surfaces were depressed by pumping during the mining operation, as shown in Figure 9.10(b). However, after the mine had ceased working·and been back-filled and pumping had been stopped, both aquifers were unconfined and a water table rebound occurred, shown in Figure 9.10(c). The time taken for this to occur will depend upon the rate of recharge to the system and also the nature and permeability of the back-fill material. Closure of deep mines in Yorkshire, UK, and the cessation of associated dewatering have led to concerns about the possible future pollution of groundwater and surface water resources once groundwater rebound is complete. Burke and Younger (2000) have used a computer model to predict the rate of groundwater recovery in the abandoned workings and the timing and flow rates of future surface discharges. Similar studies will be required in those mining areas which have now ceased to operate or are about to do so, which have been subject

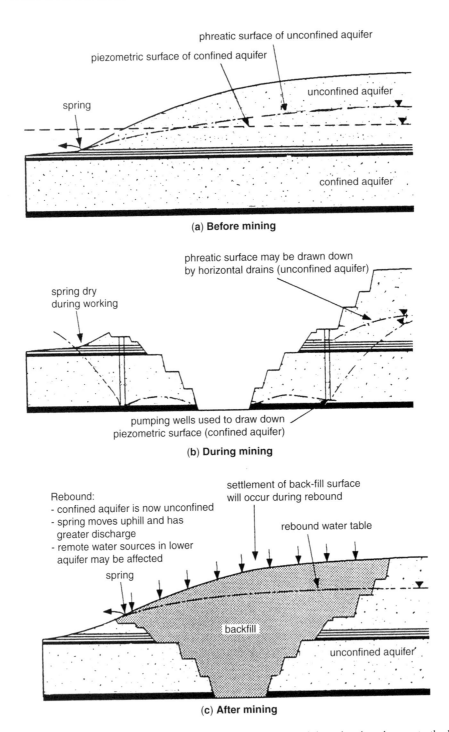

Figure 9.10   Rebound effect on the water table after cessation of opencast mining, showing changes to the local aquifer system. From Clarke (1995) with permission of IEA Coal Research

to dewatering over long periods of time. Cessation of pumping on a regional scale can mean the movement of original spring lines, and variations in the flow rates at springs and abstraction points. In areas where mining has existed for a long period of time, it is not unknown for buildings erected ostensibly on 'dry' ground during the pumping period to be flooded at a later date when pumping has ceased and the water table has reached equilibrium at a higher elevation than that maintained during mining operations.

10

Geology and Coal Mining

10.1 INTRODUCTION

The mining of coal and the methods used are well accounted for in the literature. Ward (1984) and Hartman (1992) give details of the technology used for both underground and opencast or surface mining. Only a brief outline is given here to highlight the importance of geological knowledge in mine planning and in mining operations.

The ability to mine coal is the direct result of the evaluation of geological data collected during the exploration phase of the project. The provision of values for coal reserves, coal qualities, and detailed geological conditions is essential for mine planning and process design.

Coal mining takes place in the geological medium; consequently geological conditions have a profound influence throughout the life of a mining operation. In this context, coal mining geology is a concentrated extension of the exploration process (outlined in Chapter 6), in that greater detail is required covering a small area. In addition, all the geological support studies assume a greater significance, e.g. engineering geology and geotechnics, hydrogeology, coal analysis and detailed geophysics.

The ultimate aim is to maintain the coal mine's production; one of the main reasons for failure is unreliable coal reserve estimates or unforeseen structural limitations.

The modern mining process itself is a highly mechanised operation and the selection of the mine layout to optimise equipment use is based on the mine design to which geology is a major input. The most cost effective mine operation will influence the financial return made from the mine and, in the development stage, encourage investment from interested parties.

The mining of coal has taken place in Europe for over 700 years and in most other major coal producing countries for over 150 years, using a number of methods to extract coal from the ground. In modern-day mining there are two basic methods of mining coal: first,

underground mining, where coal seams occur at depth beneath the ground surface, and are accessed by tunnels and/or shafts; and second, opencast or surface mining, where coal seams are close to the ground surface and can be accessed by direct excavation of the land surface.

Coal mining developed as a result of the Industrial Revolution, when the demand for coal for industrial use, power and shipping was suddenly accelerated. Those areas where this phenomenon first took place were Europe, North America and the former USSR. Coals were mined by underground methods as the coals in these areas were situated at depth. Despite the fact that underground mining is practised worldwide and that the bulk of the world's coal resources lie at depths only mineable by underground methods, the modern trend has been to increase the mining of coal by opencast methods. This is due to an increase in geological knowledge, cheaper methods of operation and the ability to utilise coals of all ranks and qualities in the industrial process.

10.2 UNDERGROUND MINING

To develop an underground coal mine, four main operations have to be carried out: a shaft sinking and/or an adit drivage to reach the target coal seam(s) beneath the surface; the drivage of underground roadways either within the coal seam or to access a coal seam; the excavation of working faces in order to extract coal; and provision for temporary underground storage of materials.

Once a mine design plan is complete, the first obstacle to be overcome is to access the target coal seam. This may be strongly influenced by surface topography: if the coal seam is exposed on a hillside, an adit or tunnel may be driven directly into the coal (Figure 10.1(a)); if the coal seam is at depth beneath the hillside, an inclined adit may be driven down at an angle to intersect the coal seam (Figure 10.1(b)). In areas where the topography is either flat or is a valley bottom, and the coal seams are at considerable depth, a vertical shaft is sunk to access the coal (Figure 10.1(c)). The cost of sinking

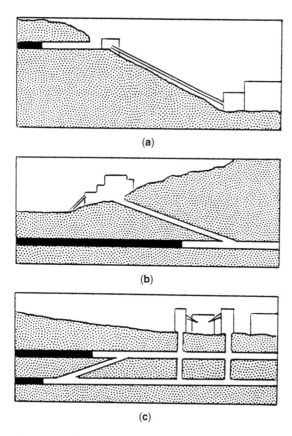

Figure 10.1 Methods of entry for underground mines. (**a**) Inseam adit; (**b**) Inclined shaft; (**c**) Vertical shafts. Adapted from Barton (1974)

the other is used for expelling the returned air, and is the 'upcast shaft'. The downcast shaft is also used for coal haulage and other access. The upcast shaft's primary function is for ventilation but also is an emergency exit. Shafts and adits are the most important capital item when opening underground mines. They provide all services for underground operations. These include fresh air, transportation of equipment and supplies, personnel traffic, power, communications, water supply and drainage, and not least the transportation and removal of coal from the mine. Depending on the depth of the mine, shaft sinking may take up to 60% of the development time (Unrug 1992). Adits are chosen where possible in preference because of the lower cost and shorter construction time. In a number of mines, a combination of shaft and adit is used. The shaft diameter and hoisting depths need to be selected to accommodate the maximum design use of the mine. It is better to overdesign in the first stage of a mine than face a bottleneck in future years. In using adits equipped with belt transporters and using a maximum slope of 15.5°, the coal transportation system is uninterrupted all the way to the surface. However, there are economic limits to the adit length, and shafts can be less expensive for depths exceeding 350 m. Roadways in the mine may be excavated within the coal seam, as is the case in some mines in USA and China; this makes development quicker and yields a coal tonnage at the same time, but is only possible when the coal is 2–3 m thick and has a strong roof.

shafts to the level of the coal seam (or just below, to provide a water drainage sump) is often the largest single cost in developing an underground mine. Sinking vertical shafts over 700 m deep may take two years or more; the deepest shafts in the UK (comparable to coal mines anywhere) are approximately 1000 m deep. Additional costs are incurred when the shafts are sunk through porous sandstones or running sands; in such cases, the ground is first solidified by injecting cement through vertical boreholes (cementation method) or by circulating saline water at temperatures below the freezing point of freshwater, through vertical boreholes to freeze the ground before shaft sinking (freezing method). In this case, the shaft walls are often reinforced by cast iron tubing to prevent water inflows when sinking has been completed and after the ice mass has thawed. Underground mines must have at least two points of entry: one is used for the intake of fresh air to the ventilation system, known as the 'downcast shaft';

10.2.1 Geological factors

Ideally, mine access should be central to the planned extraction area. In siting a shaft, all geological details of rock types, structure, and hydrogeology should be determined. In the case of adits into hillsides, hazardous sites where rock falls, landslides or flooding can occur should be avoided.

Geological investigations for underground mines must include: the identification of any mining hazards or breaks in coal seam(s) continuity, such as faults, igneous intrusions, washouts and seam splitting; any areas containing these features should be identified on the mine plan; the drawing up of plans showing thickness variations in the target seam(s) and thickness of strata between coal seams, together with coal quality trends obtained from borehole and mine sampling; and the geotechnical characteristics of the roof and floor strata for all seams that are planned to be extracted. The geotechnical behaviour of the roof and floor strata will influence the type of support to be used, e.g. whether roof bolting will be possible by having a strong cohesive rock above the coal.

The incidence of faults in coal seams has a significant effect on the selection of mining systems and on the productivity. Major faults with throws greater than 20 m very often delineate mine boundaries, or may cut out the coal entirely. Major faults are often associated with numerous minor faults, running roughly parallel. The effect of these faults is far more serious in underground mining than in surface mining operations. A fault of 1–5 m may be inconsequential in a surface mine but be a serious impediment in underground mining; for example, a fault with a displacement of 2 m can 'lose' a coal seam of 1.5 m. When this occurs, a new coal face has to be established at the new level after roadways or connections are made between the two seam levels. This may take several days or weeks, with a consequent loss in production. The problem is compounded if a series of faults is encountered and may result in the abandonment of working the seam in the problem area. Highly faulted coalfields often reflect a structural and metamorphic history which has increased the rank of the coal, as seen in the anthracite coalfields in South Wales, UK, North Vietnam, southwest China and western Siberia. Igneous intrusions are found in a number of coalfields throughout the world. Where the intrusions are vertical, they are known as dykes and can be of various widths. The rock is often doleritic and the effect is relatively local. Dykes are difficult to locate underground and have a direct effect on coal quality; the hot molten rock reduces the volatile content and 'cinders' the coal seam. Igneous intrusions in the horizontal plane above the coal seam or cutting through the coal seams are known as sills. The effect of sills on coal workings can be more subtle and dangerous. Sills are often several metres in thickness, very hard and competent; such a roof may not 'cave' or subside regularly in longwall workings and thereby cause problems. In room and pillar workings, the strong roof has led to the design of undersized pillars for roof support, which have eventually collapsed. In the 1960s, in a mine in South Africa, the whole of the underground workings collapsed virtually instantaneously when all the undersized pillars collapsed under a massive dolerite sill, with a catastrophic loss of life. Research following the disaster has led to empirical tables being available to mining engineers designing room and pillar workings for any combination of coal seam depth and thickness.

Other studies include the hydrogeological regime of the planned mine area and whether the mine is likely to be gassy. These studies will be part of the mine planning stage but will also extend as part of an ongoing programme during mine development and then continue during coal mining to ensure a continuing accessible coal reserve. The understanding of the geological parameters is essential to ensure successful high productivity coal mining.

Technical development in underground mining has concentrated principally on coal mining by two systems, longwall mining and room and pillar mining. Both methods have developed to meet the criteria of reductions in the work force and increased productivity, leading to reduced operating costs. The preference for either system relates to a number of factors: depth of mining, geological conditions, size of reserve area, availability of equipment, mining regulations and, most importantly, availability of investment capital.

Suitable geological and geotechnical conditions are necessary for high productivity mining, whichever of the two methods is chosen. If this is not the case, even with the best equipment and large reserves the mine will prove uneconomic to mine. Geotechnical conditions need to be such that development roadways and mining faces can be opened up rapidly.

10.2.2 Mining methods

10.2.2.1 Longwall mining

Modern longwall coal mining has developed over the last 50 years. It has a simple system layout and is designed to be fully automatic and to provide continuous production. The planned area to be mined is divided into a series of elongate panels accessed from an entry roadway. Mining involves the removal of coal from a single face representing the width of the panel. The workable coal seam thickness is usually 1.5 m up to 4 m, depending on the size of the hydraulic supports. The coal is mined by either longwall advance, where the face is moved forwards into the coal away from the entry roadway (Figure 10.2(a)), or by longwall retreat, where drivages are made around the edges of the selected panel area, and the face is then worked back towards the entry roadway. The working area is protected by movable hydraulic supports and overhead shields, which, as the equipment advances, protects the workforce from the roof which collapses behind as the support is removed. Figure 10.2(b) shows the basic layout of the longwall retreat working model. In both cases the coal is cut by either a rotary drum shearer (Figure 10.3(a)) or a coal plow (Figure 10.3(b)); these are attached to the front of the roof supports and run along a flexible chain conveyor. The use of coal plows in preference to shearers is more effective in thin coals as they can cut different areas of the face at different cutting depths.

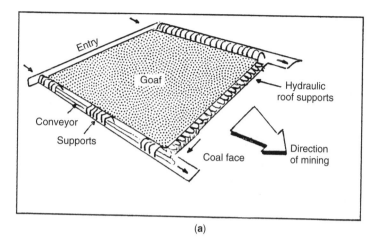

(a)

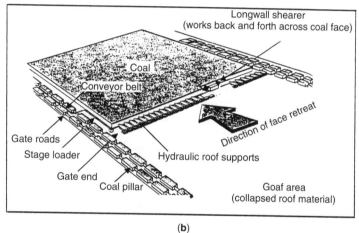

(b)

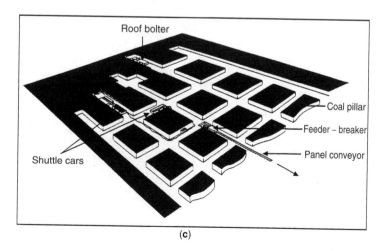

(c)

Figure 10.2 Methods of mining in underground mines. (**a**) Longwall advance mining; (**b**) Longwall retreat mining; (**c**) Room and pillar mining. (a) Adapted from Ward (1984) with permission of Blackwell Scientific Publications, (b) and (c) from Hunt and Bigby (1999) reproduced by permission of Rock Mechanics Technology Ltd

(a)

(b)

Figure 10.3   Longwall mining equipment. (**a**) Rotary drum shearer, Huabei mine, PRC (photograph by Dargo Associates Ltd); (**b**) Coal plow, Friedrich–Heinrich mine, Germany. Reproduced by permission of DBT

Longwall advance is used to mine thinner seams or while a retreat panel is being established. Longwall retreat is the more widely used method, with the advantages of collapsing the roof towards the main entry, which must be kept open. Then when the panel is completely mined, the equipment can be transported more easily to the next panel. Another advantage is that by driving access tunnels along the sides of the panel, any change in the predicted geology can be ascertained. This can limit the mineable length of the panel if the coal seam is severely dislocated by previously undetected faulting.

The panel width and length will depend on the geological conditions and capacities of the transportation, ventilation and power equipment that can be supplied and installed. In the USA, panel width is usually 120–293 m; in the UK it is 200–250 m, and panel lengths can be 600–4000 m. If the panel width is, for example, 50 m and the length less than 500 m, then this is referred to as shortwall mining, and in application it is intermediate between longwall and room and pillar mining.

From the point of view of economics, increasing panel width reduces the number of panels in a mine reserve area, which results in the reduction in development costs for panel entry drivages, an increase in the recovery level of coal due to fewer pillars, and an increase in the production of coal. However, if the panel width exceeds 300 m, further increases have less effect on coal production, as the coal may have to be moved over longer distances. Increased panel width also increases the roof exposure time, creating the potential for roof fall between the face and the overhead shields.

Longwall mining of seams thicker than the height of the available supports will mean either leaving a portion of the seam behind, or by mining the whole seam by two longwall faces progressively staggered, as shown in Figure 10.4(a); alternatively, one longwall face can be used and the remaining coal above is collapsed and collected through sliding gates at the back of the powered supports (Figure 10.4(b)). Problems arise when the coal seam attenuates or splits. Because the face is set at a fixed height, any change to the seam configuration could mean cutting noncoal material with the coal. This is not desirable, particularly if the saleable product is run-of-mine (ROM) coal. Also the shearer can emit sparks if a quartzose sandstone is encountered; this is to be avoided in gassy mines. Figure 10.5 shows three

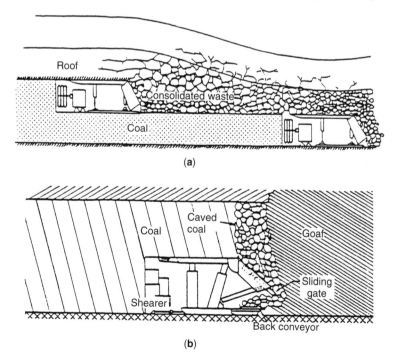

Figure 10.4 Methods for longwall mining in thick seams. (**a**) Multiple slicing, using two longwall systems; (**b**) Single longwall system combined with sub-level caving. From Ward (1984), by permission of Blackwell Scientific Publications

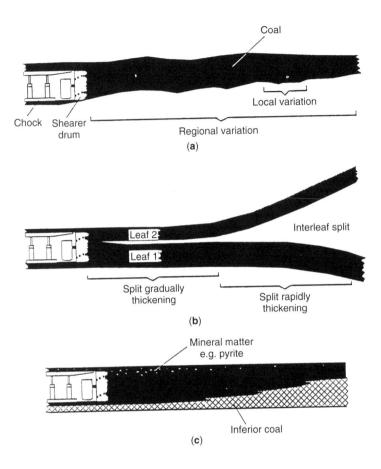

Figure 10.5   Effects of changes in coal seam development on longwall mining (**a**) Local thickness variation affects working of individual faces; (**b**) Coal seam splits affect quality of face product, rapid thickening may terminate mining; (**c**) Change in mineral content affects face product and increase wear on mining equipment. From Fulton *et al* (1995) with permission of The Geological Society

possible instances of coal seam change in front of a longwall face; other changes may be small scale faulting or washouts. Because these features are extremely local, the exploration geology may not have detected such changes. The use of in-seam seismic techniques has helped to minimise the loss of longwall faces by identifying conditions in advance of the longwall face, and so reduce failure rates.

Longwall mining is used extensively in the USA and western Europe, and is now becoming established in Australia, South Africa, China, India and the CIS.

10.2.2.2 Room and pillar mining

The room and pillar method is a type of open stoping used in near horizontal strata in reasonably competent rock. The roof is supported by pillars and coal is extracted from square or rectangular shaped rooms or entries in the coal seam, leaving coal between the entries as pillars to support the roof (Figure 10.2(c)). The pillars are usually arranged in a regular pattern to simplify planning and operation. The rooms or entries are normally around 5 m wide, and the roof is supported by either steel or timber beams or by long metal rock bolts.

Coal is extracted by drilling and blasting the coal face, a system called 'conventional mining', or by mechanically cutting and loading the coal using a 'continuous mining' system.

In conventional mining, a mobile loader collects the broken coal after blasting and loads to a conveyor or shuttle car to be transferred to the main mine haulage system. In continuous mining, a single machine with a cutting head cuts the coal without requiring blasting (Figure 10.6). This machine also collects and moves the coal to a shuttle car or directly on to a conveyor for transfer to the main mine haulage system. The continuous miner provides a higher production rate than conventional mining, and is the most widely used method in modern room and pillar coal mines. Room and pillar mining is most suitable for thick shallow seams with strong roof and floor strata. The thickness of the seam is critical, usually seam thicknesses of 1–3 m are worked; however, the largest continuous miners can cut seams up to 4.8 m.

In shallow mines, up to 60% of the seam may be mined without pillar extraction, but as the depth of the coal increases, larger pillars are necessary and the percentage recovery is reduced. Once a district has been developed and access through it is no longer required, coal is then taken from the pillar areas. The roof is allowed to collapse into the abandoned area. Pillar extraction requires little additional equipment movement such as conveyors, ventilation and electrical cables.

In order to operate a productive mine, it is important that the prevailing geological conditions facilitate rapid development in seam roadways with the most economic level of support. The ideal is for the mine to have small pillars and good roadway conditions. In modern room and pillar mines, roof support is by roof bolting with or without additional support, such as mesh or bench bars placed behind the anchor-bearing plates into which the bolt is threaded. Several types of bolt can be used; the column resin roof bolt is used to spread the bearing load over the whole length of the roof bolt. Together with the mechanical point anchor bolt, these are the most commonly used roof support. Bolts are usually 1.0–1.8 m long, of which 40–50% should penetrate a geotechnically sound anchor rock. Bolt spacing varies according to roof stability, but is based on a 1 m square grid. The best development rates are where roof bolting is undertaken in good roof conditions, i.e. the continuous

EIMCO DASH THREE MINER IN OPERATION AT A MINE IN ALABAMA

Figure 10.6 EIMCO Dash Three Continuous Miner in operation in Alabama, USA. Photograph by courtesy of EIMCO, Bluefield, WV, USA

miner can advance 2–5 m without stopping to roof bolt, or alternatively, as with the most modern continuous miners, coal cutting can proceed simultaneously with on-board roof bolting.

Mechanised room and pillar mining has long been established in the USA, Australia and South Africa, but it has rarely been used in Europe. In the past in the UK, underground coal workings which extended offshore were worked by room and pillar method as longwall shearers had high levels of vibration. Room and pillar mining is also widespread in Indian and Chinese coal mines, with varying levels of mechanisation. In northern China, new fully mechanised underground mines use a combination of longwall and room and pillar. In India, mines operate modern longwall faces in one part of a mine, and conventional room and pillar mining in another.

10.2.2.3 Stress fields

The choice of mining support systems will be determined by the type of rock strata and the loads acting on it, i.e. the stress field. The nature of the coal-bearing sequence is a complex mixture of mudstone, siltstone, sandstone and coal, with occasional limestone or igneous intrusions, all of which are subject to varying amounts of stress. This is borne out by the presence of interbed shearing and major and minor structural discontinuities.

In shallow mines (<200 m deep) and in good mining conditions, the superincumbent strata do not impose strata control problems. However, in deeper mines strata control problems are not uncommon. The vertical component of stress is dependent on the depth of mining. In the UK, Germany and Poland, mining depths are greater than 600 m, whilst in Australia, South Africa and the USA they are less than 300 m (Hunt and Bigby 1999). At greater depths mining becomes progressively more difficult and expensive. Larger areas of coal (pillars) have to be left between longwall panels or in room and pillar districts. There is also the increased possibility that rock bursts may occur.

Horizontal stresses can be up to three times greater than vertical stress, and the deeper the mine the greater the horizontal stress. The horizontal components of stress are the result of the regional tectonic framework. From studies of plate tectonics, geophysicists have created a world stress map which can be used to indicate regional stress fields. In the UK and Europe, a major horizontal stress component exists, oriented northwest–southeast, whilst in Australia, horizontal stress orientations in the southern coalfields of New South Wales can vary from north–south in some areas to east–west in

others. In the eastern USA, the greatest horizontal stress is east–northeast, whilst in the western USA, horizontal stress directions have a wide variation in direction across the region (Figure 10.7). It is essential that the direction of maximum horizontal stress is determined at an early stage, as it has a profound effect on the orientation of longwall panels and road entry directions in room and pillar mines. In northern England, UK, longwall panels aligned east–west were unsuccessful, but when changed to a north–south orientation exhibited greatly improved conditions. Here, horizontal stress is redirected about the goaf rather than wholly transferred through cracked and caved ground. Vertical stresses are redistributed within the solid coal pillars and within the goaf depending on extraction geometry. The stress distribution along one such longwall retreat panel is shown in Figure 10.8. The stress fields influencing a roadway can therefore vary over time due to mining activity and can be directly related to the geometry of the mining layout (Siddall and Gale 1992).

Another method of horizontal stress relief is the creation of 'sacrificial roadways'. In a UK mine, a longwall panel was aligned 110° to the assumed major horizontal stress and suffered from severe floor heave and poor roof conditions, but once a new face line was driven, leaving a 4.5 m pillar between, conditions improved with little roof movement and floor lift. It could be seen that the failed roadway provided stress relief to the new roadway (Siddall and Gale 1992). In Australia, the selection of longwall panel locations was severely affected by a change to the orientation of the principal horizontal stress field. The change from north–south to east–west caused severe damage in the installation roadways and existing longwall panels.

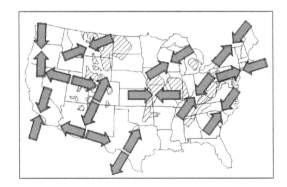

Figure 10.7 Principal directions of horizontal stress in the USA. In northeastern USA, where most of the underground mining takes place, the stress field is orientated in an east–northeast direction. From Mark (2001) with permission

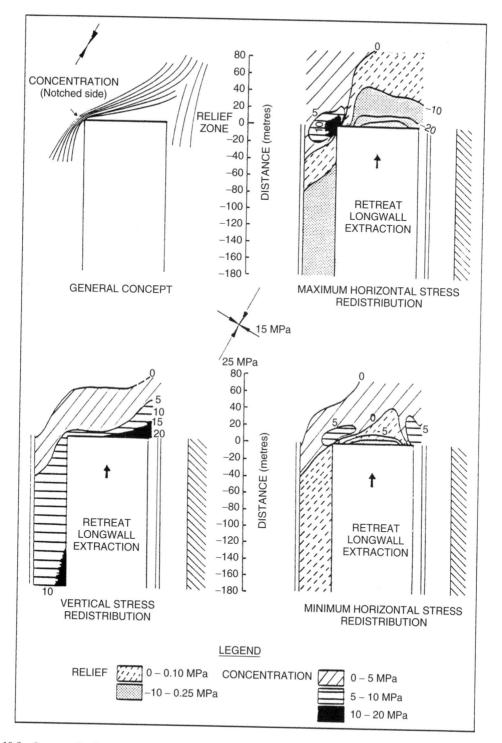

Figure 10.8 Stress redistribution about a retreating longwall panel. From Siddall and Gale (1992). Reproduced by permission of The Institution of Mining and Metallurgy

This was relieved by driving a stress relief roadway parallel to the proposed installation roadway. During the drivage bad roof conditions exhibiting shear failures were encountered due to the high lateral stress fields. This roadway was caved, and when the new installation roadway was driven, excellent roof conditions prevailed. The roof remained intact during the whole longwall installation period, proving the success of the technique. In recent years, horizontal stress problems have been recognised in a number of mines in the USA and South Africa, which has increased the importance of identifying the horizontal stress field as early as possible. Figure 10.9 shows an optimised longwall layout and working sequence for high horizontal stress conditions (Hunt and Bigby 1999).

The stress field is also affected by multiseam mining, where workings are closely spaced, and seams currently worked may overlie or underlie seams already worked out. Interaction effects may be severe and act as a constraint on further development.

In Australia, South Africa and the USA, room and pillar mining has become fully mechanised with the use of continuous miners and roof bolting machines. The ability to rapidly develop in-seam roadways with limited support requires suitable geotechnical conditions. Rock stress conditions are therefore important, and the orientation of the planned room and pillar district should take the direction of horizontal stress into consideration. It may be necessary to change the configuration of the pillars to rectangular or even diamond-shaped rather than remain square, in order to maintain favourable roadway orientation (Figure 10.10). The selection of pillar size is determined by the depth of working (vertical stress) and the position of shear zones around the coal seam.

Roof conditions may determine the minimum pillar size rather than pillar strength (Hunt and Bigby 1999).

Horizontal stress can be measured by drilling into the seam roof and installing a measuring device, and then drilling a larger hole around the first one. The second hole relieves the stress and allows the rock to expand, the amount of expansion allowing the original stress field to be measured. Other methods include pressurising a drill hole with fluid until the rock fractures; the fracture pressure and fracture orientation is then measured. Indications of horizontal stress can be simply observed by checking shifts in bolt holes, fixed equipment etc.

The use of downhole geophysics has led to the realisation that horizontal stress regimes may be recognised, and their orientations measured from the nature of associated breakouts. Breakouts are indicated by increases in borehole diameter along one axis. Boreholes elongate in a direction perpendicular to the maximum horizontal stress orientation, and are measured using x-axis and y-axis calipers together with borehole verticality data. Figure 10.11 is a borehole breakout log showing minimum and maximum calipers. In recent years, the use of acoustic scanning tools has produced high resolution formation images in boreholes, and breakouts can be identified and plotted using this technique (see Chapter 8 and Figure 8.24). In the UK, an average breakout orientation of 54°/234° has been identified in conjunction with minimum stress orientation measurements from other techniques such as hydrofracturing and overcoring (Brereton and Evans 1987).

Computer modelling is now established as a major tool in mine design. It is now possible to model the predicted and actual behaviour of coal-bearing strata, to assess both longwall panel and roadway orientation, as well as identify support requirements and optimum

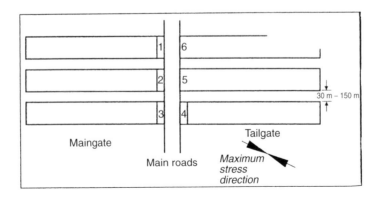

Figure 10.9 Optimised longwall layout and working sequence for high horizontal stress conditions. From Hunt and Bigby (1999). Reproduced by permission of Rock Mechanics Technology Ltd

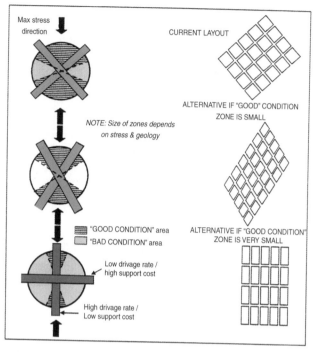

Figure 10.10 Possible layouts to minimise horizontal stress effects for a room and pillar district. From Hunt and Bigby (1999). Reproduced by permission of Rock Mechanics Technology Ltd

pillar size. This improves the accuracy of interpreting coal mine geotechnical properties, which will reduce the risk factor in underground coal mining, with obvious commercial advantages.

10.2.2.4 Strata and air temperature

In underground mines, the virgin rock temperature increases with depth, which can make mining conditions uncomfortable. A typical temperature gradient is 30 °C per 1000 m depth (as in the UK). In deep coal mines the temperature is kept at reasonable levels by increased air flows (ventilation) of colder air from the surface. As underground workings get further from the ventilation shafts there is more time for heat transfer from the surrounding strata into the airways, so that the longer the air has to travel to the coal face, the nearer the air temperature will approach that of the virgin rock temperature. Deep mines may employ booster fans underground to increase the speed of ventilating air and reduce the temperature increase.

10.2.2.5 Production

A comparison of production between longwall and room and pillar operations shows the highest face production

is achieved in longwall mines, e.g. 1.0–2.5 Mtpa, dependent on seam thickness, compared to 0.25–1.0 Mtpa for continuous miners.

Typical continuous miner section production per shift for one continuous miner and 2–3 shuttle cars is: 300–500 t for 1.5–2.0 m thick coal seam, 500–700 t for 2.0–2.5 m thick coal seam and 700–900 t for 2.5–3.0 m thick coal seam. Production will increase if two continuous miners are used per section with 3–4 shuttle cars.

In the USA longwall production has increased from 5000 t/employee-year in 1989 to over 10 000 t/employee-year in 1999 (Sabo 2000), with a ten-hour shift producing up to 25 000 t in large longwall systems.

In Australia, the overall productivity of longwall mines (including surface and coal preparation plants) averages 9000 t/employee-year compared to 6000 t/employee-year for nonlongwall mines.

Where high production is essential to remain competitive, e.g. in the USA and Australia (where wages are a high proportion of the total cost), where good management and planning prevail, together with skilled operators in good mining conditions, then a modern longwall represents the best model, i.e. outputs of +2 Mtpa with less than 200 employees underground. Longwall mining is able to work under a greater range of geological

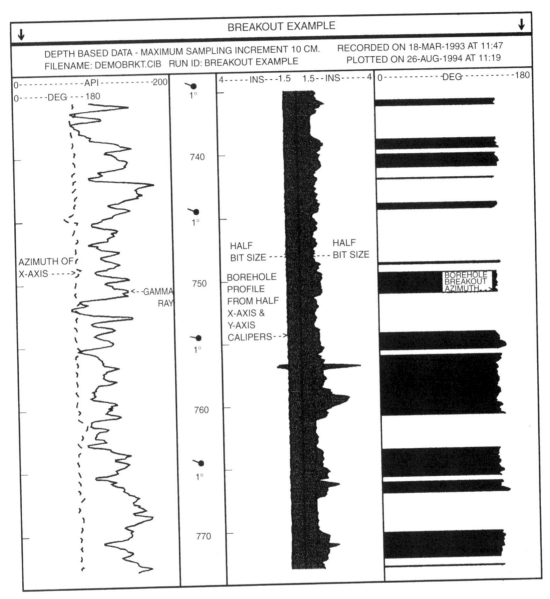

Figure 10.11 Borehole breakout log showing minimum and maximum calipers. Breakout is identified by rock spalling or an increase in borehole diameter. Reproduced by permission of Reeves Oilfield Services Ltd

conditions and is preferred for deeper mining, for thinner coal seams and for poor roof conditions. However, any interruption to production can cause serious problems. The relocation of longwall equipment to a new panel or to overcome an unforeseen obstacle is a high cost process.

Room and pillar mining is more flexible than longwall mining in meeting variable underground conditions. It is

less capital intensive and the risk of losing the working faces is a lot lower. Planning and design is simpler than for longwall mining, and training and organisation is also simpler. Because of the depth limitations, room and pillar mines are not generally considered for depths beyond 200–300 m.

Apart from geological considerations, finance can influence which mining method is selected. To equip a heavy duty longwall face with an extraction height of 2.0–2.5 m will cost £7.5–10.0 million ($15 million), whereas a room and pillar continuous mining system is nearer £3.5 million ($5 million). For the investment in coal mining in developing countries, room and pillar mines present fewer risks than longwall systems.

10.3 OPENCAST OR SURFACE MINING

Surface mining, also referred to as opencast, open pit or open cut mining, describes the accessing of a coal seam or seams from the ground surface by excavating all of the material above, between, and including the coal seam(s).

Surface mining has a number of advantages over underground mining: a higher degree of geological certainty, lower capital costs, lower operating costs and a safer mining environment for personnel. The major disadvantages are the restriction of depth due to cost and geotechnical limits, surface and groundwater influences, the direct effect of climate and the commitment to restore the land to meet environmental requirements.

Ward (1984) divides surface mining into two types:

- Strip mining: the material above the coal, known as overburden, is excavated and deposited in one operation adjacent to the working face. This method is usually employed along the outcrop of a coal seam or a number of seams. The strip mine will extend along the strike for long distances, but only a short distance down dip.
- Opencast mining: the overburden is taken away from the working face and dumped elsewhere on the mine property. Opencast mining is the best suited for mining thick seams or a series of seams. The mine configuration is less elongate and is sometimes referred to as 'area' mining.

The majority of new coal mine development in the world is for surface mines, and both black and brown coals are mined worldwide by this method. The large volume of black coal exported from Colombia, Indonesia and Venezuela is from large surface mines, and the large black coal mining operations in the Powder River and Green River Basins in the USA are also surface mining operations. Brown coal mines are virtually all surface mines, many of which are large scale operations, as at Belchatow in Poland and in the Latrobe Valley in Victoria, Australia.

10.3.1 Geological factors

A surface mine will be designed, first, by assessing the basic geological data of the area, together with geographical and economic constraints. This will determine whether further investigation is warranted, or should cease. Second, reserves will be assessed from more detailed geological data followed by hydrogeological and geotechnical studies, to test the viability of a mine. From this, a decision is made on whether to commit finance and other resources to develop a mine design. Third, the refinement of the geological data and selection of the mining method is completed and a final mine design is produced. From this point, construction of the mine can begin.

As can be seen, each stage is based on an increase in the amount of geological knowledge. The first stage consists of exploration work with some limited drilling, so that the thickness, mining depth, and extent of the target coal seam(s) is known, together with the structural framework of the mine area, which includes identifying all major faults and changes in dip or strike of the strata. Samples for coal quality will be taken from outcrops and boreholes. Again, if the geological conditions are unfavourable or the coal quality is unsuitable, the project will be terminated at the end of the first stage.

During the second stage, a planned drilling programme will identify the weathering and hardness of the overburden, changes in coal seam thickness, seam splitting, washouts, small scale faulting and the nature of the noncoal interbeds between coal seams. The groundwater regime should be ascertained and all water levels known; a flow net plan should be plotted for the proposed mine site. Additional drilling will be carried out to collect coal samples for more detailed analysis and cores taken to determine the geotechnical nature of the overburden.

All of these data will be fed into the computer mine software and the resulting mine design will be assessed for technical and economic viability. Once this is completed, the selection of the mining method and equipment will be made.

10.3.2 Mining equipment

In surface mining, the excavation of overburden is by dragline, electric or hydraulic shovels or bucketwheel excavators (BWEs). Coal excavation is usually by shovel (black or brown coal) and BWEs (brown coal).

1. Dragline. The large walking dragline is a mainstay for large surface coal mines, particularly in the USA. Draglines are used to remove overburden in both brown (lignite) and black (bituminous) coal surface mines, and can move large amounts of overburden at a lower cost than other mining techniques. More

than 90% of all overburden removal by draglines is handled by large machines with 30 m³ and larger buckets (Gilewicz 1999). In the USA, a single dragline of this size can move 7 million bcm per year, and the largest draglines (60–90 m³) can move over 20 million bcm per year, and can excavate down to coal seams 45 m below the surface (Pippenger 1998). Such equipment is only usable in the largest mines; elsewhere dragline capacity may be lower (10–20 m³) when the mining operation is smaller, e.g. in India.

As the coal reserves in surface mines have become deeper and less accessible, walking draglines have been designed with longer booms and with extended digging and dumping ranges. Draglines currently in use with such modifications are shown in Table 10.1.

Draglines mounted on crawlers are smaller capacity; the largest with a 69 m boom and 17 m³ bucket is operating in New South Wales, Australia.

Figure 10.12 shows a large dragline removing overburden and dumping adjacent to the excavation area. The average cost of a walking dragline with a 30 m³ bucket and 90 m boom is $10.5 million, and with a 60 m³ bucket and 104 m boom is around $22 million. A crawler dragline with 7.6 m³ bucket and 61 m boom is $3.2 million. All of these are major capital items to equip a surface coal mine.

2. Powered shovels. The modern trend in surface mining is one of large scale operations that produce in excess of 1.0 Mtpa. Most of these operations utilise large loading equipment together with large capacity trucks; this style of operation is referred to as 'truck and shovel'. There are three types of loading equipment: electric mining shovels, hydraulic excavators, available as either front shovel or 'backhoe' type, and wheel loaders.

Wherever overburden cannot be economically handled by draglines or BWEs, it has been traditionally removed by electric mining shovels working in tandem with large trucks. In recent years, these have been challenged by the developing hydraulic excavator industry. The number of electric shovels has declined as unit size has increased, and as smaller models in smaller mines are replaced by competitive products. However, electric shovels remain the dominant loaders when size is considered. The average electric shovel is nearly 40% larger than the average hydraulic excavator, and 50% larger than the typical wheel loader.

The USA is currently the largest market for all three types of loader, with large electric shovels and large trucks (220+ t) dominating the larger coal mines.

The manufacture of such equipment is very specialised, and there are three major manufacturers of electric shovels, eight principal makers of hydraulic excavators, and three producers of wheel loaders. Table 10.2 (after Gilewicz 1999) shows the size and capacity of a selection of electric and hydraulic excavators and wheel loaders that are currently in use.

Surface coal mines producing 10 Mtpa or more will probably select the biggest trucks and shovels, as is the case in the USA and Australia. Smaller mines will require smaller units, as used in the UK and India. Other factors influencing the choice of equipment can be the necessity for mobility in an area, or electric power may be impractical in another.

For loading coal, it is usual to use smaller size hydraulic excavators with either a front bucket or 'backhoe' configuration; these load coal into either road trucks or dump trucks.

Hydraulic excavators are preferred in Europe, whereas wheel loaders are rarely used. This is in contrast to Australia, where hydraulic excavators and wheel loaders dominate mining. In India, coal mining has relied on smaller machines, chiefly because surface mining has only recently begun to increase in scale. The older machines are electric; these are being replaced by hydraulic models. The new coal exporting countries of Colombia, Indonesia and Venezuela use mainly hydraulic excavators, and South Africa has changed to similar equipment in recent years. China, the USA and the CIS utilise both large and small electric shovels and are the chief manufacturers.

Table 10.3 shows the world distribution of loading equipment (after Gilewicz 1999), and Figures 10.13(a), (b) and (c) show examples of all three kinds of equipment.

In considering electric shovels and hydraulic excavators, haul or dump trucks are utilised for both overburden and coal removal. For overburden, large dump

Table 10.1 Dragline capacities currently in use. Adapted from Grimshaw (1997)

Operating area	Boom length (m)	Bucket capacity (m³)
Callide, Australia	128	100
Fording, Canada	128	100
Thunder Basin, USA	110	122
N. England, UK	100	57

Figure 10.12   Dragline removing overburden in opencast mine, USA courtesy of Bucyrus International Inc. reproduced by permission of *World Coal*, Palladian Publications Ltd

trucks (over 90 t) are used in the larger mining operations to complement the larger size shovels. Dump truck sizes range from 35 to 100 t for smaller shovels and 100 to 220 t for the large shovels (Figure 10.14); most are rear dump models, but bottom dump trucks, usually 100–190 t, are also used.

To equip a surface mine with a truck and shovel fleet is a major expense: the capital cost for electric shovels is between $2 and 7 million, for hydraulic excavators $0.5–5.0 million, wheel loaders $0.1–2.0 million and dump trucks $0.3–3.5 million. Depending on the mine size, stripping ratio and production scheduling, the capitalisation is a major item, but has the advantage of flexibility in that the capital cost may be phased in as the mine increases in size over time. Most truck and shovel equipment has a mine life of around seven years, and capital must be available to replace worn-out equipment at several stages throughout the life of the mine.

3. Bucketwheel excavators (BWEs). In large scale surface mines, with thick coal seam sections, as found in the large lignite mines in eastern Europe and Australia, BWEs (or dredgers) are used. These machines consist of a boom with a rotating wheel at one end

Table 10.2 Size and capacity of a selection of electric and hydraulic excavators and wheel loaders currently in use. Adapted from Gilewicz (1999), with permission

	Make	Bucket capacity (m³)	Examples of where in use
Electric shovel	Bucyrus 495B	30–60	USA
	Marion 351M	44–63	USA
	P&H 4100A	43	USA/Australia
	P&H 2800XPB	25	Venezuela/Canada
Hydraulic shovel			
	CAT5230	15–20	USA/Canada
	O&K RH120C	11	USA
	O&K RH170	20	USA
	O&K RH400	45	USA
	Liebherr R994	20	Australia
	Liebherr R996	30	Australia
	Hitachi EX2500	14	USA
	Komatsu H485S	33	USA
Wheel loaders			
	CAT994	19	USA
	JCB456	2–5	Europe/USA
	Komatsu WA900	13	USA

Table 10.3 Worldwide distribution of large loading equipment operating at major opencast coal mines. From Gilewicz (1999), with permission

Country/ region	No. of units, all types	Electric shovels	Hydraulic excavators	Wheel loaders
USA	561	121	178	262
Europe	277	99	160	18
Australia	233	43	62	128
India	125	120	5	0
Canada	147	65	13	69
Colombia	51	12	25	14
China	44	24	4	16
South Africa	47	22	20	5
Russia	26	25	1	0
Indonesia	37	0	34	3
Rest of World	84	20	39	25
World units	1 632	551	541	540
m³	27 081	11 495	8 159	7 427
average size m³	16.6	20.9	15.1	13.8

around which a series of buckets or scoops with a cutting edge can excavate relatively soft lignite or soft overburden. The excavated material is fed on to a series of conveyors which then load on to a main belt conveyor or into trams for transport out of the mine. The normal capacity of these machines is 420–2300 m³ per hour (1000–3000 yd³) and they are particularly effective in excavating soft overburden in flat lying strata. Figure 10.15(a) shows a BWE cutting overburden in a mine in Bosnia-Herzegovina and Figure 10.15(b) shows a large BWE in operation in the Ekibastuz mine in Kazakhstan. Disadvantages are the inability to cut hard overburden or overburden containing boulders or large consolidated rock masses which typify glacial deposits. The system of fixed conveyors make the use of BWEs less flexible than truck and shovel operations, and BWEs are not suitable in small, confined mines.

The cost of BWEs varies according to capacity: a 420 m³/hour machine has a capital cost of around $1 million (not including ancillary equipment), and a 1550 m³/hour machine a cost of $4 million. As large mining operations may have between two and four BWEs operating, they make up a very large initial capital cost, and ongoing maintenance costs. They do have, however, a long mine life.

10.3.3 Surface mining methods

The method of mining and the equipment used in surface mining is dependent upon the size, configuration and depth of the planned mine, together with the ability to excavate hard or soft strata to access the coal seam(s). Surface mines are typically up to 50 m wide and the deepest level of excavation up to 80 m. Working faces can have an angle of 50°–90°, and mine batters and spoil tips have angles that range from 30° to 45°, the lower angles being most common.

A brief outline is given of the principal mining methods currently used in surface mining.

1. Strip mining usually begins close to where the coal seam crops out at the surface. If there is a significant weathering profile, then the initial box cut may be located down dip to expose the coal seam. The overburden is excavated directly or, if hard, is blasted before excavation. Overburden removal is by means of large electric or hydraulic shovels, and/or a dragline. The shovel stands on the top of the coal and excavates overburden from the highwall, while the dragline is situated on the top of the overburden and excavates down to the coal. The working face is advanced along the strike of the coal; this leaves the coal seam exposed in the floor of the pit. The coal, which may or may not need to be blasted, is

then excavated by a smaller shovel and loaded into trucks (Figure 10.16). The overburden from the first box cut is placed up dip or below the outcrop of the coal. Once the first cut has been completed, the second strip of overburden is removed down dip and parallel to the first. The overburden from the second strip is placed in the area left after removing the coal from the first strip, and successive strips are

(a)

(b)

Figure 10.13   (a) Electric shovel removing overburden in central India. This type of shovel is still commonly in use, but has been superseded by larger capacity models in the larger mining operations in USA and Australia. Photograph by courtesy of Dargo Associates Ltd; (b) Hydraulic shovel loading overburden in Spain. Modern shovel capacities are $20-30\,m^3$. Photograph by courtesy of Liebherr Mining Equipment Co.; (c) Wheel loader removing overburden. Photograph by courtesy of *World Coal*, Palladian Publications

(c)

Figure 10.13   (*continued*)

Figure 10.14   Large dump truck (280 t) being loaded at Fording coal mine, Canada. Photograph by courtesy of Komatsu Mining Systems

(a)

(b)

Figure 10.15   (a) Bucketwheel excavator (BWE) cutting overburden in Gacko lignite Mine, Bosnia-Herzegovina. Photograph by courtesy of Dargo Associates Ltd; (b) Large BWE and conveyor system in operation in Ekibastuz opencast mine, Kazakhstan. Photograph by courtesy of Dargo Associates Ltd

Figure 10.16   Coal being loaded into trucks by hydraulic excavators, Western Coalfields, India. Photograph by courtesy of Dargo Associates Ltd

cut in this manner. Excavation is continued until the thickness of overburden becomes too great, because the stripping ratio is too high, and/or the excavation equipment has reached its maximum working depth. With this method of mining, land restoration begins early on in the mining schedule: the spoil area is landscaped, the topsoil is then replaced and prepared for appropriate land use. Such large scale strip mines are operating in the Powder River Basin, Wyoming, USA; e.g. Jacobs Ranch mine produces 13 Mtpa of subbituminous, low sulphur coal for electricity generation (Hartman 1992).

2. Opencast mining is more complex, particularly when a very thick coal is to be extracted, or a series of coals are targeted in one mine. In these circumstances, a series of benches will be developed and coal extracted from each bench; this can be on a very large scale as at El Cerrejon Norte in Colombia (Figure 10.17(a)) or be of more modest size, as seen at Pljevlja mine in Montenegro (Figure 10.17(b)). The use of explosives may be required to break up resistant rock in the overburden. As the mine develops, benches are constructed at succeeding lower levels. This means that all noncoal material, i.e. overburden and interburden, has to be removed

and dumped away from the working bench areas. To achieve this, electric and hydraulic shovels with truck fleets will be used, particularly when the overburden is hard and requires blasting, when there are restrictions of space for equipment and when the pit reaches lower and lower levels. Where large volumes of relatively soft overburden or thick brown coal (lignite) has to be excavated, BWEs may be used. If the pit is relatively shallow and wide, BWEs with their associated conveyor systems may be most appropriate. Black coal may be blasted, if necessary, and shovelled directly into waiting trucks by small shovels or wheel loaders. Figure 10.18 illustrates such an operation at Kalinga mine in Orissa, India. Brown coal will either be cut by BWEs and conveyored, or shovelled directly into trucks. In the case of exposed horizontal brown or soft coal, it can be cut and loaded by means of a surface miner such as a Wirtgen machine. This is used to strip off a thin layer of coal and to load simultaneously into a truck (Figure 10.19). This method of coal stripping has been used for selective mining of coals to leave noncoal partings out of the run-of-mine product. A Wirtgen 3500 machine produces around 500 bcm/hour. Overburden is removed to a

(a)

(b)

Figure 10.17 (**a**) Large scale benched mining operation, El Cerrejon Norte, Colombia. Photograph by courtesy of Dargo Associates Ltd; (**b**) Smaller scale benched mining operation, Pljevlja brown coal mine, Montenegro. Photograph by courtesy of Dargo Associates Ltd

Figure 10.18 Truck and shovel operation, Kalinga mine, Orissa, India. Photograph by courtesy of Dargo Associates Ltd

Figure 10.19 Wirtgen surface strip miner in operation in Gacko mine, Bosnia-Herzegovina. Photograph by courtesy of Dargo Associates Ltd

designated dumping area and remains there until a void area in the mine is available. This is then filled in and the land restored. Coal may be taken to a loading area for transport away from the mine, or stockpiled for use in the area adjacent to the mine. Examples of large black coal opencast mines are El Cerrejon Norte in Colombia, Guasare in Venezuela, Grootegeluk in South Africa as well as numerous similar operations in Australia and the USA. Smaller black coal opencast mines can be found in East Kalimantan, Indonesia, in India and the UK. Brown coal mines dominate central and eastern Europe, which produce 50% of the world's total; there are also significant producers in the Ekibastuz Coalfield, the CIS, in the Latrobe Valley, Victoria, Australia, at Belchatow, Poland and in the Big Brown mine in Texas, USA.

Contour mining is carried out in rugged topography, characterised by hill, ridges and V-shaped valleys. The coal is extracted by removing the soil and rock overlying the coal, and is often referred to as the mountain removal method. The location is then restored to its approximate original contour. This procedure is followed along the outcrop of the coal seam as successive cuts are made. Contour mining has been practised successfully in the Appalachian coalfield, eastern USA, particularly in Kentucky and West Virginia.

3. Highwall mining is a remotely controlled mining method which extracts coal from the base of an exposed highwall, usually in a series of parallel entries driven to a shallow depth within the coal (Shen and Fama 2001). This method enables coal to be mined that otherwise would remain in the ground. The arrival at the final highwall position may be due to an uneconomic stripping ratio, or be in an area of the mine that had effectively sterilised the coal due to overlying mine infrastructure. Highwall mining is reliant upon the self-supporting capacity of the ground because there will be no artificial support in the entries. It is essential that the nature of the highwall and the ground behind it is fully understood, so that a mining layout can be designed. Two types of highwall mining systems are used. The continuous highwall mining system, which uses a continuous miner to cut rectangular entries approximately 3.5 m wide (Figure 10.20), and the auger system, which creates individual or twin circular holes of various diameters depending on whether a single or twin auger system is used. The continuous highwall mining system has been more widely used than the auger system because of its higher productivity and recovery rate. However, the auger system can better tolerate changing and difficult geological conditions, such as unstable seam roof and floor and

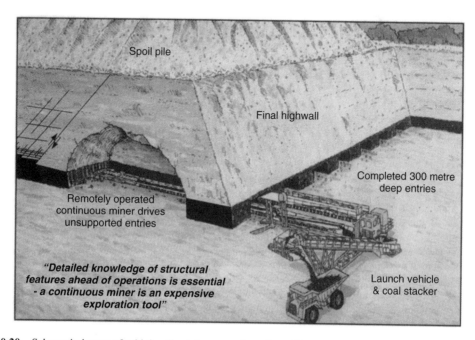

Figure 10.20 Schematic image of a highwall mining operation. From Shen and Fama (2001) with permission

Figure 10.21   Clean-cut entries into the highwall at Appalachian Fuels' K2 mine in Kentucky. From Walker (2001). Reproduced with permission of World Coal, Palladian Publications Ltd

seam thickness variations. In Australia, the continuous highwall miners have reached 500 m penetration and 124 000 tpm, the auger system reaching 200 m penetration and 60 000 tpm (Shen and Fama 2001). In the USA, auger systems have produced over 16 Mt of coal from the Appalachian coalfields from coal seams ranging from 0.6 to 4.9 m in thickness (Walker 1997), as shown in Figure 10.21. The final highwall can be 40–60 m deep with a face angle of 70°. One of the chief concerns is the stability of the highwall face. Instability can result from subsidence, discontinuities in the rock, either parallel to the face or intersecting at an angle to produce rock wedges, and from fracturing due to previous blasting. Both highwall mining systems have a protective shield above the working area, but this is only effective against small rock falls. In highwall mining, pillars are left between the entries; the success of the technique depends upon the stability of the pillars, because if they collapse during mining this may lead to the loss of mining equipment. It is essential that the unsupported spans are sufficiently stable to remain for days or even weeks. This is particularly true for continuous highwall mining with wide spans. Auger mining creates an arched roof with an effective span of around 1.0 m. Highwall mining is still a relatively new concept, having begun commercially in the USA in the 1970s and in Australia in the 1980s. It is now being introduced in South Africa, and is a means of accessing hitherto considered unmineable coal reserves in opencast mines.

10.3.4 Production

A major part of world coal production comes from surface coal mines, Gilewicz (1999) quotes 1.3 billion tonnes produced from the world's major surface mines. Surface black coal mines produce on average 1.0–15.0 Mtpa, whilst lignite mines can be much larger, producing up to 40 Mtpa. In surface mining, the capital cost savings when compared to underground mining are offset by the higher rehabilitation costs incurred when land restoration is required.

Surface mining is the preferred mining option where environmentally possible, and many of the major coal producing countries have invested heavily in surface mines. This trend is likely to continue in the future.

11

Coal as an Alternative Energy Source

11.1 INTRODUCTION

The essential property that distinguishes coal from other rock types is that it is a combustible material. In the normal course of events, coal is burnt to provide warmth as a domestic fuel, to generate electricity as a power station feed stock or as a part of the industrial process to create products such as steel and cement.

Coal, however, is more versatile than this and has been, and still is, able to provide alternative forms of energy. This may be from its by-products such as gas, or through chemical treatment to become liquid fuel, and by *in situ* combustion to convert coal to liquid and gaseous products.

The development of these energy alternatives is important, particularly in those areas where coals are too deep for exploitation, or where underground mining has ceased for economic reasons. Those coalfield areas once thought to be exhausted can still provide large amounts of energy through the use of modern technology. In addition, the understanding of the origins of oil and natural gas shows coal to be a contributory source rock.

Although the bulk of coal utilisation is, and will continue to be, by direct handling and combustion, the alternative uses of energy from coal are increasing in significance, and are being developed in all the major coal producing countries.

11.2 GAS IN COAL

Bituminous coals contain a number of gases including methane, carbon dioxide, carbon monoxide, nitrogen and ethane. The amount of gas retained and held by a coal depends on various factors, such as pressure, temperature, pyrite content and the structure of the coal. Fresh coal contains more gas than coal which has been subject to oxidation. Large volumes of gas can be accommodated on the internal surfaces of the coal as a result of adsorption. It is released by the removal of pressure, usually by mining or drilling. The gas may migrate into associated strata such as porous sandstones, which release the gas into openings such as boreholes and mine excavations.

The association of gases with coal has been a constant problem in mine workings since underground coal mining first began. In underground workings, methane is released from coal exposed at the coal face, plus the broken coal being transported through the mine. Methane is a flammable gas and is explosive between a lower limit of $c.5\%$ and an upper limit of 15% when mixed with fresh air. This highly combustible gas is known as 'firedamp'. The faster the coal is mined, the larger the amount of methane released into the workings, so that it is essential that an adequate ventilation system is in operation. A danger is that of methane collecting in roof pockets and in the upper parts of 'manholes' or cuts in the roadway sidewalls where the rock sequence may still be exposed.

The safety lamp invented by Sir Humphry Davy in 1815 was the greatest single invention in the cause of safety, since it enabled coal miners to measure the concentration of methane in the mine ventilation system. The safety lamp could detect firedamp levels as low as 1.25% on a lowered flame. Since that time, the statutory maximum limit for methane content has been 1.25% for the use of electrical power. The use of locomotives and shotfiring, i.e. using explosives underground, must be discontinued if the methane exceeds this limit. At 2% methane, labour must be withdrawn from the workings until the methane content is diluted to within the statutory limit. Other coal mining countries used the same method of detection as the UK until hand-held monitors and continuous recording monitors were introduced in the 1970s and 1980s, so that gas emanating from the coal face can be monitored by keeping a methanometer in close proximity to working personnel. This allowed much lower concentrations of firedamp to be detected, and therefore allowed lower statutory limits to be introduced.

In the UK, the current statutory limits are 0.25% of methane in air for air entering the working area and 0.5% for methane in air for air returning from the workings.

In France, the maximum percentage of firedamp is 1.0–1.5%, and in Germany 1.0% is the normal limit, but has been increased to 1.5% in certain longwall installations.

In Australia, intake airways are kept to below 0.25% methane, and up to 2.0% in return airways. Above 1.25%, electrical power must be switched off, and persons are not allowed to travel in roadways with 2.0% methane. Continuous mining equipment may be required by the Inspectorate to be equipped with automatic methane monitors, which emit audible signals at 1.0% at the cutting head, and the power is automatically tripped at 2.0%. Similarly on longwall faces power is cut off at 1.25% methane.

In New Zealand, the limit for methane in the general body of the air in a coal mine is 1.25%. An inspector can call for a ventilation survey of the mine if this figure is exceeded.

In the USA, electrical shutdowns are required at 1.0%, and labour is withdrawn at 1.5%.

Carbon dioxide is more common in brown coal than in bituminous coal workings. However, bituminous coals which have a high pyrite content contain higher amounts of carbon dioxide, due to the fact that coals rich in pyrite absorb more oxygen when moist, and this absorbed oxygen produces not only water by combination with hydrogen, but also carbon dioxide by combination with carbon. Carbon dioxide, also known as 'blackdamp', is a colourless gas and is heavier than air. It therefore tends to accumulate in the lower parts of mine workings.

Carbon monoxide originates from the incomplete oxidation of coal, especially after methane explosions. The gas is combustible and poisonous.

Only a small proportion of the nitrogen found in coal gases has its origin in the nitrogen present in the coal material; the bulk of the nitrogen originates from the surrounding air.

Free hydrogen occurs in small amounts associated with methane, but is not usually found in any great amounts.

Ethane is more prominent in gases derived from oxidised coals; cannel coal contains ethane in its pore structure.

Radon is a naturally occurring radioactive gas, and as such is distinguished from the other gases present in coal; it does have significance in posing a health hazard to humans (see Chapter 12).

The methane content of the coal can, however, be regarded as a significant source of energy and is the subject of a large amount of research and development.

Methane is usually referred to as coal-bed methane (CBM) in the literature.

11.2.1 Coalbed methane (CBM)

11.2.1.1 CBM generation

The process by which plant material is progressively altered through peat, lignite, subbituminous, bituminous to anthracite coal is termed 'coalification' (see Chapter 4). As the organic material is altered through the effects of temperature and pressure, both physical and chemical changes take place. Diagenetic change occurs up to the lignite/subbituminous boundary, depending on time–temperature relationships. Above subbituminous rank, changes can be equated to metamorphic alteration.

The major products of the coalification process are CBM (CH_4), carbon dioxide (CO_2), nitrogen (N_2) and water (H_2O). CBM is generated in two ways: first, during the early stages of coalification at temperatures below 50 °C; this is biogenic methane and is formed by decomposition of the organic material, and where biological activity induces reducing conditions which remove oxygen and sulphate. Where subsidence and burial are rapid, biogenic CBM may be trapped in shallow gas reservoirs (Rightmire 1984).

Second, CBM is generated by means of catagenesis, the process by which organic material is altered as a result of the effect of increasing temperature. CBM generated at temperatures in excess of 50 °C will be due to this process and is referred to as thermogenic methane. The relative volumes of CBM generated by biogenic and thermogenic mechanisms are shown in Figure 11.1.

During coalification, more than twice as much CO_2 as CH_4 is generated up to the high volatile bituminous–medium volatile bituminous coal boundary. CBM volumes generated increase rapidly above this point, with the CBM generation peak occurring at about 150 °C, or at the medium volatile bituminous–low volatile bituminous coal boundary (Figure 11.1). The two gases associated with CBM are CO_2 and N_2; the latter is only found as a minor constituent of thermogenic gas as it migrates readily from the system due to its small molecular size. CO_2 is a principal constituent of early thermogenic gas (Figure 11.1), but is only a relatively minor and extremely variable constituent of the gas produced at high temperatures; it is highly soluble in water, and this facilitates its mobility from the system. Analysis of gas produced from coal beds shows that 95% is CBM, <3% of CO_2 and N_2, and trace amounts of higher hydrocarbons such as ethane, propane etc.

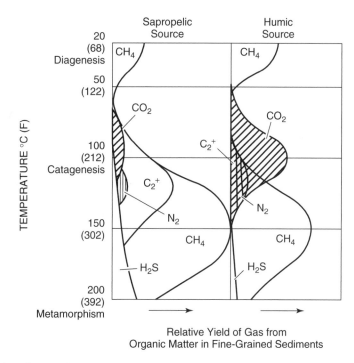

Figure 11.1 Generation of gases with depth, C_{2+} represents hydrocarbons heavier than CH_4 in gas phase. N_2 is generated initially as NH_3. From Rightmire (1984), based on Hunt (1979)

11.2.1.2 CBM retention

CBM is retained in coals in three ways: first, as a free gas within the pore space or fractures in the coal; second, as adsorbed molecules on the organic surface of the coal; and third, dissolved in groundwater within the coal. Porosity in coals occurs as fracture porosity and matrix porosity. The latter is more significant when considering the CBM retention potential of coals.

Gas generated in excess of that which can be adsorbed on the coal surfaces will be 'free' gas within the porosity of the coal, most notably in the fracture porosity. This gas is available to be dissolved in groundwater moving through the coal. The CBM saturation of large volumes of water can remove large volumes of gas from the coal seam(s) which will be lost to the system and possibly vented to the atmosphere. The fracture porosity in coal is primarily due to the formation of fractures called cleat. Cleat is a joint or set of joints perpendicular to the top and bottom of the coal seam. Usually there are two cleat sets developed in an orthogonal pattern. Cleat is a major control on the directional permeability of coals.

The primary mechanism of CBM retention in coal beds is adsorption on the coal surface within the matrix pore structure. A number of fractures influence the ability to utilise pore surface for CBM adsorption. The size of the aperture accessing the pore surfaces, the variation in moisture content and/or the degree of coalification may alter the surface areas of a coal. Surface area is also related to carbon content; studies have shown that coals with carbon contents of <76% and >83% generally have surface areas <1 m²/g, whereas coals within that range have areas >10 m²/g. An exception to this is anthracite with >92% carbon, which also has high areas of 5–8 m²/g (Gan *et al* 1972). Studies of open pore distribution in the coal rank series have been carried out to support this (Table 11.1). Porosity may also depend on the maceral content in high volatile bituminous coals. Vitrinite has fine porosity, pores are 20–200 Å diameter, and inertinite is the most porous, with pore diameters of 50–500 Å (Rightmire 1984). The adsorptive capacity of coal appears to increase with increasing rank. The maximum amounts of CBM that can be adsorbed on to the internal surfaces of coals according to coal rank and depth are shown in Figure 11.2.

11.2.1.3 CBM production

The production of CBM from underground sources is either by draining old and current mine workings or by

Table 11.1 Gross open-pore distribution in coals. Based on Gan *et al* (1972), reproduced with permission from Elsevier Science

Rank	Porosity distribution (%)			
	C (% d.a.f.)	<12 Å	12–300 Å	>300 Å
Anthracite	90.8	75.0	13.1	11.9
Low vol. bit.	89.5	73.0	nil	27.0
Med vol. bit.	88.3	61.9	nil	38.1
High vol. A bit.	83.8	48.5	nil	52.0
High vol. B bit.	81.3	29.9	45.1	25.0
High vol. C bit.	79.9	47.0	32.5	20.5
High vol. C bit.	77.2	41.8	38.6	19.6
High vol. B bit.	76.5	66.7	12.4	20.9
High vol. C bit.	75.5	30.2	52.6	17.2
Lignite	71.7	19.3	3.5	77.2
Lignite	71.2	40.9	nil	59.1
Lignite	63.3	12.3	nil	87.7

production from wells sunk into virgin or unmined coal seams (Figure 11.3).

CBM mine drainage

In many coal producing countries, underground mining operations are becoming deeper and deeper due to the exhaustion of shallower seams and improvements in mining technology. The increasing depth of coal seams usually equates with higher CBM content; this puts pressure on mine ventilation systems which have to be increased, as worldwide safety standards require that mines cease operation when methane-in-air levels exceed a predetermined percentage. Lost production due to excessive methane-in-air content can have serious economic implications. When faced with this phenomenon, mines should recover and use the CBM. In the USA, those mines with high gas contents vent up to 75% of all gas liberated. Worldwide, mines vent 1.23×10^{12} ft^3 (35×10^9 m^3) to the atmosphere, and use around 0.07×10^{12} ft^3 (2×10^9 m^3) (Schultz 1997). (For all unit conversions see Appendix 5.) Depending on the drainage method selected, a mine can remove between 20% and 70% of CBM in a coal seam. This will provide significant relief to the ventilation system, as well as producing gas of suitable quality for the commercial market.

Drainage methods include vertical wells (vertical premine), gob wells (vertical gob), long hole horizontal boreholes and horizontal and cross measure boreholes (Figure 11.4). The vertical wells are drilled from the surface to drain gas from coal situated in advance of the mine workings; these wells produce almost pure CBM. Similarly, gob wells are drilled from the surface to drain gas from the mined-out areas of the mine, such as the mined-out voids behind longwall faces or disused or collapsed areas of room and pillar districts. Gob wells produce a CBM/air mixture. Horizontal and cross measure boreholes are drilled within the mine to drain CBM from seams in production and from surrounding gob areas. Again, the holes in-seam produce pure CBM and the holes in the gob areas produce a CBM/air mixture. Significant advances in directional drilling systems have been made in recent years which have led to a larger range of CBM drainage options in gassy mining operations. Long, directionally steered in-seam boreholes can effectively reduce *in situ* CBM contents of large coal volumes in advance of mining in low to high permeability coals, as well as to drain faults and fissures containing free CBM. As well as the coal currently mined, coals both above and below this seam can be drained, and gob gas can be drained from mined areas that lie above future longwall panels in a lower seam.

For example, in Australia, in the Sydney Basin, CBM drainage has been carried out by drilling a series of boreholes both vertically from the surface into the coal, and by drilling horizontal boreholes into the seam ahead of the working panels. In Figure 11.5 a pattern of horizontal boreholes has been drilled into the panels ahead of heading B, each borehole ranging from 50 to 100 mm in diameter. Figure 11.6 shows a comparison of the resultant gas emissions in headings A and B; the gas emissions in heading B are reduced to less than half those of the undrained heading A, demonstrating the effectiveness of this method.

In the UK, conventional methane drainage has concentrated on boreholes which give access to methane held in gas sands and carbonaceous material above and below the coal seam being extracted. Trials have been carried out in known gassy mines to improve the accessing of methane from the coal itself. Mixed results were obtained due to technical and geological difficulties in drilling near-horizontal wells in the confined space underground. Following the trials, a comprehensive underground drilling programme was considered impossible due to the technical difficulties and prohibitive cost. Studies were also carried out into improving ventilation systems in rapidly advancing drivages, i.e. where continuous miners are in operation. It was shown that approximately 20% of the gas quantity ultimately released from the roadway sides enters the workings during the first hour after the

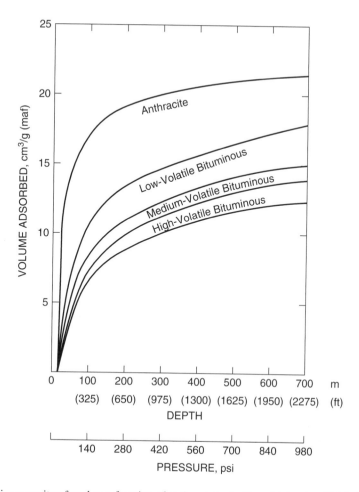

Figure 11.2   Adsorptive capacity of coal as a function of rank and depth. From Rightmire (1984), based on Kim (1977)

coal is cut. After this initial high flow, the coal continued to degas for a period up to seven months (IMC 1997).

As well as the presence of gas in the operating mines, the problem of surface gas emissions associated with coal mining has greatly increased in the UK, due to the rapid rundown of the UK coal industry. The large volumes of gas in abandoned mines cause not only surface emissions, but gas from abandoned workings migrates into adjacent operating mines, resulting in high gas levels, particularly during falls in barometric pressure (IMC 1997). Operating mines have exercised direct control of mine gas by means of ventilation and water pumping, often over large areas; however, these methods cannot be continued and maintained for abandoned workings indefinitely. The redevelopment of mine sites adds to the number of locations where

mine gas emissions may collect to form hazards. Such redevelopment may also disturb the naturally occurring gas seals such as glacial clays, and so exacerbate the problem. So in the future, the major concentration of effort is likely to be the drainage of gas from these old working areas.

Where CBM drainage has been successfully established, a mine can utilise its CBM on site or transport it to a dedicated user. In the USA, a number of mines sell their CBM through existing gas pipelines; this is dependent on the quality of the recovered CBM meeting set standards and that the production, processing and transportation are competitive with other gas sources. The quality control of the CBM is more applicable to the gob gas rather than pre-mine gas recovery. In the UK, the production of CBM from abandoned underground mines has a potentially bright future, and licences have

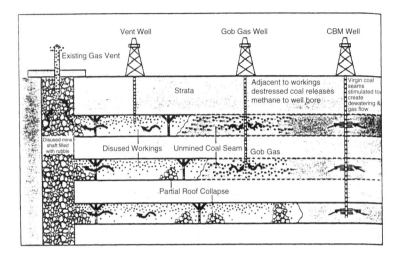

Figure 11.3   Diagrammatic view of production methods for abstraction of coal mine methane. From Garratt (2001) with permission of World Coal, Palladian Publications Ltd

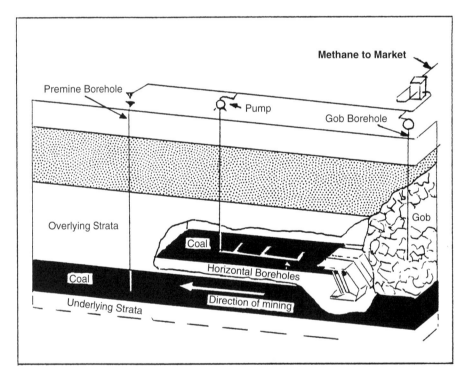

Figure 11.4   Methane extraction from active mine workings. Adapted from Schultz (1997) with permission of World Coal, Palladian Publications Ltd

been issued for selected coalfield areas to recover CBM and therefore avoid venting CBM to the atmosphere. This could mean the capture of 250 000–600 000 ft^3/d from abandoned workings (Garratt 2001).

In China, the amount of CBM drained from mines is increasing. In 1996, 600×10^6 m^3 of CBM was drained, of which 76% was utilised by the mines. Up to 1996, CBM drainage efficiency was only about 20% compared

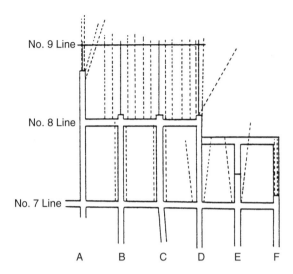

Figure 11.5 Pattern of drainage holes ahead of a panel front, Metropolitan Colliery, Sydney Basin, Australia. From Hargreaves and Lunarzewski (1985)

to 40% in the USA, but this is improving with increased efficiency and experience.

CBM well production methods

The production of CBM from surface wells penetrating virgin or unmined coal has experienced phenomenal growth worldwide over the past ten years. Studies of the major coal-bearing basins in the world suggest that more than 50% of the estimated *in situ* CBM resources is found in coals at depths below 1500 m (5000 ft). Drilling in deep low permeability reservoirs has demonstrated that open fractures can exist at depths of 2000–3000 m (7000–10 000 ft) (Myal and Frohne 1991). Of major concern are, first, the effect of horizontal and vertical stress components on deep lying coal beds; here tests have been shown that CBM can be produced at economic rates from coals below 1500 m under low to moderate stress conditions (Murray 1996). The second concern is the effect of gas and water saturation of coal on CBM production; the ideal would be 'dry' coal (no mobile water and free gas in the cleats and fractures) and the rapid desorption of CBM as the formation pressure is lowered, combined with low stress conditions. It has been demonstrated that CBM wells have not deteriorated over time. In the USA, in the San Juan Basin, New Mexico, a CBM well drilled in 1953 has produced 150–180 million ft^3/d (4.2–5.1 Mm^3/d) for 30 years. This well has shown that CBM under conditions of favourable geology and reservoir characteristics can produce economically over a long period of time (Murray 1996).

There are a number of methods by which CBM is produced from wells. The standard means for CBM production is by reservoir pressure depletion (Figure 11.7(a)). Reservoir pressure is reduced by dewatering the coal

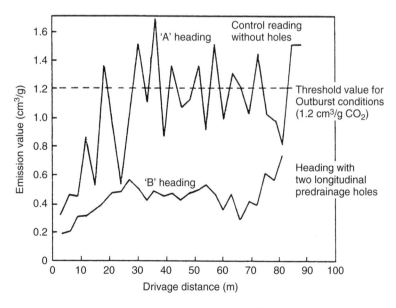

Figure 11.6 Reduction of gassiness resulting from pre-drainage. From Hargreaves and Lunarzewski (1985)

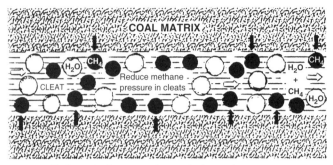

• Reduce cleat pressure by producing water
• Methane desorbs from matrix and diffuses to cleats
• Methane and water flow to wellbore

(a)

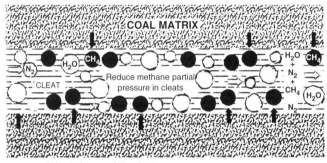

• Inject nitrogen into cleats
• Keep total cleat pressure high
• Reduce partial pressure of methane
• Methane desorbs from matrix and diffuses to cleats
• Methane, nitrogen and water flow to wellbore

(b)

Figure 11.7   (**a**) Coalbed gas recovery by reservoir pressure depletion. From Murray (1996), based on Puri and Yee (1990); (**b**) Enhanced coalbed gas recovery by use of nitrogen injection. From Murray (1996), based on Puri and Yee (1990)

bed; gas then desorbs from the matrix and micropores of the coal by a process of diffusion. The desorbed gas then flows to the well bore via the coal cleat and fracture system, along with any groundwater still present in the fractures. However, this method does not recover much more than 50% of the CBM in place. To overcome such low recovery rates, a series of experiments has been carried out in the USA on reservoir enhancement. One method is the injection of an inert gas such as nitrogen into the coal; this lowers the partial pressure in the coal, which allows a greater percentage of CBM to be recovered (Figure 11.7(b)). Alternatively, CO_2 is injected in similar fashion into the coal bed reservoir to release a greater percentage of CBM. This has been used successfully in the San Juan Basin, USA, and has been

tested in Alberta, Canada. The use of dynamic openhole cavity completion techniques has also been developed in the USA. In this type of operation, perforated casing is run to total well depth and the target coal beds are subjected to various types of fracture inducement or cavity completion. This is designed to create numerous fractures of varying orientation, linking the reservoir to the well. Tests have shown that adequate reservoir permeability, reservoir overpressuring and thermal maturity to at least high volatile A bituminous rank are required for successful cavity completions. This technique has been pioneered in the San Juan Basin and helped to make it the most prolific CBM producing basin in the world (Murray 1996). The USA leads the field in such experimentation, driven by the fact that CBM has become an increasingly important energy resource available to large energy consumers.

World CBM production

The greatest percentage of world CBM resources is thought to concentrate in twelve countries. An estimate of the coal and CBM resources of these countries is shown in Table 11.2, with Russia, China, the USA, Canada and Australia having the highest CBM potential (Murray 2000).

In the USA in 1997, 1.1 Tft^3 of CBM was produced, of which 90% came from basins in the Rocky Mountains region. The San Juan Basin, Colorado–New Mexico, produced 95% of this total, and there are now more than 8000 wells producing CBM in the USA (Murray 2000). The increasing importance commercially of CBM is borne out by the fact that in 1997, the CBM industry grossed more than \$2 billion in annual revenues (Stevens 1999). The distribution in the USA of CBM resources and their potential is shown in Figure 11.8, and the CBM resources for the major coalfield basins is given in Table 11.3. In the western USA, the San Juan Basin in Colorado–New Mexico, the Uinta Basin, Utah, and the Powder River Basin, Wyoming, have all increased CBM production. In the eastern USA, the Black Warrior Basin, Alabama is the second largest CBM producer in the USA, with an annual production of 110 billion ft^3, and the Central Appalachian Basin produces over 40 billion ft^3 per year. Other areas of development are the Gulf Coast region and southern

Table 11.2 Major coal and coalbed methane resources in the world. From Murray (1996)

Country	Coal resource (10^9 tonnes)	Methane resource (TCF,[a] in place)
Russia	6500	600–4000
China	4000	1060–2800
USA	3970	275–650
Canada	7000	300–4260
Australia	1700	300–500
Germany	320	100
UK	190	60
Kazakhstan	170	40
Poland	160	100
India	160	30
Southern Africa[b]	150	40
Ukraine	140	60
Total	**24460**	**2976–12640**

[a] TCF = trillion cubic feet.
[b] Includes South Africa, Zimbabwe and Botswana.

Alaska, the latter having estimates of 245 Tft^3 CBM *in situ* (Murray 2000).

In China, it is estimated that there are vast CBM resources, up to 1000–2000 Tft^3. The CBM resources with the highest potential are found in northeastern China, in the Ordos, Qinshui and Huabei Basins (Figure 11.9). Geological conditions, with the exception of the Ordos Basin, are structurally more complex than the commercial areas in the USA, and become increasingly complex towards the east. The Ordos Basin has coals at 300–1500 m with simple structure, and has estimated CBM resources of 50–100 Tft^3. The Qinshui Basin also has a moderately shallow perimeter suitable for CBM development, but there are numerous structural elements present. The Huabei (North China) Basin comprises a number of discrete fault bounded coalfields containing coals of bituminous rank, and this region has a large number of gassy coal mines. Texaco have drilled wells in the Huaibei region and obtained a gas content of 12 m^3/t at 610 m depth (Stevens 1999). The North China Bureau of Petroleum Geology has carried out CBM exploration at the Liulin CBM pilot site in Shanxi Province. Three coal beds with a combined thickness of 8–10 m have been selected, at a depth of 340–400 m, dipping at 3–8°. The coals contain 20% ash and 80% vitrinite and the CH$_4$ content ranges from 10 m^3/t to 20 m^3/t and the estimated resource is 1.6×10^8 m^3/km^2 (Xiaodong and Shengli 1997). Both Phillips Petroleum and Lowell Petroleum have also carried out tests in the Liulin and Hedong region of the eastern Ordos Basin, and Arco have also drilled a number of test wells in the eastern Ordos Basin and have obtained favourable results (Stevens 1999). The results of the Liulin tests indicate that China has coal-bearing areas with high CBM potential comparable to those in the USA. Although historically a number of technical and logistical problems have been encountered, the potential in China for CBM production on a commercial scale is still great, and advances can be expected to exploit this energy resource.

The CIS is the third largest producer of coal behind China and the USA. In 1990, an estimated 8×10^9 m^3 was vented to the atmosphere, of which 90% was liberated by underground mining operations primarily located in Russia, Ukraine and to a lesser extent in Kazakhstan (Figure 11.10). Hard coal production is falling in the CIS; in Russia, a 28% decrease between 1988 and 1993 from 274 Mt to 198 Mt, and in Ukraine, a 20% reduction in 1993–94 to 93 Mt (Marshall *et al* 1996). There are opportunities to recover and utilise CBM in conjunction with coal mining in Russia and Ukraine, as well as the development of the CBM

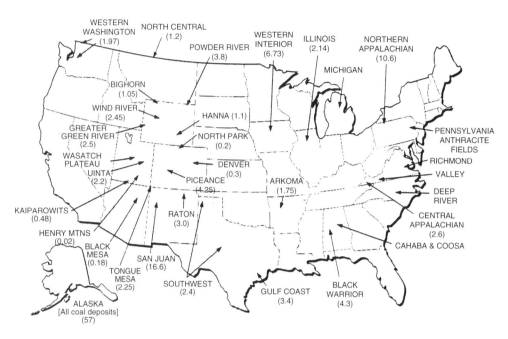

Figure 11.8 Coalbed gas resources of the USA (in trillion cubic feet); includes probable, possible and speculative potentially producible supply. From Murray (1996) with permission of The Geological Society

Table 11.3 Estimates of coalbed methane potential in the major coalfield basins of the USA. Based on data in Rightmire (1984)

Region	Area underlain by coal group (sq miles)	CBM resource 10^{12} ft
Appalachians	22 850	10–48
Warrior Basin	35 000	$7–10^3$
Illinois Basin	53 000	5.2–21.2
Arkoma Basin	13 488	1.4
San Juan Basin	7 500	31
Piceance Basin	6 680	60
Uinta Basin	14 450	0.8–4.6
Green River Basin	21 000	0.2–30
Wind River Basin	1 500	2.2
Powder River Basin	25 800	30

resource independently from mining. In Russia, in the Kuznetsk Basin, all of the underground coal mines are gassy, more than $1.2 \times 10^9 \, \text{m}^3$ was liberated in 1991 (Marshall *et al* 1996), whilst only 17% was removed by mine drainage systems. The CBM resources in the Kuznetsk Basin are estimated to range from 190 to $340 \times 10^9 \, \text{m}^3$, of which $70–120 \times 10^9 \, \text{m}^3$ are associated with virgin coal reserves. Further investigation may well

prove a much greater volume is present in the Basin. On the west Russian and south Ukraine border, the Donetsk Basin has CBM resources of $430–790 \times 10^9 \, \text{m}^3$ of which $100–200 \times 10^9 \, \text{m}^3$ are in virgin coal reserves. These figures may be also underestimated; however, together with the other black coal basins in CIS, the total CBM resource is extremely large. Until recently, the CIS has lacked the technology and, more significantly, the capital, to develop further CBM production. It is clear that CBM usage would benefit the industrialised regions by replacing brown coal and low quality black coal combustion in terms of air quality. It is expected that CBM production will become a significant contributor to the CIS energy needs.

In Australia, the identification of CBM resources has been and is being assessed, notably in the Gunnedah Basin, Gloucester Basin and Clarence–Moreton Basin. In the Gunnedah Basin, exploration has identified over $35 \times 10^{12} \, \text{ft}^3$ of CBM *in situ*, over an area of $9500 \, \text{km}^2$. In the Gloucester Basin, exploration has shown that favourable geological conditions exist for CBM exploitation. Coal seam permeabilities are high, seams are higher rank and have low *in situ* stress. Conditions in the Clarence–Moreton Basin suggest that there is high potential for CBM but as yet the Basin has not been extensively explored. There have been low levels of exploration in the Surat, Sydney and Darling Basins,

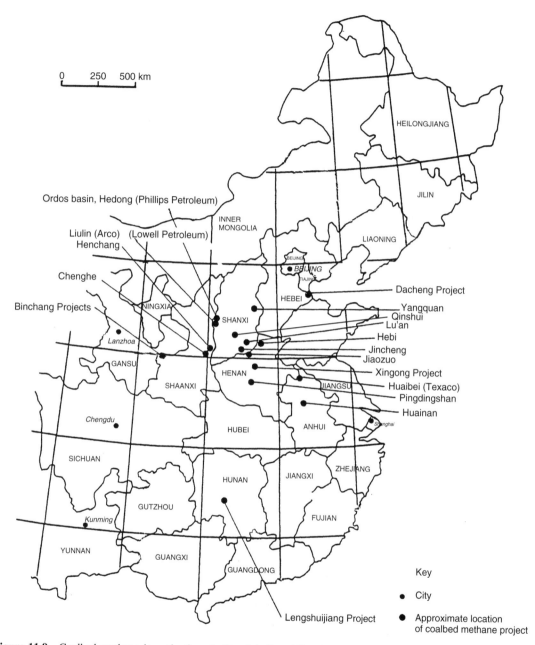

Figure 11.9 Coalbed methane investigations in People's Republic of China, including areas of western oil company participation. Reproduced by permission of Dargo Associates Ltd

but a new approach to gas exploration and open access to markets should encourage further exploration for gas in these areas.

In India, tests have been carried out in the Jharia Coalfield, Bihar, where recoverable resources are estimated to be 0.25–0.5×10^{12} m^3 (at a 50% recovery rate), and a limited amount of exploration has taken place in the Raniganj Coalfield, West Bengal. Current estimations suggest a conservative resource of 2×10^{12} m^3 of CBM is present in the coal basins of India (Bhaskaran and Singh 2000). Further extensive exploration for CBM in India will give

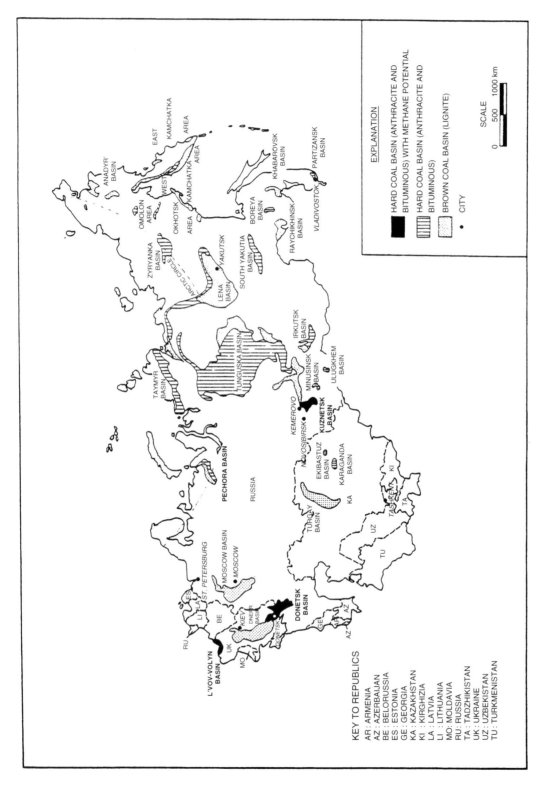

Figure 11.10 Major coal basins of the former Soviet Union, including those with the highest coalbed methane potential. From Marshall *et al* (1996), with permission of Marshall, Pilcher and Bibler and The Geological Society

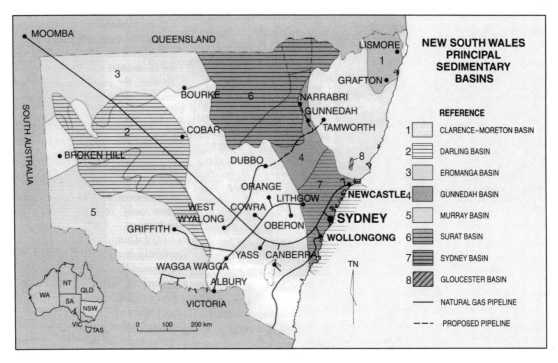

Figure 11.11 Principal sedimentary basins with oil and gas potential, in New South Wales, Australia. Reproduced with permission of the New South Wales (Australia) Department of Mineral Resources

a more accurate reflection of the country's CBM producing potential.

In Poland, CBM exploration has been carried out in the Lower and Upper Silesian Basins, the latter being the main focus of interest. An estimated resource of $200-400 \times 10^9 \, m^3$ has been given for the Upper Silesian Basin (Kotas 1994). With the closure of the Lower Silesian coal mines, and rationalisation of the mines in the Upper Silesian Basin, production of CBM will be critical to the maintenance of energy supplies in these industrialised areas.

In the UK, the last few years have seen a significant increase in exploration for CBM. Initial targeting of CBM prospects has been governed by the extent of underground mine workings, the depth of the target sequence (preferably <1500 m), the volumes of coal present *in situ*, and the coal rank and measured gas content (Bailey *et al* 1995). The coalfield areas of the UK (Figure 11.12) vary in rank, in confining pressure, and in gas content; this, with low coal seam permeability, will constrain CBM production in many parts of the UK (Creedy 1999). The South Wales Coalfield has high rank coal combined with significant confining pressure, and other areas have lower rank coal preserved at greater depth. These areas appear the most attractive for CBM

exploration. CBM development is still at an early stage in the UK; however, a number of licences have been granted for both conventional CBM wells and also for gob gas and mine drainage wells. CBM capture from former mine workings will have the added advantage of reducing the amount of CBM currently escaping to the atmosphere. In Scotland, a pilot electricity generation scheme is planned, using CBM from four wells drilled in unusually permeable virgin seams.

11.3 UNDERGROUND COAL GASIFICATION

The concept of the underground gasification of coal was first envisaged by Sir William Siemens (1868) and Mendelev (1888). A patent was issued in the UK in 1909, but there was no follow-up. It was not until the 1930s that tests were done in the former USSR, where a number of field stations were established for the purpose of developing a workable underground gasification technology. This led to the establishment of a number of large industrial installations which have supplied a low calorific value (CV) gas to power stations and other industrial consumers. Further experiments were conducted in the USA and Europe after the Second

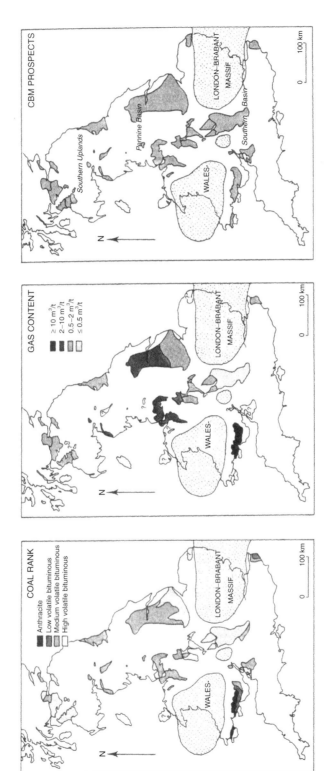

Figure 11.12   Rank map of carboniferous coals, map of hypothetical gas content of carboniferous coals and the principal coalbed methane prospects in the UK. From Bailey *et al* (1995). IPR/26-3C British Geological Survey. © NERC. All rights reserved

World War, but were discontinued when the availability of low cost natural gas removed the incentive for further underground coal gasification development work. More recently, successful demonstrations in the USA have been carried out at depths of 250 m; there are considerable potential reserves at this order of depth. Trials have also taken place in Europe, but at depths of 500 m as the coals are typically at deeper levels. These trials are designed to enable the technology to access unmined deep coals (1000+ m), such as those in Belgium and the UK.

11.3.1 Underground coal gasification (UCG) technology

The chemistry involved in UCG is complex, but essentially the process is a combination of three reactions, as outlined by ETSU (1993).

1. The producer gas reaction: when oxygen (O_2) is passed over/through hot coal it combines with carbon (C) to form carbon monoxide (CO), which is a flammable gas. This is an exothermic reaction (i.e. it produces heat). If too much oxygen is supplied, then carbon dioxide (CO_2) is produced, which is inert; therefore careful regulation is required.
2. The water gas reaction: when water or steam (H_2O) is passed over/through heated coal, the oxygen in it combines with carbon to form CO, and releases the hydrogen (H_2) content. This mixture of CO and H_2 produces a high CV gas. However, this reaction is endothermic (i.e. requires the addition of heat to sustain it). The conventional gasification method is to alternate the two reactions. However, in UCG, the two would be carried out simultaneously, the reaction temperature being regulated by adjusting the O_2/H_2O ratio.
3. The methane synthesis reaction: at high pressures achievable in the gasification of deep coal, H_2 combines with C to form methane (CH_4). This reaction is beneficial in that it increases the CV of the product, and since it is exothermic, it reduces the amount of input O_2 required to gasify a unit mass of coal.

UCG is coal gasification conducted *in situ*, so it does not require mining, gasification reactors, or ash disposal. It does, however, require some coal seam preparation before gasification can proceed. The permeability of the coal between boreholes (wells) must be enhanced; this is normally done through reverse combustion linking, or directionally drilled linking, or a combination of both.

In reverse combustion linking, two wells are drilled 30–100 m apart into the coal seam, and coal at the base of one well is ignited (the ignition well), and high pressure air or steam is injected down the other well (the injection well). Air permeates through the coal from the injection well and intersects the combustion zone at the base of the ignition well. This zone will expand and follow the air source back to the injection well, creating a high permeability channel between the two wells. In directionally drilled linking, a borehole is drilled between the vertically drilled well pair. This requires directional drilling techniques whereby an initially vertical borehole can be deflected at an angle to coincide with the dip of the coal seam. This is the preferred method for establishing the link because the inclined borehole can be placed near to the base of the coal seam, thereby increasing resource recovery. It is usual for reverse combustion linking to then be used after directional drilling to complete the mechanical link. To further enhance gas circulation through the coal seam, it may undergo 'hydrofraccing', a process that applies pulsating hydraulic pressure to produce fracturing of the coal seam.

Other configurations have been developed and tested. In steeply dipping coal seams, an inclined production well is drilled near to the base of the coal seam, and an inclined injection well is then drilled below the coal seam and only enters the seam at the area to be gasified where it intersects the production well (Figure 11.13).

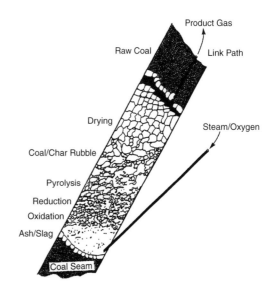

Figure 11.13 Typical configuration for steeply dipping bed gasification (SDB). From Oliver (1991). Reproduced with permission

Another method, developed by the Lawrence Livermore National Laboratory in the USA, is termed the controlled retracting injection point (CRIP). A horizontal injection well is drilled at the base of the coal seam into which is also drilled a vertical and a horizontal production well. The vertical production well is used during process start-up, after which the horizontal production well is used. A liner is inserted in the injection well and a mobile igniter–burner is placed inside the liner. The igniter–burner can be retracted and used to burn off the liner, exposing unburnt coal to the process (Figure 11.14). A third vertical borehole can be drilled, offset by at least 50 m from the line of the two original boreholes. Water and oxygen can be injected via this borehole to achieve a link between this and the initial reaction area, which should have increased the natural permeability of the coal seam (Figure 11.15). This will allow a lateral expansion of the reaction zone, and the process can be repeated to further increase the area of production (Oliver and Dana 1991). Observations made during tests near Hanna, Wyoming in 1987 suggested that groundwater has a profound influence on the UCG system. Not only is the volume of water used substantial, but the resulting impacts on the hydrogeological system in the fractured coal medium were felt over a substantial area. During the test it was estimated that 2% of the water available in the coal seam was consumed. This substantially lowered the hydraulic head on the site and also affected head beyond the site area. The relationship of hydrogeologic conditions under high and low pressure operating conditions are seen in Figure 11.16 (Beaver *et al* 1991). When the cavity pressure was less than

the hydrostatic pressure, groundwater moved towards the cavity, where it was converted to steam, resulting in a loss of head. High pressure conditions at the gasification centre initiated movement of product gas up dip, and the gas, being less dense than water, displaced the water at the top of the coal seam, and as migration progressed, gas was detected in the surrounding monitoring wells.

Therefore, in order to prepare an area for UCG, a detailed knowledge of the geology and hydrogeology is required. An example of the desired underground gasification site characteristics are given in Table 11.4 (Oliver and Dana 1991). These include the depth, thickness, quality, structural condition and hydrogeological character of the target coal seam, together with knowledge of the properties of the overlying strata. Vertical boreholes are drilled in the conventional way, but curved or 'deviated' boreholes require detailed knowledge of the depth and dip of the coal seam. Because of the problems associated with drilling the more difficult directed injection boreholes, it is preferable, where possible, to drill it first and adjust the position of the production well(s) to match rather than the other way around.

As described above, the first objective of UCG is to achieve ignition of the coal by producing a gasified channel by high pressure injection; the second phase is to extend the area of combustion by drilling additional injection wells. The final phase of the UCG operation is to extinguish the fire. This is done by injecting nitrogen (N_2) into the reaction area, and then after a time lapse of several days, the underground cavities are filled with water.

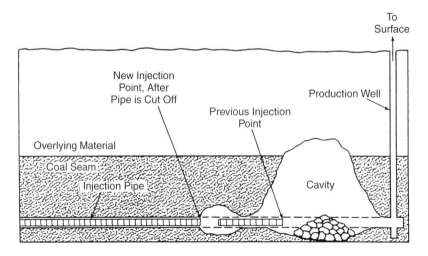

Figure 11.14 Typical configuration for controlled retracting injection point (CRIP) gasification. From Oliver (1991). Reproduced with permission

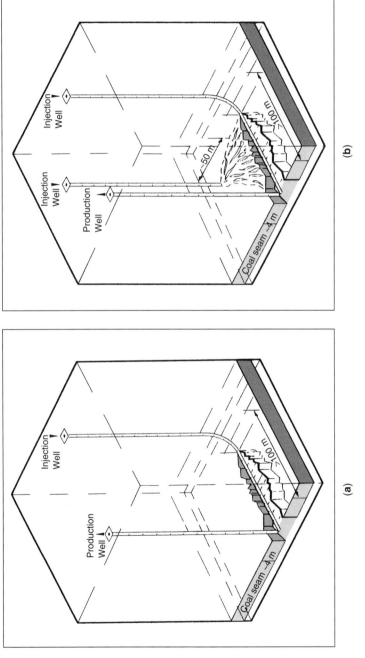

Figure 11.15 (**a**) Injection and production wells using the CRIP technique; (**b**) Addition of an offset well designed to extend the combustion zone laterally. From ETSU Report 1993, with permission of DTI Cleaner Coal Programme

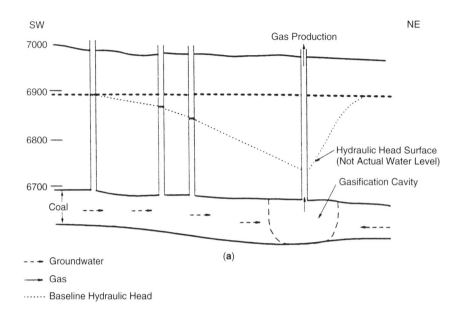

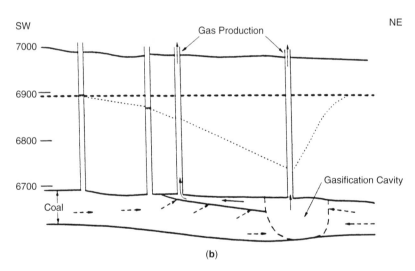

Figure 11.16   Schematic diagram of hydrogeologic relationships under high and low pressure operating conditions during the Rocky Mountain 1 UCG test. From Beaver *et al* (1991), with permission. (**a**) Under typical operating conditions (cavity pressure less than hydrostatic), groundwater moves toward the cavity where it is consumed in the process and converted to steam, resulting in loss of head; (**b**) Under elevated operating pressures, product gas migrated updip along fractures and was detected in groundwater monitoring wells in the southwest part of the site

The high cost associated with the UCG field trials requires an efficient control and monitoring system. This will include the development of a series of boreholes which will indicate the extent of the burn zone by detecting any temperature rise, by the low flow of gas as the burn zone nears the monitoring boreholes, and when 'breakthrough' occurs.

11.3.2 Development of underground coal gasification

In the former USSR, in the 1950s a number of UCG stations were built in Tula, Yushno–Abinsk, Shatsky and Angren; all were designed to burn lignite. In addition, UCG installations were constructed

Table 11.4 Desired underground coal gasification site characteristics. From Oliver and Dana (1991)

Parameter	Value
Coal thickness (m)	1.5–15
Thickness variation (% seam thickness)	<25
Dip[a] (degrees)	0–70
Dip variation[b] (degrees/31 m)	<2
Single parting thickness (m)	<1
Total partings thickness (% of seam thickness)	<20
Fault displacement (% of seam thickness)	<25
Fault density (no. of faults/31 m)	<1
Coal rank[c]	Bituminous
Coal moisture (wt%)	<15
Coal sulphur (wt%)	<1
Overburden thickness (m)	92–460
Thickness of consolidated overburden[d] (m)	>15
Coal seam permeability (md)	50–150
Immediate overburden permeability[e] (md)	<5
Distance to nearest overlying water-bearing unit (m)	>31
Coal aquifer characteristic	Confined
Nearest producing well completed in coal seam (km)	>1.6
Available coal resources[f] (10^6 m^3)	15.4

[a] Depends on technology.
[b] If using directionally drilled process wells.
[c] If bituminous, free-swelling index should be low.
[d] Thickness should be directly above or near top of coal seam.
[e] 15 m (50 ft) directly overlying coal seam.
[f] For 20-year-long plant operation.

Table 11.5 Underground coal gasification production in the former USSR. Based on Douchanov and Minkova (1997) with permission of The Geological Society

Installation	Production, 10^6 m^3 gas/year	Calorific value, kcal/m^3
Tula (Russia)	400	750–850
Yushno–Abinsk (Kuznetsk Basin)	100	1000
Shatsky (Russia)	200	800
Kamenskaya (Donets basin)	730	900
Lisichansk (Donets Basin)	120	850

at Lisichansk and Kamanskaya in bituminous and anthracite coals. These installations have produced gas at varying CVs (Table 11.5), but only the stations at Yushno–Abinsk and Angren remained in operation after 1980 (Douchanov and Minkova 1997).

In tests in the USA, at the Energy Center in Wyoming, hydrofracturing of coal at 120 m depth enhanced the permeability, and in 1973 produced 0.24×10^6 m^3 gas/day. The Lawrence Livermore Laboratory in Wyoming has produced gas at 5×10^4 m^3/day from coal seams 150–900 m deep, and the Morgantown Energy Center has produced gas at 0.1×10^6 m^3/day with a CV of 1100 kcal/m^3 from coal seams 275 m deep (Douchanov and Minkova 1997).

At the same time tests have been carried out in Europe, in the UK, France, the Czech Republic, Italy, Hungary and Spain. The coal seams in Europe are characterised by their greater depth and relatively thin development. This means that to achieve the successful development of any UCG operation, a number of technical difficulties will have to be overcome, such as the effective linking in the coal seam, the control of the gasification front and overall control of the multistage gasification process.

In 1988, six member states of the European Union formed a European Working Group (EWG) on UCG. To demonstrate the commercial feasibility of UCG, field trials were carried out, together with the development of a semi-commercial plant. The trials were carried out in the Teruel region of Spain at El Tremedal, and two coal seams 1.9–7.0 m in thickness were targeted at a depth of 600 m. The coal was subbituminous in rank and had low permeability, and the test area was at least 200 m from any significant faults. As demonstrated in the USA, the test used the CRIP method to control the enlargement of the gasifier; this was seen to be successful (Figure 11.15). The influx of groundwater was sufficient to meet the requirements of the chemical reactions of gasification, but during the test was uncontrolled, which created the risk of quenching the reactor. The trial also experienced significant gas losses from the gasification zone to the surrounding strata; this was partly due to permeable strata above the coal, and a working pressure above hydrostatic. However, overall the trial was a success and demonstrated that the production gas had a quality and heating value consistent with the theoretical estimates and would appear suitable for industrial use (Green 1999).

UCG has been known and understood as an alternative method of capturing energy from coal by burning *in situ*. However, efforts to establish UCG as a viable alternative to conventional methods such as mining have been hampered by the high cost of development and the technical difficulties experienced in directional drilling. The latter has now largely been overcome, particularly by significant developments in directional drilling in the petroleum industry. Although a number of countries would like to benefit by obtaining energy from their deep lying coal resources, the question of the cost of producing this energy by UCG still places it in a minor role as an energy provider from coal sources.

11.4 COAL AS A LIQUID FUEL

The world's coal resources are greater than the known oil resources; because of this, it is likely that the liquefaction of coal will be necessary to provide synthetic fuel as a substitute for crude oil once oil sources begin to run out.

The conversion of coal to oil has been developed commercially since the 1920s, whenever oil supplies became unavailable. This has usually been due to physical and technical production constraints or for political reasons. Commercial production of coal-derived synthetic liquid fuels is still limited, the prime reason being the high cost of current coal-to-oil processes. Future crises in the oil industry may stimulate further development of coal-to-oil production.

11.4.1 Coal liquefaction technology

There are various methods of coal liquefaction, the main problem being the deficiency of hydrogen in coal compared to liquid fuels. This can be overcome by adding hydrogen to the coal by a number of processes:

1. Direct liquefaction by hydrogenation.
2. Indirect liquefaction by the Fischer–Tropsche synthesis.
3. Removal of part of the carbon content from the coal by pyrolysis.

The three processes differ in technology and in the yield of liquid and solid products.

The most direct method is by hydrogenation; to overcome the hydrogen deficiency, the H/C ratio is increased by adding a hydrogen donor. Coal is dispersed in a thermally stable 'solvent' and/or 'hydrogen donor' and passed into a pressurised autothermal reactor at temperatures between 400 and 500 °C. If additional hydrogen is not supplied to the reactor, the hydrogen-depleted solvent oil is itself rehydrogenated at a later stage. The reaction products are filtered and distilled to separate the solvent from the coal extract which is subjected to vacuum distillation to produce distillate oil (Taylor *et al* 1998).

Experiments have shown that liquid yields equivalent to 4 bbl/t dry coal have been obtained by this method

(Taylor *et al* 1998). In practice, these figures would be lower, but still considerably higher than those from coal without a hydrogen additive.

The Fischer–Tropsche synthesis was developed in Germany in the 1920s, and formed the basis for the production of oil from coal in Sasolburg, South Africa, the name SASOL having become synonymous with the process. The development of the SASOL plants was motivated by South Africa's long political isolation and attendant oil embargo. This overrode any financial considerations and up to 60% of transportation fuel was supplied in this way.

The Fischer–Tropsche synthesis involves the gasification of the coal, carried out in a Lurgi gasifier to produce the synthesis gases carbon monoxide, hydrogen and methane. The methane is treated in a gas reformer and synthesised in a Kellog reactor. The carbon monoxide and hydrogen are subjected to fixed-bed Fischer–Tropsche synthesis by passing them through ovens containing circulating water with an iron or cobalt catalyst. Gasoline and other products can then be obtained by cracking the resulting synthetic crude oil. Around 34 Mtpa of coal is used by Sasolburg in plants designed to produce 50 000 bbl/d of gasoline and other products for chemical feedstocks from the processing of 30 000 t/d of coal (Sage and Payne 1999).

Pyrolysis involves the heating of pulverised coal extremely rapidly in a vacuum, known as 'flash pyrolysis'. The feed coal passes through a plastic stage during which the macerals soften and decompose into gas, char and tarry liquids. The tars are hydrogenated to produce heavy or light oil as required.

11.4.2 Coal properties for liquefaction

Coal quality requirements vary according to the method used for coal liquefaction. In both pyrolysis and hydrogenation, the use of low rank coals with high hydrogen contents enhances the liquid yields. The required high H/C ratio is closely linked to rank and petrographic composition; the latter requires a high proportion of reactive components in the coal, such as vitrinite and liptinite. Liptinite remains highly reactive over a large range of rank whereas vitrinite first increases then decreases with increasing rank. Inertinite shows varying degrees of reactivity in coal liquefaction (Taylor *et al* 1998).

In hydrogenation, the use of coals containing high amounts of oxygen, nitrogen and sulphur means that hydrogen is consumed in the removal of these heteroatoms. Because the supply of hydrogen is expensive, this constitutes a financial loss in the process. Inorganic impurities such as the mineral matter content of the coal have been found to influence the liquefaction behaviour of coals. A high ash content can lower reaction throughout and increase problems of solid and liquid separation, and may deactivate any catalyst. Some inorganic material can have catalytic effects of their own, e.g. pyrite has favourable catalytic properties (Taylor *et al* 1998).

The gasification processes are comparatively insensitive to coal properties, and can utilise coals which would be unsuitable for other processes. For example, the Fischer–Tropsche synthesis used for SASOL in South Africa uses inertinite-rich, high volatile bituminous coal with a high ash content, a typical Gondwana-type coal.

Taylor *et al* (1998) summarize those characteristics favourable for coal hydrogenation as:

Vitrinite reflectance	<0.8%
H/C atomic ratio	>0.75%
Vitrinite + liptinite	>60%
Volatile matter (d.a.f.)	>35%
Low concentration of heteroatoms	

11.4.3 Future development of coal liquefaction

A large number of studies have been carried out to produce liquid hydrocarbons from coal. This has been triggered in recent times by the oil crises in the 1970s and early 1980s; since that time much of the development has been put on hold.

The front runners in coal liquefaction development have been the USA, South Africa, Germany, Japan, and the UK. Other major coal producers have the potential to consider upgrading coal to supplement their own oil reserves and/or to reduce their oil imports, e.g. China, India and Poland. Additional smaller coal producers such as Indonesia, Turkey, Greece, Romania and Spain could consider coal liquefaction if the coal market price and the cost of coal liquefaction were to be such that the process was economically viable.

In the UK, research and development into coal liquefaction has concentrated on the direct liquefaction process known as Liquid Solvent Extraction (LSE), developed by British Coal. To date no process has been demonstrated at a commercial scale (Robinson 1994). The model plant is designed to produce approximately 50 000 bbl/d of liquid transport fuels from 17 000 t/d of coal. In the first stage, bituminous coal is digested in a hydrogen-donating solvent in the absence of hydrogen, at a low pressure. The resulting mixture is filtered to produce a low ash extract solution. In the second stage, the extract solution is catalytically hydrocracked in the presence of hydrogen to upgrade the products (Barraza *et al* 1997).

In the USA, a variation on the pyrolysis technique is the International Liquids from Coal (LFC) process (Weber and Knottnerus 2000). Its development has been influenced by the electricity generators in the USA changing from high sulphur coals from eastern USA to low sulphur coals from western USA. This change is in order to meet the sulphur dioxide emission standards set up by the US Clean Air Act. However, because the western coals are high in moisture content, and therefore expensive to transport, the LFC process was designed to overcome these problems. The coals used are from the Powder River Basin; they are high moisture coals (25–32%) with relatively low heating values (7900–8800 Btu/lb). The coal is dried to almost zero moisture content; it is then mildly pyrolysed and approximately 60% of the volatile matter and most of the organic sulphur is removed. The coal char is then cooled and has controlled amounts of moisture and oxygen added to produce a stable solid fuel. The volatile matter driven off during pyrolysis is partially condensed in a multi-step operation to produce crude hydrocarbon liquid. It has been shown that this process can produce approximately 0.5 t of solid fuel and 0.5 bbl of crude hydrocarbon liquid from each short ton of raw coal feed. The liquid is low sulphur heavy liquid hydrocarbon, which can be further processed to produce other chemical and industrial products.

In Yugoslavia, studies into the conversion of low rank brown coals into liquid products have been undertaken by using the direct catalytic hydrogenation process. Results indicated that the yield of particular liquid products varied markedly depending on temperature and residence time. A high degree of conversion (84%) was observed; this was confirmed by petrographic analysis which showed that there was no unreacted coal in the solid residues. It was also noted that the petrographic composition of the residues depended on the reaction conditions (Aleksic *et al* 1997).

The greatest drawback to coal liquefaction has been the high cost of production, particularly when oil, natural gas and coal prices have been low. It has always been cheaper to obtain coal supplies from either indigenous sources or as imports than to invest in coal liquefaction plants.

In the future, the greater availability of coal, combined with low coal prices and rising natural gas and oil prices, could provide the ideal economic scenario for future development of coal liquefaction technology.

11.5 COAL AS AN OIL-PRONE SOURCE ROCK

In addition to the treatment of coal to produce alternative sources of energy, geological processes acting upon coal-bearing sequences have produced hydrocarbon reserves, which have been preserved as oil, condensate, and wet and dry gas.

Coal-bearing sequences are essentially nonmarine in nature and have been estimated to account for less than 10% of the world's oil, and much of this nonmarine contribution is derived from lacustrine source rocks, accounting for 85–95% of the oil in areas such as Brazil, China and Indonesia (Fleet and Scott 1994). However, in spite of being a minor contributor to the world's oil resources, oil-prone coal sequences are considered as significant oil-source rocks in southeast Asia and Australasia, whereas the role of coal sequences in providing oil in North America and western Europe is much more debatable. The identification of oil derived from coals has important implications; recognition of such oils in a basin can indicate the presence of coals not previously identified.

11.5.1 Suitability of coal as an oil-source rock

In order to comprehend how coals or coal-bearing sequences have the ability to expel petroleum in the liquid phase, it is necessary to understand the quantity and quality of liquids and gases which coal-bearing sequences can expel in response to their thermal and structural history. Laboratory techniques have been used to try to distinguish oil-prone coals from other coals; these indicate that oil-prone coals are richer in hydrogen relative to carbon. It is generally accepted that sediments must contain moderate to high concentrations of hydrogen-rich kerogens in order to have significant oil-source potential. The progenitors of hydrogen-rich kerogens are derived from vascular plants, lacustrine algae, photosynthetic bacteria and in-sediment bacteria. To have oil potential there has to be a combination of sedimentary and environmental processes which has enhanced the production and preservation of the organic constituents (Thompson *et al* 1994).

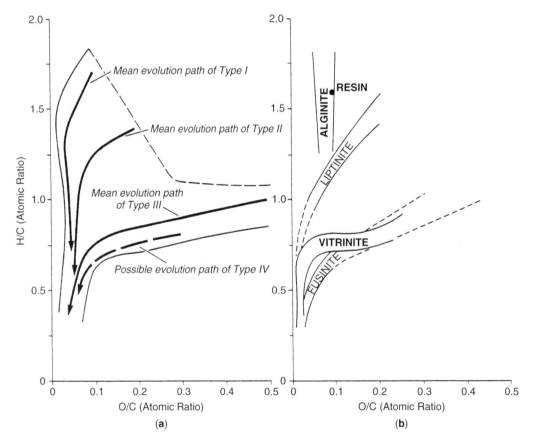

Figure 11.17 Comparison of kerogen types, evolution paths and petrographic components of coal based on atomic ratios. From Powell and Boreham (1994), with permission of Geoscience Australia

The composition of macerals reflects the original composition of the plant precursors, despite having been substantially modified during the biochemical and early thermal stages of coalification. Liptinite (or exinite) is richer in hydrogen than vitrinite, which in turn is richer than inertinite. These maceral groups occupy different coalification pathways with thermal maturation, and are similar to the pathways for Types I–IV organic matter (kerogen) defined for sedimentary rocks in general (Figure 11.17).

Studies of organic matter or kerogen types suggest that in order for a source rock to have hydrocarbon potential, 10–20% of its organic matter must equate with Type I organic matter, or 20–30% must equate with Type II organic matter. The bulk H/C ratios would therefore be in the range 0.8–0.9, or Hydrogen Indices in Rock–Eval analysis would be above 220–300 mg HC/gC before oil expulsion is considered (Powell and Boreham 1994).

The use of the petrographic composition of coal as an indicator can have limitations, and it is possibly the association of macerals, the microlithotype, rather than the macerals themselves that controls the expulsion of liquid petroleum (Fleet and Scott 1994). Clues to this type of source can be high wax and low sulphur content in the oil (Hedberg 1968). Biomarker molecules in the oils, which can be linked to land plant communities, can add further evidence to an origin for the oil. The character of the vegetation component in coal deposition will have changed through geological time as plant communities have evolved, being influenced by climate and environmental change. The Jurassic coals of Australia have a dominance of conifers in swamp floras, whereas the late Cretaceous coals contain angiosperm flora. Both of these provide an abundant amount of potentially oil-prone material which has been preserved as exinite. The Tertiary oil-prone coal sequences of southeast Asia have resulted from deposition of oil-prone detritus in coastal plain environments under wet tropical conditions (Fleet and Scott 1994). These two sets of conditions have led to the suggestion that the significant oil-prone coal-bearing sequences are restricted to late Jurassic–Tertiary basins

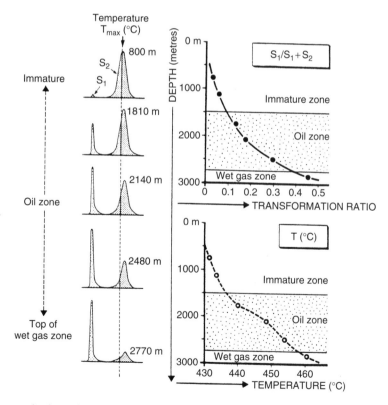

Figure 11.18 Characterization of source rock maturity by pyrolysis methods. Transformation ratio and/or peak temperature T_{max} may be used as indicators of thermal evolution. From Taylor *et al* (1998), with permission of Gebruder Borntraeger

of Australasia and the tropical Tertiary basins of southeast Asia (Macgregor 1994).

Because hydrogen is the significant factor in the generation of hydrocarbons from sedimentary organic matter, it is suggested that the hydrocarbon potential of terrigenous organic material may be expressed as a ratio of hydrogen-poor and hydrogen-rich components. Rock–Eval pyrolysis records the release of hydrocarbons and CO_2 with increasing temperature, and determines the temperature of maximum hydrocarbon generation (T_{max}). Free hydrocarbons (S_1) already present in the rock are liberated at low temperatures, whereas newly generated hydrocarbons (S_2) are given off at higher temperatures. The transformation ratio given as S_1: ($S_1 + S_2$), and T_{max} both increase with increasing maturity, as shown in Figure 11.18, which illustrates the characterisation of source rock maturity by pyrolysis methods (Taylor *et al* 1998). The ratio S_2/TOC (total organic carbon, expressed in weight %) or Hydrogen Index correlates with the atomic H/C ratio measured by elemental analysis on kerogen.

Isotopes can be used to characterise the total carbon or bulk fraction of a kerogen, oil or gas. Isotope analysis is considered most useful for characterising gases, and is therefore important in studying the petroleum

generated from coal-bearing sequences, which, although containing liquid products, also contain a high proportion of gas.

11.5.2 Coal-sourced oil and gas occurrences

The principal coal-bearing sequence sourced oil basins are listed in Table 11.6. The Gippsland Basin in Australia and the Kutei Basin in East Kalimantan, Indonesia are the largest, most fully documented and least disputed cases of coal-bearing sequence sourced oil provinces (Macgregor 1994). Some 80% of Australian oil is attributed to a coal-bearing sequence source, namely from the Gippsland Basin. The Cooper, Eromanger and Taranaki Basins also contribute hydrocarbons, chiefly as large amounts of gas. In southeast Asia, the proportion of oil reserves sourced from coal-bearing sequences is estimated at 10–30%. Other significant contributions of coal sourced are from mid-Jurassic coals in China and Egypt. Minor reserves of oil may be present sourced from Tertiary coals in Venezuela. Dry gas-prone coals become increasingly significant with increasing geological age. This is represented by the European Westphalian coals, which seem to have expelled only dry gas, although these coals are well

Table 11.6 Examples of case studies of petroleum systems derived from terrigenous sediments, i.e. coal or terrestrially sourced organic matter. Part based on Powell and Boreham 1994, reproduced with permission from Geoscience Australia

Country Basin/province	Source rock		Reference
	Age	Hydrogen indices[a]	
Australia			
Gippsland Basin	Late Cretaceous–Tertiary	200–350	Moore *et al* (1992)
Cooper/Eromanga Basin	Permian	150–300	Vincent *et al* (1985)
	Jurassic	200–400	
Bowen/Surat Basin	Permian	150–250	Boreham (unpublished)
Canada			
Beaufort-Mackenzie Basin	Eocene–Paleocene	130–250	Issler and Snowdon (1990)
China			
Turpan Basin	Jurassic	200–500	Zhao *et al* (1997)
Indonesia			
Ardjuna Sub-basin, Java	Late Oligocene	250–400	Noble *et al* (1991)
Kutei Basin, Kalimantan	Middle Miocene	200–350	Durand and Oudin (1979)
New Zealand			
Taranaki Basin	Late Cretaceous–Tertiary	230–360	Curry *et al* (1994)
Nigeria			
Niger Delta	Late Cretaceous–Tertiary	<200	Bustin (1988)
Norway			
Haltenbanken area, North Sea	Jurassic	275	Forbes *et al* (1991)

[a]Hydrogen indices are measured on immature samples.

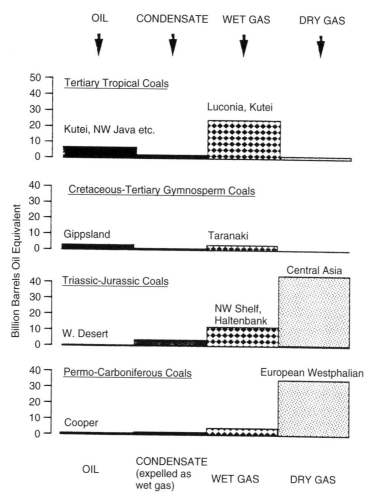

Figure 11.19   Hydrocarbon reserves tied to coal-bearing source sequences by age. Volumes of different hydrocarbon phases are plotted in billion barrels oil equivalent. Oil-prone coals plot to the left; gas-prone coals plot to the right. In pre-Cretaceous times gas-prone coals predominate; later times see the appearance of significant oil-prone coals MacGregor (1994) with permission of The Geological Society

within the oil-producing window. Figure 11.19 shows the hydrocarbon reserves in relation to geological age, and in particular, the change from gas-prone coals in pre-Cretaceous times followed thereafter by oil-prone coals (Macgregor 1994).

The Cooper Basin is a Permo-Triassic intercratonic basin situated in south central Australia. During the Permian the basin was filled by a series of lacustrine and fluvial deposits; the coals are characterised by high concentrations of inertinite and low levels of liptinite, a common feature of Gondwana coals. However, in spite of this, the H/C ratio of the inertinite is 0.5 and the noninertinitic material is 1.0, and the Hydrogen Index

values for coals in the Cooper Basin range from 116 to 300 mg HC/gC. These coals are considered to be derived from terrestrial organic matter co-deposited with algal derived organic material (Curry *et al* 1994, Powell and Boreham 1994).

In northwest China, a thick sequence of Jurassic sediments is located in a number of tectonic basins along the flanks of the Tianshan–Qilan Mountains; one such basin is the Turpan Depression, which contains 4500 m of sediments. The sequence is characterised by rhythmic transgressions and regressions of lake and swamp facies over a large area. Deltaic sand bodies are associated with the lacustrine–swamp transition zone and act as hydrocarbon reservoirs. The organic material is primarily composed of liptinite and vitrinite, classified as Type I and Type II organic matter. The Hydrogen Index has a range of 200–500 mg HC/gC, and the depth of hydrocarbon maturation and expulsion is given as 3000–4000 m (Zhao *et al* 1997). These hydrocarbons from coal-bearing source rocks are of major importance in China, and large reserves of oil and gas have been identified.

It is significant that coal-bearing sources have not been considered for any of the world's 30 largest oil provinces, and that less than 1% of the world's known oil reserves is sourced from coal-bearing sequences (Macgregor 1994).

There are numerous occurrences of coal worldwide that are not tied to either oil or gas reserves due to the geological history of the deposits. For example, the Tertiary coals in the Philippines are similar to those in nearby Indonesia, yet no significant oil discoveries have been made other than those clearly derived from marine source rocks. Clearly, the oil potential of coals must vary across the region.

Coal-bearing source rock sequences are more confined in space and geological time than other oil-source rocks. Botanical controls and the environment will define the likelihood of oil availability within any coal-bearing sequence. Lacustrine margin coals appear to be more favourable than marine margin coals (Macgregor 1994).

The study and understanding of coal-bearing sequences as source rocks for oil and gas is still in its infancy, but it is clear that such studies will need to consider the sedimentological, paleobotanical and geochemical characteristics of the coals in each individual sequence, as well as identifying suitable reservoir rocks. Those areas containing coal but hitherto considered unprospective may be reassessed in the future due to the increasing need for hydrocarbon reserves.

12

Coal and the Environment

12.1 INTRODUCTION

In the last 30 years, public awareness has increased regarding local, national and international environmental issues. This has resulted in the concentration of political attention by means of statutory regulations covering the majority of industries in every industrialised nation. Developing nations are being asked to conform to environmental standards conceived in the industrialised nations without which, aid, financial support and trading facilities will not be forthcoming. However, there is a concern that environmental standards can be imposed without due consideration of scientific and technical evidence together with the economic welfare of the community.

It is clear that no industry has attracted greater attention than the coal industry. The mining and use of coal remains an emotive issue with environmentalists and their political supporters despite the tremendous improvements in mine rehabilitation and coal-fired power station emissions. Historically, the coal industry had left a legacy of both land and atmospheric pollution; indeed, the traditional image of dirty mines and industrial chimneys belching smoke is still too easily evoked in irrational discussions about the coal industry, even though it bears no relation to the modern coal industry. The media are sympathetic to the environmentalists and have depicted the mining and use of coal as a threat to human health which must therefore be strongly regulated. Such regulations serve to increase the cost of coal production and use, but in spite of this, coal remains a low cost, abundant and secure source of energy. In addition, coal is a democratic fuel in that it is widespread globally and therefore less sensitive to political instability.

Because of the close attention given to it by environmentalists, the coal industry has had to address numerous environmental issues. This has included remedying previous environmental damage and preventing future occurrences. Planned current and future environmental regulation as a result of international agreements will ensure that the coal industry will continue to improve both its working practice and its public profile.

However, it is undeniable that the mining and use of coal does have pronounced effects on the environment, and are principal causes of environmental concern. Figure 12.1 is a well known summary of the effects of the use of coal on the environment. These have a direct influence upon the geological investigation and exploration for coal. Coals that are environmentally disadvantaged, e.g. high sulphur coals, are unlikely to be the prime target of mining companies, who know their limitations in the coal sales market.

12.2 COAL MINING

The increasingly complex regulatory regimes imposed by governments have brought environmental planning to the forefront in the mine planning and development process. This has resulted in changes in the methods of working, in types and utilisation of equipment and in coal preparation techniques. In addition, changes in mine planning and operation have had to be developed, but at the same time both complying with the regulations and remaining cost effective and competitive.

Once the results of the exploration phase are known, and these indicate that a viable mining operation is possible, it is normal practice to prepare a preliminary environmental report. This report, plus the final details of the coal mine planning, will serve as support documents in the preparation of the official Environmental Impact Assessment (EIA). All potential mine developments require an EIA in order to obtain the necessary legal permits and concessions in order to mine coal. Although each mine can have different environmental effects, there are a number of factors which strongly influence the environment. Not only will the mine be assessed but also the effect on the surrounding landscape, water courses, and native flora and fauna, as well as social effects on the local community.

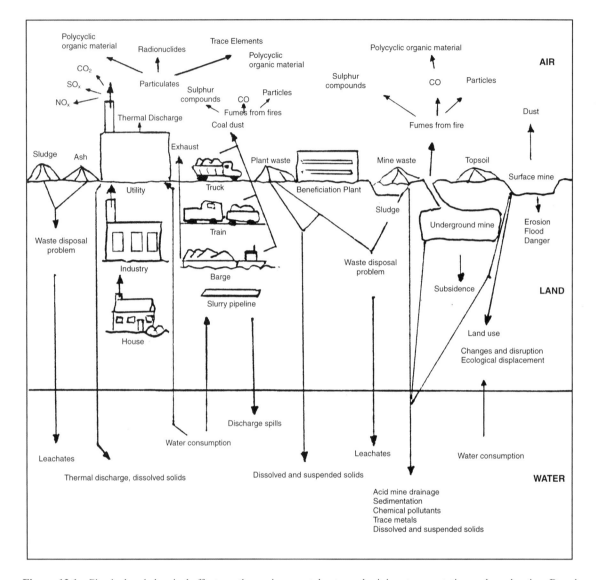

Figure 12.1 Physical and chemical effects on the environment due to coal mining, transportation and combustion. Based on numerous sources

12.2.1 Watercourse diversions

During surface mining operations, it is possible that certain drainage divisions may be required, particularly for the larger scale operations. All surface water originating upstream of a mining site should be diverted around the excavation and spoil areas to avoid contamination of the water and to reduce other problems within the pit. Diversions should be hydraulically efficient, and designed and constructed to control erosion and sediment load. Many countries have published regulations to establish design and performance standards for such diversions. It is necessary to determine the design of the channel section which will carry the diverted flow, and also to ensure that the flow velocities will remain below that which can be tolerated by the chosen channel design.

In the case of perennial streams, increased flow velocities and sediment loads in channel diversions may not be conducive to freshwater fauna and flora.

The diversion ideally should include shallows and deeps and some meander pattern to suit the local ecological regime.

12.2.2 Contamination of mine waters

One of the most serious effects of underground coal mining has been the escape of polluted water from both old and current mine workings. Before the days of environmental regulation, such acid and alkaline mine waters had been allowed to pollute waterways and surrounding land, rendering the area unusable and sterile. Old industrialised countries such as in western and eastern Europe and the USA exhibited industrial wastelands as a consequence of coal mining. Such practices are now long gone, but the potential for underground mine waters to escape and enter the surface water regime is still a real one in some locations.

Acid mine drainage is the principal cause of contaminated water arising from coal mining. It results from the exposure of sulphide minerals, particularly pyrite, to water and oxygen during and after mining or in piles of mine waste. Many underground mines have to pump to remove water from the mine workings. This can be a major problem in old mining areas where old mining districts are often connected underground. Water entering the workings from near surface aquifers is usually of reasonable quality, but mine workings in deep seams are likely to encounter more saline waters. The minerals iron pyrite and marcasite (both FeS_2) are commonly present in coals and coal-bearing sequences, and these are reactive to atmospheric oxygen. The initial products of oxidation are ferrous and ferric sulphates, sulphuric acid and hydrated ferric oxide. With the exception of ferric oxide, these products are soluble in water, and in turn react with clays and carbonate minerals to form aluminium, calcium, magnesium and other sulphates. Ferruginous waters that flow in the presence of air in mine workings precipitate ferric oxide; this produces the extensive red/orange staining of walls and equipment that characterises many underground workings.

Water flooding into abandoned mine areas containing large quantities of sulphide minerals will rapidly become contaminated. Problems arise when mining ceases and pumping is stopped; then all the connected workings become flooded, the water level will rise using old shafts and workings as conduits, and can result in mine water discharge into the surface water regime. The initial breakout of water is the most acidic, and contains the largest quantities of iron and other dissolved metals; this is due to the fact that the greatest potential for the oxidation of pyrite is in a humid atmosphere where there is free oxygen, as in the case of old workings. However, this is greatly reduced in saturated or flooded conditions when the presence of free oxygen is removed. Consequently, once waters flowing through the workings have flushed out all the oxidised material, little additional oxidation occurs and the contamination of the mine water decreases with time, provided the water levels in the workings do not fluctuate and allow oxidation to recommence.

In the eastern USA, acid mine drainage associated with coal mining has caused severe problems. Anthracite mines in eastern Pennsylvania were abandoned and allowed to flood, and in the 1960s the initial water discharge had a pH of 3.3–5.6. As water continued to flush through the mine workings, the pH improved to 5.8–6.2 by the late 1970s. The cessation of mining and the circulation of groundwater has led to the improvement in the quality of mine water discharge in the region. In the active mines, improvement of water quality has been achieved by chemical neutralisation of the mine water before discharge. The estimated cost of such chemical treatment of mine waters in the USA has been given as >$1 million per day (Clarke 1995).

Table 12.1 shows the relationship between depth from the surface and the concentration of selected dissolved compounds found in underground mines. Elements such as sodium, calcium and chloride increase in concentration with increasing depth, whereas sulphate and hydrogen carbonate compounds decrease with depth. Waters from shallow workings contain sulphates, chlorides, bicarbonates, calcium, magnesium and sodium salts.

Table 12.1   Changes in groundwater quality with depth. Reproduced with permission of Dargo Associates Ltd

Depth from surface (m)	Concentrations of dissolved compounds (mg/l)									
	Na^+	Ca^{2+}	Mg^{2+}	Ba^{2+}	Sr^{2+}	Nh_4^+	Mn^{2+}	Cl^-	SO_4^{2-}	HCO_3^-
30	40	60	40	<2	<1	<0.1	<0.1	50	200	200
300	10 000	800	260	60	25	12	0.3	18 000	<5	200
900	41 500	11 700	2 000	550	400	70	3.0	90 000	<5	80

Waters from slightly deeper workings become more heavily mineralised with calcium and magnesium salts, and at great depths, concentrations of barium, strontium and ammonium chlorides are characteristic. Saline waters from deep coal mines contain high amounts of chlorides, e.g. 61 240 mg/l Cl^- in waters from deep coal workings in Nottinghamshire, UK (Downing *et al* 1970), together with high amounts of ammoniacal nitrogen. Other contaminated mine waters produced within deeper workings may also contain diffused methane gas. Such saline waters can be harmful to crops, and cause corrosion to metallic machinery, and have proved to be a problem in a number of areas. In the upper Silesian Basin, Poland, salinity increases with depth, and chemical analyses have shown salinity levels of more than 250 000 mg/l (Clarke 1995). In addition, Polish saline waters contain natural radioactive isotopes, mainly ^{226}Ra from the uranium series and ^{228}Ra from the thorium series. Up to 40% of the total amount of radium remains in the ground but up to 225 Mega Becquerels (MBq) of ^{226}Ra and 400 MBq of ^{228}Ra are released daily into rivers, along with other mine effluents. To counteract this, technical measures such as induced precipitation in gob areas has been undertaken in several mines, and the results have shown that the total amount of radium released to the surface waters has diminished by 60% in the last 10 years (Chalupnik *et al* 2001).

The management of saline waters from coal mines operating in arid regions can pose difficulties. In the Hunter Valley, New South Wales, Australia, coal seams carry saline groundwater which drains naturally into the Hunter River. During normal rainfall years, the mine waters are diluted by better quality surface flows, but during extended drought conditions, the saline groundwater makes up a larger proportion of the river recharge water. During such periods, it is essential that carefully controlled discharge of saline mine waters is maintained.

Table 12.2 shows some selected analyses of saline and acidic waters from deep coal workings in the UK. Of note is the acidic iron-rich nature of the initial mine drainage water from Bentinck Mine compared to the pumped water after equilibrium is reached from Moorgreen and Pye Hill mines (after Banks *et al* 1997).

In open pit mines the exposure of rock (with its content of sulphide minerals) to the atmosphere and the hydrological cycle can produce acidic mine waters. Piles of removed overburden and interburden, whether as infill or as spoil heaps, together with all surface mine waste and spoil heaps from underground workings, can produce contaminated water.

In new mine development, a detailed hydrogeological investigation is essential to understand the movement of groundwater around and within any proposed mine workings. In underground mines this is never as straightforward as in open pit mines. The transmissivity of water at changing depths and structural intensities make it difficult to anticipate groundwater behaviour. Nevertheless such studies will assist in the planning of groundwater removal from the mine (see Section 9.5) and reduce the potential for groundwater contamination.

Table 12.2 Selected analyses (in mg/l) of deep waters from coal measures strata in the UK, illustrating compositions of saline brines. Note the acidic iron-rich nature of the first drainage (i.e. nonequilibrium) water from Bentinck Colliery when compared to the pumped (equilibrium) saline water from the nearby Moorgreen and Pye Hill Collieries. Adapted from Banks *et al* (1997) reproduced by permission of the Geological Society

Source	Eakring 8 Crawshaw sandstone	Glentworth 5 Lower coal measures sandstones	Plungar 4 Crawshaw sandstone	Moorgreen Piper Colliery Pumped water	Pye Hill No. 2 Colliery Pumped water	Bentinck Colliery Initial drainage water
Na^+	8 079	7 005	7 900	–	–	–
K^+	96	9	31	–	–	–
Ca^{2+}	792	1 552	822	–	–	–
Mg^{2+}	218	192	556	–	–	–
Cl^-	14 555	11 786	14 910	3 600–10 800	1 100–3 900	31 400
SO_4^{2-}	nil	2 718	342	–	–	–
HCO_3	73	549	220	–	–	–
TDS	23 776	23 532	24 669	–	–	–
PH	7.1	7.4	7.8	6.9–7.9	7.3–8.0	5.7
Fe (total)	–	–	–	<0.1–7	1–9	150

The problem of contaminated mine drainage is best dealt with by preventing polluted drainage from occurring, or by collecting and treating it before it is discharged. In most cases, the reprocessing of spoil heaps is too expensive, and the removal of underground sources of contamination is not feasible other than to flood completely the area from which the contamination is generated. If the source cannot be removed, then the alternative is to keep water away from spoil heaps and within the mine. A number of barrier methods is used, particularly in relation to spoil heaps. Compaction and revegetation are two ways of inhibiting water passage through spoil heaps, but more effective isolation of spoil material from percolating waters is to use low permeability barriers, such as clay or plastic membranes (Clarke 1995). In the USA, a common method used to isolate pyritic spoil from groundwater flow, percolating surface waters and oxygenation is the *high and dry placement* method. Figure 12.2 shows the segregation of acid-prone material to reduce exposure to water and oxygen. The acid-prone material overlies porous material which allows groundwater to pass through, and is itself compacted to restrict surface water infiltration. This reduces the contamination but may not achieve the levels demanded by modern legislation. Also the method is mainly applicable to spoil in open pit operations. Spoil from underground mining is much more difficult to assess and it is difficult to segregate. Clay and bentonite-rich liners have been used as clay caps to prevent water entering the spoil, but have been prone to cracking in dry conditions, and they may also be attacked and breached by certain mineral-rich mine waters. Plastic membranes can be used, preferably in ongoing mining operations where they can be integrated into the reclamation programme;

however, the high cost of their use is likely to restrict them to the most acid producing sites. Effective prevention of rainwater inflow is achieved by using asphalt or concrete caps to cover spoil materials.

An interesting development has been the use of bactericides. Bacteria, principally *Thiobacillus ferrooxidans*, catalyse the oxidation of pyrite, greatly increasing the rate of reaction. The use of bactericides in the form of surfactants has proved effective in the treatment of acid mine drainage. Commercially produced, slow-release surfactants are now available in the form of spray or pellets and have an effective life of 2–7 years (Clarke 1995), and have been applied to sites in Pennsylvania and Kentucky, USA. The cost of bactericide is offset by savings in reductions in the amount of topsoil required or in the use of other remedial methods.

There has been recent research into the construction of wetlands as an effective method for the treatment of drainage from abandoned mines. The treatment of mine drainage requires the use of aerobic wetland processes; these are designed to encourage the oxidation process and are consequently of shallow depth (0.3 m). This will remove iron in the wetland by the precipitation of ferric hydroxide which in turn lowers the pH of the water which will reduce the oxidation rate. This is then compensated by growing plants such as reeds that pass oxygen through their root systems, causing aeration of the substrate. Wetland treatment studies are being carried out in Kentucky, USA and in South Wales, UK (Robinson 1998).

12.2.3 Other water pollution

Pollution of surface waters can occur from the use of drilling muds and additives. In both greenfield and

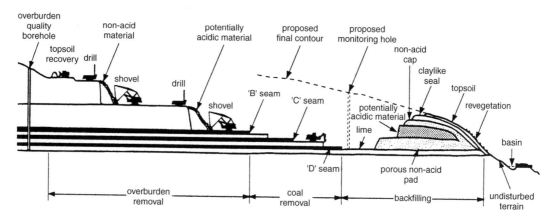

Figure 12.2 Selective handling and placement of mine spoil to prevent the formation of acid mine drainage at a mine in west Virginia, USA. From Clarke (1995), reproduced with permission of IEA

developed areas, drilling programmes must avoid the pollution of streams and rivers by drilling fluids being allowed to flow into them. Discolouration of the water, whilst not necessarily toxic, is not desired by urban and rural peoples alike. The building of a sealed circulation pit and the monitoring of flow rates into and out of a borehole should prevent this situation. Similarly in wells used for abstraction of drinking water, as well as surrounding streams, the leakage of diesel, kerosene and other industrial fluids must be avoided.

12.2.4 Runoff, erosion and sedimentation

Runoff results from precipitation and is the major cause of erosion in mining areas, particularly in regions of concentrated heavy rainfall as is the case in tropical countries. Attempts to combat soil erosion are aimed at controlling runoff, reducing the erodibility of the soil itself and removing any sediment from the runoff that does occur.

Deforestation and the stripping of vegetation cover need to be kept to a minimum, and exposure of the required area of land should be for as short a term as possible. This requires effective mine planning and scheduling the sequential stages of vegetation removal, overburden stripping, mining and reclamation. Seasonal climatic variations may play an important role in this scheduling. Such planning should include the siting of haul roads and any banking, as these are the sites of much of the runoff and erosion. Diversion structures such as terraces and ditches can be sited to intercept runoff on long steep slopes, together with keeping topsoil loose to aid infiltration, and by using new vegetation types to stabilise slopes.

Concave slopes are least affected by erosion, yield the least sediment, and change shape more slowly than other profiles. Convex slopes erode most rapidly, yield the most sediment and change shape quickly. Uniform and complex slopes are affected to an intermediate degree, but can still be severely eroded in a single storm. It is therefore recommended that slopes should be produced with as low a gradient as possible and be concave where possible.

The loss of soil and land due to erosion can result in the degradation of streams and lakes as a result of increased sediment loads. As most runoff and erosion prevention measures are not 100% successful, all runoff originating within the mined area should be routed through a sedimentation pond, the primary purpose of which is to trap sediment movement from the mined area. Suspended sediment concentrations in waters draining from surface mining areas can be very high, with concentrations of 10 000 mg/l up to 100 000 mg/l. Such a sedimentation pond should be of sufficient size to store the sediment load without having the need for frequent removal of settled material, and to be of a size so that inflowing water has a sufficiently long detention period and low velocity to allow suspended sediment to settle out.

12.2.5 Spoil dumping

Environmentally, the dumping of spoil material is considered one of the least desirable surface manifestations of coal mining. Historically the dumping of spoil was not regulated and old underground mining areas were easily identified by the characteristic skyline of conical and elongate spoil tips with their attendant dumping systems, usually by tram railway or overhead ropeways. Modern underground mines still have the problem of where to place rock waste, but this has been reduced by the widespread introduction of longwall mining and by repacking waste material in abandoned districts in room and pillar mines.

Apart from the undesirable visual effects, spoil heaps can be a problematical mining legacy. Spoil heaps or tips can be extremely large and be a product of a number of different mining episodes. The materials in tips can vary enormously; apart from noncoal rock waste, machinery, wood, ropes, boiler ash and general rubbish can all end up in spoil tips. For example, in the UK, at Cilfynydd in South Wales, a large tip fed by an aerial ropeway received up to 500 tons per day, and the tip was used for over 50 years (Bentley *et al* 1998). In areas of mountainous and dissected topography, tips have been constructed on steep slopes, and in some cases, across spring lines. Load pressures in these circumstances combined with high annual precipitation rates have led to tip failures. Failure of such tips was not uncommon in old mining areas, but none captured the headlines more than the failure of the Merthyr Vale Colliery spoil tip No. 7 at Aberfan in October 1966 (Figure 12.3) (Siddle *et al* 1996). Failure had been recorded in other tips at Aberfan in 1944 and 1963, but tip No. 7 is remembered for the high loss of life (144 people) caused when the tip failed. Rotational slides of spoil disintegrated to flow slides which ran downhill for 600 m. These flows released groundwater from a fault zone in the underlying sandstone which then caused a secondary debris flow (Figure 12.3). In the UK since the Aberfan disaster, improved tipping practices and rigorous inspections have ensured that no rapid tip failures have occurred since 1967.

It is a different picture in open pit operations: the final restoration of the site has to be included in the

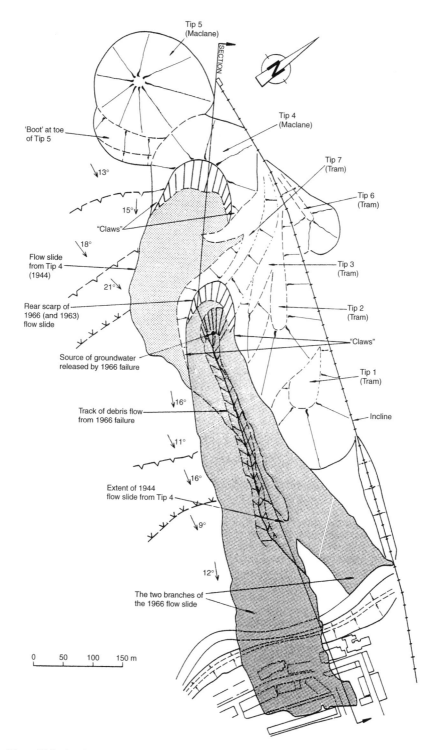

Figure 12.3 Plan of Merthyr Vale (Aberfan) colliery tip flow slides of 1944, 1963 and 1966. From Siddle *et al* (1996), with permission of The Geological Society

overall mine planning, and in the interim period when the mine is in full production, provision has to be made for dumping of topsoil and then overburden and interburden waste material. In modern-day mining, the amount of material removed is enormous, e.g. in the USA alone more than 1.5 billion m³ per year is moved by dragline; add to this figure that moved by shovels and it is clear that very large amounts of spoil are moved. Even in small open pit operations, such as in central Europe, over 7.4 million bcm of overburden will be moved to mine 1.8 Mt of coal (Figure 12.4). Certain methods of open pit mining ease the problem by back-filling the mining void as the mine progresses (see Section 10.3), and then restoring the topsoil once mining is completed. In many open pits, the overburden is tipped on to areas that will not be mined as near to the working operation as possible; long waste haulage with dump trucks is expensive and time consuming. Slope stability studies determine the size and shape of the spoil dump and the natural drainage is piped or diverted. Many mines leave the spoil dumps as a permanent feature, but modern-day restoration demands that the dumps be contoured, covered with topsoil, have adequate drainage and be revegetated with selected local plant species. Some open pit mine areas now serve as wildlife conservation areas. In areas where flooding from

rivers is a reality, a bund of overburden may be placed between the watercourse and the mine working area (Figure 12.5). Old dumps or tips where no restoration was carried out are now being retreated or removed, and some are being reworked for their discard coal content. This not only removes the dump altogether but can yield coal otherwise lost. Tip reclamation operations are well established in the USA, e.g. in southern Illinois, Indiana and western Kentucky and in the UK, e.g. in South Wales, where old tips (pre-Second World War) are recycled for coal. The percentage of coal is likely to be higher in these old tips, as coal preparation (if any) was less efficient and mining methods less precise.

Reject spoil material from coal preparation plants is disposed of by dumping or sent to landfill. Fine waste is usually disposed of as slurry in specially constructed lagoons or ponds, or else is dewatered and dumped with coarser spoil. The fine grained slurry or tailings are dumped in the lagoon or settling pond where the solids settle out; the clean water is then discharged or reused in the mine or coal preparation plant. Abandoned tailings ponds are potential sources of water pollution, as is the coarser coal waste. Rainfall and groundwater passing through spoil dumps which contain pyrite and other potential sources of contamination can generate acid mine drainage. This can be controlled by providing

Figure 12.4 Opencast mine in Bosnia-Herzegovina showing large scale overburden removal. Note small scale failures of the highwall caused by surface water action during periods of mining inactivity. Photograph by courtesy of Dargo Associates Ltd

Figure 12.5 Opencast mine in Maharastra State, India, showing protective bund of overburden to prevent flooding from nearby river. Photograph by courtesy of Dargo Associates Ltd

adequate drainage channels within and underneath spoil dumps. If the water is contaminated, it can be treated before discharge. One problem with old mine waste is that imprecise records on the location of old settling ponds can prevent modern treatment of the waste to avoid future water pollution.

Reclamation of open pit areas with back-filled spoil covered with topsoil must be accompanied by the regular testing of soil and water samples to ensure that the soil profile remains uncontaminated and suitable for revegetation. Some restored areas have failed to support revegetation because the soil has become acid-rich from waters percolating through the back-fill material. Modern mine reclamation has to assess the nature of the back-fill, its potential for contamination and the remedial measures necessary to ensure successful land reclamation.

Many mines now leave the area in better soil condition than before mining began, but it is a cost to the mining operation which has to be assessed prior to mining.

12.2.6 Spontaneous combustion

The propensity to spontaneous combustion is related to the rank, moisture content and size of the coal. In addition, mining and ventilation practices and geological conditions can also be contributory factors.

Oxygen is adsorbed on to the surface of the coal in an exothermic reaction, which is the start of oxidation (see Section 4.3.4). If the amount of free oxygen is small, the reaction is slow, with little rise in temperature, but where the quantity of oxygen passing over the coal is much larger, any heat produced will be dissipated, the temperature will not rise again and oxidation will proceed at a low level. Between these two conditions exists the situation where the quantity of oxygen is sufficient to promote oxidation but not sufficient to dissipate the heat. This increases the rate of oxidation and eventually ignition will occur. Therefore the coal's capacity to adsorb oxygen determines its propensity to spontaneous combustion. Another contributory factor to spontaneous combustion is the presence of pyrite in the coal. Pyrite can easily be oxidised, and on doing so, swells and exposes more coal surface to oxygen and therefore assist in the oxidation process. Solid coal presents less risk of spontaneous combustion, but when it is shattered by mining or broken by structural dislocation, the surface area of the coal is greatly increased.

Lower rank coals with high moisture content are most susceptible to spontaneous combustion, both in opencast

and underground mines. Stockpiles or cargoes of such coal are also vulnerable.

In underground mines, areas of coal such as in sidewalls or pillars are subject to oxidation. Regular monitoring of the mine atmosphere and exposed coal areas is essential to avoid the hazard of underground coal fires. Some mining areas have a history of coal fires, the most notorious being India and in particular the Jharia Coalfield, which contains the largest complex of surface and underground coal fires in the world. The coalfield contains 40 coal seams, of which 70% are 3.5 m or greater in thickness, and are high volatile bituminous coal of medium coking quality. Mining is by a mix of underground and opencast operations, and the first mine fire was reported in 1916 and has spread over the ensuing 85 years. At the present time, over 60 fires are still burning from 15 to 140 m below ground surface. They have produced uncontrolled subsidence, devastated land, ill health and death to the indigenous population. Throughout the 1990s up to 40 Mt of coal has been lost through fires and a further 1500 Mt isolated from further development (World Coal 1997). The big problem is how to eliminate the fires; this requires elimination of one or more of the components needed to sustain them, namely, fuel, oxygen and heat. Methods such as excavation (removes fuel), smothering (removes oxygen), quenching (lowers fuel temperature) are currently being applied. Of these, only excavation is the certain means to extinguish the fire; however, in the case of Jharia, this would require 500 million m³ of material from 60 open pit excavations, which apart from the prohibitive cost would take a very long time. This option is therefore not viable except on a case-by-case basis (Michalski and Gray 1997). Whilst Jharia is an extreme case, underground fires occur elsewhere, e.g. in underground brown coal mines in Turkey. All are hazardous, producing danger to life and loss of revenue.

In open pit mines, the problem is less easy to isolate, and fires start either at the coal face, coal stockpiles or in out-of-pit dumps where enough combustible material is present.

Coals can also catch fire at outcrop, either by spontaneous combustion or as a result of events such as forest fires; examples of such fires are shown in Figure 12.6. In the case of stockpiles, remedies are available: first, avoid having large ranges of coal particle size which allows a higher rate of oxidation; second, the stockpile can be compacted to exclude oxygen from circulating and therefore reduce the oxidation rate, and the stockpile can be sealed with a layer of clay or bitumen. It is important to prevent the practice of spraying water on stockpiles to ostensibly cool them down; this has the opposite effect of causing increases in temperature as the water reacts with the coal surfaces and can even start fires. The stockpile temperature should be regularly tested to ensure no 'hot spots' develop; this is also true for coals prone to spontaneous combustion during transport, especially on oceangoing vessels.

12.2.7 Dust suppression

In underground mines, dust control is maintained by good ventilation systems and water sprays at the coal face where coal is being cut. The quantity of air and the distance from the end of the air line to the coal face are critical. Because of leakage through and around working districts, more air must be forced into a mine than would otherwise be used for ventilation. Such leakage adds to the mine's operating costs, so efficient sealing of mine areas is necessary. Water spray systems using nonclogging water nozzles are fitted as standard on power shearers and continuous mining equipment. Coal travelling on conveyors is damped down to further reduce dust in the mine atmosphere. Cutting rock produces more dust than cutting coal, so in development areas where rock and coal have to be removed, face-mask respirators may be worn.

In open pit operations, the use of wider haul roads and larger haul trucks has put greater emphasis on road conditions and dust control. If not controlled, excessive dust can raise equipment operation and maintenance costs and shorten the life of vehicle components and systems. Heavy dust is also a sign that the haul road surface is degrading. Data collection on air quality is performed by using dust deposition gauges and operated throughout the life of the mine. Studies of dust levels can provide valuable information in the future planning of the mine. Water may appear the cheapest form of dust control but can contribute to road surface deterioration. Dust can be controlled by applications of diluted suppressant which can enhance water penetration into the road surface, lengthening the time it takes for the water to evaporate. A major factor is the nature of the climate; for example, in areas with a definite rainfall season, such as the monsoon season in India, dust is a problem for the dry period. In other areas, like Indonesia, a tropical rainfall climate of hot sun and frequent heavy showers produces a dust/mud cycle, both of which can inhibit traffickability within the mine.

Dust suppression is also necessary in coal preparation plants, particularly where the coal is only crushed and screened, and in coal loading facilities. Automatic coal loading is easier to control for dust emissions, but wagon

(a)   (b)

Figure 12.6 (**a**) Coal burning as a result of ignition by forest fires in Kalimantan, Indonesia. Photograph by LPT, with permission of Dargo Associates Ltd; (**b**) Spontaneous combustion of coal in the highwall of an Indian mine. Photograph by courtesy of Dargo Associates Ltd

loading in rail yards can be a problem in dry windy conditions. The usual method is to damp down the coal, but this can create a problem if the coal is prone to spontaneous combustion.

Environmental legislation, particularly in areas with indigenous populations, means that mining companies are under pressure to minimise dust pollution, and most mines use dust suppressants to combat it.

12.2.8 Subsidence

Subsidence is a consequence of underground mining. It may be localised or extend over large areas, and it may be immediate or delayed for many years. When a cavity is created underground, the stress field in the surrounding strata is disturbed. These stress changes produce deformation and displacement of the strata, the scale of which is dependent upon the magnitude of the stress and the cavity dimensions. Over a period of time, mine roof and sidewall supports deteriorate and can result in instability. Roof collapse induces strata above to move into the void; these movements emerge at the ground surface and appear as a depression or a series

of depressions. In mines, when the void left by coal extraction is of larger size, the collapsed strata fall into the excavation; this process continues until a height of three to six times the mined seam thickness is reached. When the cavity is filled with broken rock, the debris offers some support to the adjacent strata. As these strata settle or sag, bed separation may occur because of the tendency for lower strata to subside more than the higher beds (Figure 12.7). Overall, as the strata settle or subside, they sag rather than break and produce a dish- or trough-shaped depression on the ground surface (Figure 12.8), known as trough or sag subsidence. The *critical width* of the workings is the minimum width that needs to be mined before the maximum possible subsidence is observed at the centre of the trough. If the mined area is less than critical, it is termed *subcritical*, and the amount of subsidence that occurs will be less than the maximum. If a *supercritical* width is extracted, the central portion of the trough will attain maximum subsidence, and a flat-bottomed depression will be produced. Such flat-bottomed depressions are a feature in many of the coalfields in China, where in some circumstances they have been flooded and used

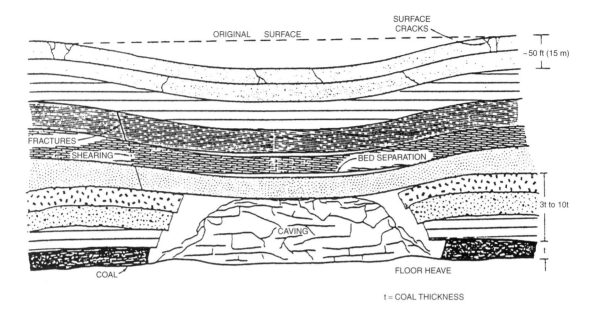

Figure 12.7   Strata disturbance and subsidence caused by mining. From Hartman (1992), reproduced with permission of SME, www.smenet.org

for fish farming (Figure 12.9). Trough subsidence may be due to mining at any depth, the overall movements of the ground around the mine cavity are shown in Figure 12.10. The direction of motion can be seen to not only be vertically downward but also horizontal and, in some locations, upward (Singh 1992).

Old shallow mines have produced small surface subsidence features but modern longwall mining, the collapse of pillars in room and pillar mines and the large network of underground roadways can produce widespread subsidence, and all countries with a long history of underground mining have experienced subsidence problems. Subsidence has caused the collapse and distortion of buildings, roads and railways and is an ongoing legacy in areas where coal mining has long ceased.

Subsidence can be reduced substantially if large pillars of coal are left in place, or if rock waste from other mining districts is packed into the mined void. Where full coal extraction takes place, as is the case in longwall mining, the maximum subsidence is usually around 80% of the thickness of coal removed (Ward 1984).

In both underground and opencast mines subsidence can be a result of the lowering of the water table as part of a mine dewatering scheme. This has a small effect on agricultural land but can seriously affect buildings and conduits close to the mine. Usually such subsidence is caused by the compaction of sands and gravels in superficial deposits such as river or glacial deposits

overlying the coal-bearing strata. In central Europe, the dewatering of large areas in order to mine shallow brown coal deposits has produced such subsidence. The land surface will not return to its original level once mining has ceased, even though the water table will return to a higher level.

As well as subsidence, a legacy of old mining areas is the numerous old shafts used to access the coal. Many of these are not shown on modern plans and their locations are largely now unknown. For example, in the UK this has become a problem in the old coalfield areas, where a number of shafts have been built over and have subsequently opened up as the capping material has deteriorated over time, exacerbated by the weight of material above and/or by the action of groundwater erosion. Figure 12.11 illustrates such a shaft which has opened up in a housing complex in Scotland, UK.

Mining subsidence and deterioration in the capping of old mine shafts can cause the problem of gas seepage from old workings, particularly methane, which is lighter than air. As shown in Figure 12.12, gas from old mine workings is able to migrate upwards via fractures caused by rock collapse into voids left between coal pillars. Gas will also migrate upwards in old shafts and, if these are inadequately sealed, will reach the surface and escape. In Figure 12.12, buildings at the surface will be vulnerable to gas invasion. House A is protected by an underlying layer of clay which prevents

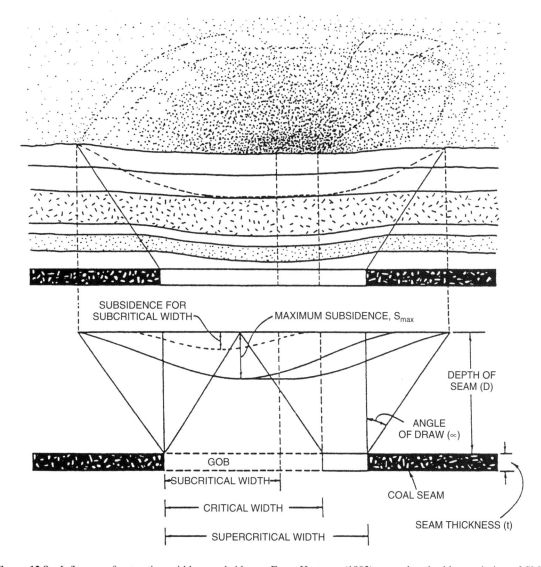

Figure 12.8 Influence of extraction width on subsidence. From Hartman (1992), reproduced with permission of SME www.smenet.org

gas from reaching the surface, whereas house B has no protection. Fatalities have occurred from gas seepage from old mine workings, and as a result, careful investigations should now be made before buildings are erected in old mining areas.

12.3 COAL USE

Coal is a versatile fuel, and has long been used for heating, industrial processes and in power generation. Internationally traded coals have predominantly been for use in coke making in the iron and steel industry, and in the electricity generation sector. This is borne out by the fact that coal provides around 23% of global primary energy needs and generates about 38% of the world's electricity, generating some 4800 TWh (WCI 2001). In addition, 17% (600 Mt) of the world's total black coal production is currently utilised by the steel industry, 70% of which is dependent on coal. Recent years have seen an increase in international traded coal, and in particular, the trading of steam coals for electricity generation. In 1985, 74 Mt of steam coal

Figure 12.9 Subsided land overlying underground workings, People's Republic of China, now flooded and used for fish farming. Photograph by courtesy of Dargo Associates Ltd

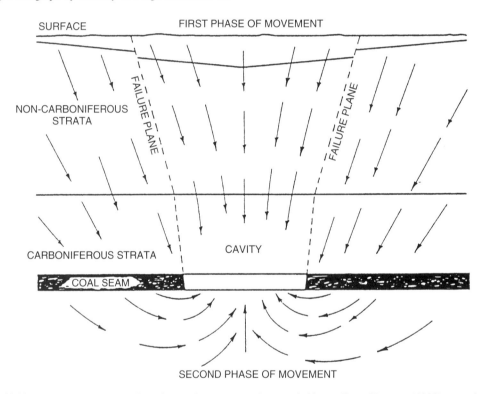

Figure 12.10 Schematic representation of ground movements due to subsidence. From Hartman (1992), reproduced with permission of SME, www.smenet.org

and 112 Mt of coking coal were traded, whereas in 2000, 381 Mt of steam coal and 192 Mt of coking coal were traded internationally (WCI 2001). By the year 2010, coal-based electricity generation is expected to reach 7400–8000 TWh (WEC 1998). Coal consumption in the iron and steel industry is moving towards the use of lower quality coking coal (soft and semi-soft) in blends with high quality coking coal. The heating market, including other areas of industry, domestic and miscellaneous consumption, will continue to contract.

Figure 12.11 Collapse of capping material over an old concealed shaft in a modern urban area, UK. Photograph by courtesy of IPR/25-36c British Geological Survey. © NERC. All rights reserved

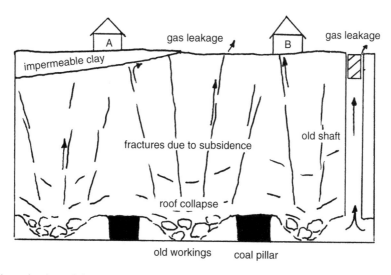

Figure 12.12 Schematic view of the upward migration of mine gas by means of fractures due to subsidence and from an inadequately sealed shaft. House A is protected by underlying impermeable clay, whereas house B is vulnerable to gas invasion. Reproduced by permission of Dargo Associates Ltd

12.3.1 Electricity generation

Although this volume is essentially concerned with coal and its character, its uses cannot be ignored, especially as the market forces of supply and demand will determine where and how much coal will be mined in the future. Because electricity generation is the biggest single user of coal, its environmental effects and technologies to control pollution are briefly discussed here.

Electricity generation is singled out as one of the largest causes of pollution of the atmosphere. The rapid growth of the demand for electricity has led to large increases in production and therefore large increases in emissions, which in turn has brought attendant environmental problems (Figure 12.13).

Nowadays boilers are designed to burn a range of coals from lignite to anthracite. The majority of stations burn coals in the middle of this range; these are selected according to their ash, sulphur, moisture and volatile matter contents, together with their heating value (calorific value), grindability and ash fusion temperature. The performance of different coals in terms of their specifications, pre-combustion, combustion and post-combustion performances is well documented elsewhere. In the context of this account, it is the result of burning coal in power plants and the direct contribution this makes to the environment in terms of waste products, both solid and gaseous, that is of concern here. Environmental legislation governs the limits of emissions and waste products from power plants and coals that may cause problems are not likely to be utilised in the future. This can have serious repercussions on coal mining. For example, the low sulphur requirements for coals to be used for power generation in the USA have meant a decline in the mining of high sulphur coals in the traditional coalfields in eastern USA, in Illinois, Indiana and western Kentucky, and the enormous expansion in mining low sulphur coal in the western coalfields of Wyoming and Colorado.

As a guide, Table 12.3 gives the coal specifications which are normally used in coal-fired boilers, although coals outside the given ranges can be burned.

In simple terms, as shown in Figure 12.14, coal is brought from the mine to the power plant, the coal having a quality determined and agreed to by both parties. It is then stockpiled and, when required, fed through the mills and then into the combustion chamber. It is at this stage that waste products are generated. Exhaust gases contain particulates, sulphur and nitrogen oxides and volatile organic compounds. Fly ash removed from the exhaust gases can make up 60–85% of the coal ash residue in pulverised coal boilers. Bottom ash includes slag and coarse, heavier particles than fly ash. The volume of solid waste may be substantially higher if environmental measures such as flue gas desulphurisation (FGD) are adopted

Figure 12.13 Modern coal-fired power station, Inner Mongolia, People's Republic of China. Photograph courtesy of Dargo Associates Ltd. Reproduced by permission of World Coal, Palladian Publications Ltd

Table 12.3 Normal range coal specifications for pf-fired boilers (BP *Coal Handbook* 1987) with permission

Parameter	Range	Comment
Total moisture	Max 15% (ar)	If high, creates handling problems. The limits are higher for lignites and low rank coals. Reduces net CV
Ash	Max 20% (ad)	If high, creates fly ash problem. Reduces net CV
Volatile matter	Min 20–25% (d.a.f.)	For conventional pf burners
Calorific value (CV)	High	Almost any CV fuel can be used; the higher the better
Sulphur	Max 0.8–1.0% (ad)	Maximum value dependent on local emission regulations
Nitrogen	Max 1.5–2.0% (d.a.f.)	Various limits apply in some countries because of NO_x emissions
Chlorine	Max 0.2–0.3% (ad)	Causes ash fouling problems in boiler
Hardgrove grindability index (HGI)	Min 45–50	Lower HGI values require larger grinding capacity and more energy
Ash fusion temperatures (AFT)	Various	Dry bottom boiler – IDT $>1200\,^{\circ}$C. Wet bottom boiler – FT $<1300\,^{\circ}$C
Maximum size	Max 40–50 mm	Dependent on capacity of grinding equipment
Fines content (−3 mm)	Max 25–30%	High fines can increase moisture content and create handling problems

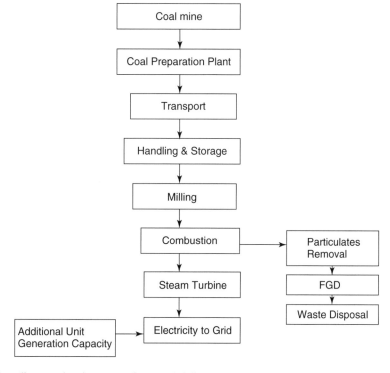

Figure 12.14 Flow diagram showing stages from coal delivery to electricity generation in a modern power station

and the residues are not reused in other industries. Steam turbines also require large quantities of water for cooling, including steam condensation. Water is also required for auxiliary station equipment, ash handling and FGD systems. Water contamination arises from demineralisation, lubricating and auxiliary fuel oils and chlorine, together with any other chemicals used to control the water quality in the cooling system, which also serves to increase the water temperature.

12.3.1.1 Emissions

In the industrialised countries, although the problem of emissions is of serious concern, the ability and desire to diminish the harmful elements in emissions is not consistent. Reasons are primarily financial, as control of harmful emissions requires the replacement or modification of existing equipment and/or the addition of new technology.

From an environmental point of view, attention is focused on emissions of particulates less than 10 microns (μm) in size, sulphur dioxide (SO_2), on nitrous oxides (NO_x), and on fly ash. In addition, carbon dioxide (CO_2) and dioxins have also attracted a great deal of attention in deciding on the effects of these pollutants on the atmosphere and global ecology.

To estimate the amount of SO_2 produced daily from the power station stack, it is necessary to know the calorific value, hydrogen and moisture content of the coal and the station heat rate. For example, a 500 Mwe plant using coal with 2.5% sulphur, 16% ash, and a calorific value of 30 MJ/kg (12 898 Btu/lb) will emit each day:

200 t SO_2, 70 t NO_2, 500 t fly ash, 500 t solid

waste, and have 17 gigawatt-hours (GWh) of

thermal discharge (World Bank Group 1998).

SO_2 is one of the principal gaseous pollutants; it can be a hazard to human health and damage both the natural and man-made environment. Total global emissions of SO_2 from the combustion of coal, oil and oil-derived fuels, refining and smelting amounts to 140 Mtpa. To control the emissions of SO_2, it may be possible to use a fuel that has a lower sulphur content, or has a sulphur content that can be removed before use. For example, the inorganic sulphur fraction in coal is usually in the form of iron pyrite (FeS_2), and this can be removed in a coal preparation plant, and has been effective in reducing sulphur content in traded coals, e.g. in South Africa. However, the organic sulphur and sulphate fractions will remain within the coal. The alternative is to remove the sulphur during use and the most efficient means of controlling SO_2 emissions is to remove the SO_2 from the flue gases before they are released to the atmosphere (see Section 12.3.1.2).

NO_x emissions are produced by the reaction of the nitrogenous compounds in the coal with oxygen. It is considered that nitrogen is released during devolatilisation and enters the gas phase as HCN or NH_3, where it reacts with air to form NO_x in a complex series of chemical reactions. High temperatures and rapid heating rates maximise the yield of the volatile nitrogen species and the presence of free oxygen favours the formation of NO_x. Among the NO_x emissions from power plants is nitrous oxide (N_2O); this gas has a significant effect on the atmosphere, it is a strong absorber of infrared radiation and is considered as a major contributor (20%) to ozone depletion.

Carbon dioxide (CO_2) is emitted as a product of coal combustion, and relates to the carbon content of the coal burned at the time. Old and inefficient plant allow higher CO_2 emissions than do modern installations.

Dioxins is the general name given to a group of some 210 species consisting of 75 polychlorinated dibenzo-para-dioxins (PCDDs) and 135 polychlorinated dibenzofurans (PCDFs). The majority of these species are considered to pose little threat to health at the levels generally found. A small group of dioxins (17) are of great concern because of their toxicity/carcinogenicity, and these are formed as unwanted by-products in some industrial processes, and combustion processes such as waste incineration. Dioxins are present in all environmental media, and there are many natural sources of dioxins, such as forest fires. Since the incomplete combustion of any organic material can result in the formation of trace hydrocarbons in flue gases, and coal contains chlorine, the combustion of coal has been implicated as a significant contributor to the release of dioxins to the atmosphere. There is little published data on dioxin emissions from coal, and current testing in the UK has so far only recorded low concentrations of dioxins in flue gases from coal-fired plant. The highest dioxin emissions were from domestic combustion appliances (Dorrington *et al* 1995).

The concentration of trace elements in ash is dependent upon particle size. Increasing concentrations are correlated with decreasing particle size. In coals from Indiana, USA, it has been demonstrated that concentrations of lead, thallium, antimony, cadmium, selenium, arsenic, zinc, nickel, chromium and sulphur were markedly increased in the size range 0.65 μm to 74 μm. In Australia, a threefold increase in concentrations of gallium, germanium, mercury and lead have been observed between the coarse ($>50 \mu$m) and fine

($<2 \mu m$) fly ash fractions. Table 12.4 shows the distribution of elements between the main coal residues in a power plant, i.e. bottom ash, fly ash and flue gas, taken from an example in the USA. The trace elements present in high percentages in the flue gas fraction can be seen to be chlorine, chromium, mercury, nickel, selenium and sulphur. Particulate emissions will contaminate surrounding soil areas and, if inhaled, have serious health effects on the local population.

To control the emission of the fine particulate fraction and its undesirable trace element content, emission level limits have been imposed by most countries. Current permitted particulate emission levels are to be substantially reduced during the current decade. For example, where permitted particulate emissions levels are now 50 mg/Nm^3 (normal cubic metre), this is planned to be reduced to 10 mg/Nm^3 after the year 2000; this means that the electrostatic precipitators will be required to collect up to 99% of particulates from flue gas.

Table 12.4 Distribution of elements among bottom ash, fly ash and flue gas. From Valkovic (1983) with permission of CRC Press LLC

Element	Bottom ash (22.2%)	Fly ash (77.1%)	Flue gas (0.7%)
Aluminium	20.5	78.8	0.7
Antimony	2.7	93.4	3.9
Arsenic	0.8	99.1	0.05
Barium	16.0	83.9	0.09
Beryllium	16.9	81.0	2.0
Boron	12.1	83.2	4.7
Cadmium	15.7	80.5	3.8
Calcium	18.5	80.7	0.8
Chlorine	16.0	3.8	80.2
Chromium	13.9	73.7	12.4
Cobalt	15.6	82.9	1.5
Copper	12.7	86.5	0.8
Fluorine	1.1	91.3	7.6
Iron	27.9	71.3	0.8
Lead	10.3	82.2	7.5
Magnesium	17.2	82.0	0.8
Manganese	17.3	81.5	1.2
Mercury	2.1	0	97.9
Molybdenum	12.8	77.8	9.4
Nickel	13.6	68.2	18.2
Selenium	1.4	60.9	27.7
Silver	3.2	95.5	1.3
Sulphur	3.4	8.8	87.8
Titanium	21.1	78.3	0.6
Uranium	18.0	80.6	1.5
Vanadium	15.3	82.3	2.4
Zinc	29.4	68.0	2.6

Power plants burning high ash coals, as is the case with most Gondwana coals, are faced with the problem of large amounts of fly ash disposal. Large plants using coal with ash contents of 50+% can produce over 1 Mt of fly ash per year. If the fly ash is suitable for industrial use then it is simply removed from site. If, however, the amount of fly ash produced far exceeds any industrial requirement, then the fly ash has to be disposed of. In the case of a mine mouth power station, and if the mine is an open pit operation, fly ash is returned and used to fill the mined-out void. Stockpiling of fly ash has proved a problem particularly in dry, windy conditions, e.g. during the dry season in India.

Where fly ash is considered for use in other industries, it has been classified based on the source coal and specified major element oxide contents. ASTM has differentiated two types of fly ash (Class F and C). Class F fly ashes are derived from anthracite and bituminous coals and should contain a minimum of 70% of $SiO_2 + Al_2O_3 + Fe_2O_3$, and Class C fly ashes are derived from lignite and subbituminous coals and should contain a minimum of 50% of $SiO_2 + Al_2O_3 + Fe_2O_3$. Both classes of fly ashes should possess their own distinctive sets of chemical, physical and engineering properties. This classification is used as the basis for selecting fly ashes as admixture in cement and concrete.

12.3.1.2 Flue gas desulphurisation (FGD)

The control of SO$_2$ emissions has centred on FGD technologies (DTI Report 2000). These are now widely used to control the emissions of sulphur dioxide (SO$_2$) and sulphur trioxide (SO$_3$). A variety of FGD processes is available; most use an alkali sorbent to recover the acidic sulphur compounds from the flue gas. The most commonly used alkaline materials are limestone – calcium carbonate, quick lime – calcium oxide, and hydrated lime – calcium hydroxide. Limestone is an abundant and relatively cheap material and both quick lime and hydrated lime are produced from limestone by heating. Other alkalis are sometimes used; these include sodium carbonate, magnesium carbonate and ammonia. The alkali used reacts with SO$_2$ in the flue gas to produce a mixture of sulphite and sulphate salts, and this reaction can take place either in bulk solution ('wet' FGD processes) or at the wetted surface of the solid alkali ('dry' and 'semi-dry' FGD processes). Selection of the FGD process is made on economic grounds, i.e. the process selected will be the one with the lowest overall through-life cost. Technical considerations include the degree of desulphurisation the process can achieve, and the flexibility of the process.

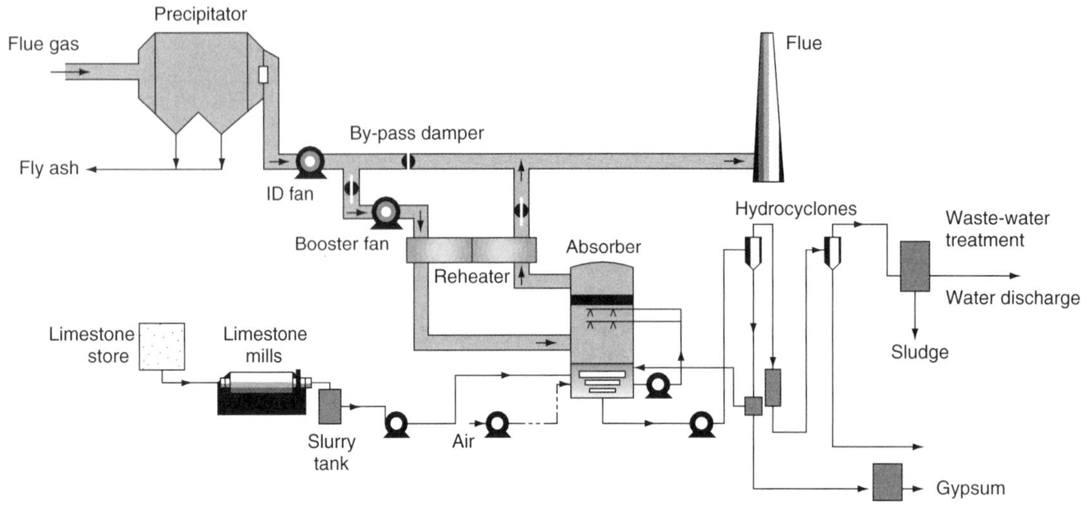

Figure 12.15   Schematic flow diagram of a limestone/gypsum FGD process, reproduced by permission of DTI Cleaner Coal Programme

The most common FGD process now being installed worldwide is the limestone–gypsum wet scrubbing process, and has evolved over 30 years. Today, a plant would be designed to achieve a high quality gypsum product, which can be used in wallboard manufacturing. There are variants in equipment arrangement, absorber type and reheat methods, according to the client's requirements. Figure 12.15 shows one of the most common limestone–gypsum plant layouts. The FGD plant is located downstream of the electrostatic precipitator so that most of the fly ash from combustion is removed before the gas reaches the FGD plant. In a coal-fired plant, fly ash removal would be 99.5%. The gas is then scrubbed with the recirculating limestone slurry to remove the required amount of SO_2. FGD plant manufacturers claim that over 95% of SO_2 can be removed with the absorber. This process also removes almost all of any hydrogen chloride (HCl) in the flue gas.

The calcium carbonate ($CaCO_3$) from the limestone reacts with the SO_2 and oxygen (O_2 from air) to produce gypsum (hydrated calcium sulphate $CaSO_4.2H_2O$), which precipitates from solution in the sump. HCl is also dissolved in the water and neutralised to produce calcium chloride solution. The gypsum slurry is extracted from the absorber sump, and treated for storage and removal to industrial users. Fresh limestone is then pumped into the absorber sump to maintain the required pH. The remaining gas is reheated and then exhausted to the stack. This process will usually offer the lowest through-life cost option for large inland coal-fired power plants with medium to high sulphur fuel, a high load factor and a long residual life.

Other wet FGD processes include seawater washing, ammonia scrubbing, and the Wellman–Lord process (using aqueous sodium sulphate solution). Semi-dry processes are the circulating fluidised bed, spray dry and duct spray dry, all producing dry powdered mixtures of calcium compounds. Dry processes inject hydrated lime or sodium bicarbonate into the furnace cavity of the boiler and absorb SO_2. The spent sorbent is extracted with the fly ash as a mixture of ash and calcium or sodium compounds.

12.3.1.3 Other emission controls

The reduction of NO_x emissions requires the fitting of NO_x reduction equipment to both existing plant (retrofitting) and to new plant. Conventional pulverised fuel (pf) burners are designed to achieve rapid intimate mixing of the fuel with the combustion air. Since the fuel devolatilises in high temperatures and air-rich conditions, the level of NO_x is high. To counteract this, low NO_x burners are used based on the physical separation of three air streams to the burner, the addition of each air stream helping to reduce NO_x production. The use of low NO_x burners can cause slagging problems in the furnace and can result in a loss of efficiency in the plant. Current research is seeking to overcome such problems and make NO_x reduction cost effective as well as environmentally desirable.

A post-combustion NO_x control method is known as selective catalytic reduction (SCR), in which ammonia is mixed with the flue gas and in the presence of a catalyst reacts with NO and NO_2 in the gas to form molecular nitrogen and water. SCR systems can be costly and difficult to retrofit, due to their physical size, but are widely installed in Japan and Germany.

Several combined SO_2 and NO_x removal systems have been developed; one such system is the SNOX process developed in Denmark. The flue gas is reheated, and then undergoes SCR, the flue gas is then further heated and a second catalytic reactor oxidises SO_2 to SO_3. The gas is then cooled to condense out the SO_3 as sulphuric acid. The condenser uses glass tubes to prevent excessive acid corrosion. SNOX units have been built in Denmark, Italy and the USA.

Reduction in CO_2 emissions at pf plants has mainly been by chemical absorption processes, and current research is seeking to improve on these methods. CO_2 emissions have been linked to reduced efficiency levels in coal-fired plants, particularly in non OECD countries. This is due to a number of factors – poor quality coal, small and ageing units, poor maintenance and obsolete technologies.

12.3.1.4 Fluidised bed combustion

In the electricity generating industry, fluidised bed combustion (FBC) in its various forms offers a technology that can be designed to burn a variety of fuels efficiently and in an environmentally friendly manner. There are two main processes in use, namely bubbling FBC (BFBC) and circulating FBC (CFBC), both of which can be either atmospheric or pressurised in operation.

Briefly, the BFBC method is when a bed of packed small particles is subjected to an upward gas flow; the bed remains static but the pressure drop across it increases in proportion to the increasing gas flow rate. When the pressure drop across the bed particles equals the weight per unit area of the bed, the bed becomes suspended. The bed is then considered to be at minimum fluidisation. Increases in the gas flow above the minimum will produce bubbles; the upwards and sideways coalescing movements of the bubbles provide intense agitation and mixing of the bed particles. This results in the bed particles transferring heat at very high rates from burning fuel to cooler surroundings. Fuel can be fed into and burned in the bubbling bed. In order to burn coal efficiently, and where limestone is used to successfully retain SO_2, the bed particles need to be controlled in the temperature range of 800–900 °C. Over

the last 30 years, BFBC technology has been shown to be well suited to the utilisation of high moisture, high ash coals as well as low volatile anthracite. In addition to the primary air source in the combustor, secondary air is introduced at several levels above the bed; this provides a balance of the combustion air to reduce NO_x levels.

Circulating FBCs are a development of the BFBC, whereby the velocity of the circulating air is increased, resulting in the particles being carried upwards away from the bed surface, and the distinctive bubbling bed disappears. The combustion chamber is then filled with a turbulent cloud of particles that no longer remain in close contact with each other. The burning particles are recovered from the air flow and fed back into the lower part of the combustion chamber. A circulating fluidised bed can sustain combustion in a similar manner to a bubbling bed, and the turbulent contact between the coal particles present and the bed solids stabilises the overall temperature. Again additional combustion air is introduced at higher levels to reduce NO_x levels.

Many CFBC units proved capable of achieving lower levels of the primary pollutants, NO_x, SO_2, CO_2 and particulates. Sorbent is added to the system to control SO_2 emissions, NO_x levels are minimised by careful bed temperature control, and particulate control systems are installed. Pressurised FBC has the advantage that the hot combustion gases leave the combustor under pressure, which if maintained, and the gases cleaned, can be fed directly into a gas turbine. Several thousand BFBC plants are in operation, predominantly in China. CFBC plants are dominant in Asia, again principally in China, with 52% of installed capacity; North America has 26% and Europe has 22%. PFBC is now expanding into the USA and Europe, with the highest level of current activity in Japan (DTI Report 2000).

12.3.2 Other major users

12.3.2.1 Iron and steel production

The potential source of pollution in the iron and steel making process relating to coal is in the production of coke. Other emissions are part of the steel making process and are not considered here.

Coal is heated in an oxygen-free environment until the bulk of the volatile constituents have been driven off. The solid residue is known as coke and its principal use is to provide heat energy and to act as a reducing agent for iron ore in the blast furnace. Coke has to be a strong material, able to withstand handling and be capable of supporting the overlying weight of coke as it moves down through the blast furnace. Coke can be produced from a single coal or a blend of selected coals. Only coals with a specific range of rank and type are capable of forming coke, and particular properties of the coal decide the nature of the coke produced (see Section 4.3.2).

Coke can be manufactured within the iron and steel plant on a large scale or produced on a small scale and transported to the plant. Traditionally coke was produced in beehive ovens by combusting covered piles of coal until all was carbonised, usually taking 3–4 days. This method is still widespread in China (Figure 12.16), where coke making is still a cottage industry. This method has been replaced by carbonising a thin vertical layer of coal in less than a day; the coke oven is a standard part of any iron and steel plant. Both methods have resulted in the venting of gas by-products to the atmosphere. This has produced high levels of local atmospheric pollution as well as raising regional CO_2 emissions.

Today, OECD countries produce about 61% of the world's steel, but emit only 34% of the CO_2 from global steel production (WCI 2000). This has been achieved by reducing fuel consumption of the blast furnace, making more efficient use of exhaust heat, increasing on-site electricity generation from waste gases, using expansion turbines, the partial substitution of coke through the injection of pulverised fuel and heavy fuel oil, and by introducing larger blast furnace units to gain from economies of scale (WCI 2000).

In the UK, in Yorkshire, the Monckton Cokeworks produces 535 t of coke and 276 000 m^3 of coke oven gas per day from 790 t of coal. A new 11 MWe combined heat and power plant (CHP) has been developed as part of an environmental clean-up and redevelopment programme. Of the coal gas produced from the coke ovens, 40% is reused to heat the coke ovens, and 60% is diverted to the new CHP plant. The CHP plant utilises the coke oven waste gas to produce 31 t/h of superheated steam; this in turn produces 11 MW, with 1.4 MW required for on-site demand and the remainder sold via the local grid. This facility is able to generate more added value from the surplus high calorific value gas released during the coke making process, with the added benefit of environmental improvement.

In Europe these methods have been implemented by closing down old style steel plants and replacing them with basic oxygen furnace (BOF) and electric arc furnace (EAF) plants. The energy consumption of basic oxygen steel making is very low compared to other production methods. In addition, it is possible to transform this method into a net supplier of energy by utilising the waste converter gases. The energy input and associated

Figure 12.16   Local coke manufacture in small 'ovens', Guizhou Province, People's Republic of China. Photograph courtesy of Dargo Associates Ltd

level of CO_2 emissions per tonne of steel produced by the international steel making industry could be reduced by over 25% if the status of the production technology used in Europe and the USA were introduced on a global scale. Currently the CO_2 emissions per tonne of steel are three times higher in China ($3.9\,t$ CO_2) compared to western Europe ($1.3\,t$ CO_2). Apart from utilising new technology, a switch to iron ores of higher quality would lower the consumption of coal and reducing agents in the blast furnaces. The main difficulty in achieving a major reduction in CO_2 emissions is a financial one; the capital expenditure necessary to achieve such reductions in the CIS and China is estimated at $20 billion and $42 billion respectively (WCI 2000).

12.3.2.2 Industrial use

Although electricity generation and iron and steel production make up the bulk of the use of coal by the industrial sector, coal is used in a number of industries for heating. The principal effect of coal on the environment is the venting of waste gases to the atmosphere. The share of coal's contribution to this has been reduced due to the fact that modern industry has made substantial reductions in the use of coal for conventional heating, having replaced it with gas or oil. However, industries such as the manufacture of cement still utilise significant quantities of coal. Cleaning of flue gas and reducing the particulate emissions are both contributing to the improvement of air quality.

12.3.2.3 Domestic use

Coal as a household fuel has almost disappeared in most well developed countries. Strict regulations on air quality in urban areas has led to the replacement of coal by gas and oil heating. The thick smogs of large cities are now a thing of the past, although photochemical smog produced by the internal combustion engine is still a reality.

In less developed countries, domestic heating using coal is still prevalent. In the CIS, China and eastern Europe coal is plentiful, oil and gas are expensive or not available, so atmospheric pollution can still reach high levels. Improvement in industrial use and the gradual replacement of coal for heating will reduce the problem, but this is likely to be a long term prospect.

12.3.3 Coal transportation

The transportation of coal is by road, rail and conveyor on land, and by barge and oceangoing vessels on water.

The effects on the environment are usually minimal, but can cause local problems.

- Road transport: coal is moved from the mine to the customer by lorry fleets. This means using public roads and this can cause problems such as wear and tear, traffic congestion and dust from coal loads. In countries such as China and India, where villages alongside main roads are numerous, coal lorries do cause degradation of roads and village streets. This, coupled with poor maintenance, produces bad road conditions and slow delivery schedules. Roads built specifically for the purpose of transporting coal do not impinge on the local transport system, e.g. in East Kalimantan, Indonesia, private coal roads transport coal to the loading areas on the major rivers or ports.
- Rail transport: the overland transport of large shipments of coal by rail is the established means throughout the world. Rail transport has little effect environmentally other than dust and noise at the loading/unloading areas. Where coal is loaded/unloaded automatically, such effects are minimised.
- Conveyor: overland conveyors are used to transport coal from the mine to the stockyard. Conveyors are usually covered and have no adverse effect on the environment.
- Water transport: coal transported by barge or oceangoing vessel has only a dust problem on loading/unloading, and some coals have the propensity for spontaneous combustion (see Section 12.2.6).

12.4 HEALTH

Worldwide coal mining remains a growth industry, and coal mining, particularly underground mining, is still perceived to be a dangerous occupation. A great amount of time and money has been invested in mining health and safety research, and the industry is much safer as a result.

Most of the problems associated with mining are common to all countries, i.e. strata collapse, fires and explosions, dust, fumes and heat, noise and water. Improvements in mining practices and in the development of equipment with high safety standards have helped to drastically reduce accidents in both underground and open pit mines. Improved ventilation and soundproofing have made underground conditions less hazardous and a good understanding of the groundwater conditions can prevent unexpected water inflows in mines. Apart from accidents, coal mining, and in particular underground mining, has historically been associated with lung diseases caused by the breathing in of coal and stone dust over a number of years. Pneumoconiosis and silicosis were the legacies of the coal miner, together with newer complaints such as vibration white finger.

Coal miners working in deep mines with high virgin strata temperatures, high use of machinery, intake air passing over machinery and standing and sprayed water can be affected by the heat and humidity. The human thermo-regulatory system tries to regulate the body temperature at 37 °C, but when working in hot and humid conditions this control is not maintained; the body temperature starts to rise and produces various physiological effects such as heat rash, fainting, heat exhaustion, cramps culminating in heat stroke when the body temperature exceeds 41 °C (Leeming and Fifoot 2001). To combat these effects, it is essential to maintain good air quality to the working areas and to prevent unwanted heat being picked up by the air stream. A well maintained ventilation system is also important to dispel diesel fumes where such equipment is in use.

In open pit mines air quality is maintained by dust control on in-pit roadways and protection is worn to exclude noise. Local populations do have problems with respiratory complaints such as asthma which they attribute to mining, and these are often used as claims for compensation.

One hazard which is now receiving greater attention is the presence or absence of radon in mine waters and mine atmospheres. Whilst harmless externally, radon gas, if breathed in on dust and moisture particles, remains in the respiratory system. The major health hazard from radon is thought to be an increased risk of lung cancer. Radon gas is soluble in water and it may be carried for great distances. When such groundwater discharges into mine workings, there is a pressure release of gas into the mine atmosphere. The occurrence of radon in coal mines is now closely monitored and hydrogeological studies are essential to anticipate whether any radon-rich water is likely to inflow into a mine.

Also associated with coal mining is the release of naturally occurring toxic organic compounds into the environment. In Yugoslavia, researchers believe that a relationship may exist between organic compounds leached by groundwater from shallow lignite deposits and a disease known as Balkan endemic nephropathy. This disease, recognised since 1956, is a progressive kidney disease that leads to death. The disease occurs in villages situated on alluvial deposits overlying Pliocene lignites. The lignites contain large amounts of organic compounds, some of which may be water soluble. Current research is studying the drinking water sources of each village, the hydrological regime and the water and lignite chemistry, to determine whether there is a definite link between the disease and the water supply (Finkelman *et al* 1991).

12.5 ENVIRONMENTAL REGULATIONS

12.5.1 Introduction

The effects of mining and utilising coal on the environment are closely monitored and regulated. Regulations and codes of practice are implemented at national, regional and at local government levels in the majority of industrialised countries.

In recent years, prominence has been given to the harmful effects of 'greenhouse gases' on the earth's atmosphere. These gases comprise carbon dioxide, methane, nitrous oxide, hydrofluorocarbons, perfluorocarbons and sulphur hexafluoride. In addition, the emission of sulphur dioxide and particulate matter has given rise to concern.

Emission regulations are many and varied all over the world; some countries do not have national emission standards, although emission limits may be issued as guidelines by health councils, e.g. Australia. In some countries, the responsibility for emission standards is divided between the State Ministry and local government, e.g. Italy, South Korea. Others have emission standards which are the sole responsibility of central government, as in the case of the People's Republic of China, France, Germany and the UK. The USA involves both federal and state governments in the regulation of air pollution. Most industrialised countries follow one of these models for air quality legislation. (See relevant literature for details of individual country emission limits, e.g. McConville 1997.)

Emissions into the atmosphere may traverse individual country boundaries, and because of the international nature of air pollution have led to a number of initiatives to control transboundary pollution. Agreements between countries have been made, e.g. between the USA and Canada, the United Nations Economic Commission for Europe Conventions (UNECE), the European Community (EC) environmental legislation, and the Kyoto Protocol. In addition, the World Bank environmental guidelines are regularly implemented when projects seek international finance.

12.5.2 UNECE conventions

The UNECE Convention of Long Range Trans-Boundary Air Pollution (LRTAP) was signed in 1979 by

Table 12.5 UNECE (1994) SO$_2$ emission standards for new coal-fired plants. From McConville (1997), with permission from IEA Coal Research

New plant type	Plant size (MWt)	Emission standard[a]
Combustion plants	50–100	2000 mg/m^3
Combustion plants	100–500	2400 − (4 × P) mg/m^3 (where P = plant size in MWt)
Combustion plants	>500	400 mg/m^3
Combustion plants; domestic high sulphur coal	100–167	800 mg/m^3 or 40% SO$_2$ removal
Combustion plants; domestic high sulphur coal	167–500	800 mg/m^3 or 15 + (0.15 × P)% removal
Combustion plants; domestic high sulphur coal	>500	800 mg/m^3 or 90% removal

[a]Emission standards figures are given in mg/m^3 on dry flue gas at 6% O$_2$ and standard temperature and pressure (0 °C (273 K), 101.3 kPa).

33 countries including the USA and Canada. Now 42 countries are party to the Convention. The Convention outlines the responsibility of governments to minimise transboundary air pollution and came into force in 1983. Two Protocols have been made under the LRTAP Convention dealing with sulphur emissions. The first, the 'Helsinki Protocol', signed in 1985, required sulphur emissions to be reduced by 30% on 1980 levels by 1993. Twenty-one countries agreed and in fact achieved a 48% reduction. A second sulphur Protocol was signed in 1994 by 27 European countries, the EC and Canada. The emission standards for SO$_2$ are given in Table 12.5. The Second Sulphur Protocol is based on a concept of critical loads. Critical loads are quantitative estimates of pollutant deposition, below which plants and ecosystems are not adversely affected. This Protocol's objective was to reduce the gap between deposition and the critical load by 60%. All signatories were allocated targets to be achieved by the year 2000; however, the Protocol comes into force only after it has been ratified by 16 countries; by 1997, only 8 countries had ratified the Protocol (McConville 1997).

The Protocol on nitrogen oxides (NO$_x$), the 'Sofia Protocol', was signed in 1988 and came into force in 1991, signed by 23 countries and ratified by 16. The Protocol requires that emissions of NO$_x$ be frozen at 1987 levels by the end of 1994 and then maintained.

A protocol on the control of volatile organic compounds (VOC) emissions was signed in 1991 and came into force in 1997. VOCs are defined as all organic compounds of anthropogenic nature, other than methane, that are capable of producing photochemical oxidants by reactions with NO$_x$ in the presence of sunlight (McConville 1997).

Protocols on persistent organic pollutants and heavy metals are under preparation.

12.5.3 European Community (EC)

The Commission of the European Communities (EC) puts forward legislation to be adopted as formal proposals by the Commissioners. EC legislation is presented either as regulations which are binding on member states or as directives which give targets to be achieved and deadlines in which to implement them; how this is done is left to the member states. Environmental legislation is in the form of directives allowing member states flexibility in trying to achieve environmental objectives.

The Directive on Controlling of Emissions from Large Combustion Plant (LCPD) 1988 sets out emission standards for particulates, SO$_2$ and NO$_x$ (see Table 12.6). The Integrated Pollution Prevention and Control Directive (IPPC) 1996 requires the introduction of an integrated environmental licensing system applied to a range of industrial processes including power plants larger than 50 MWe. This is to be applied to all new and existing plant by 2007, and member states are to implement their own schedules in order to achieve this target.

The EC has since drafted additional directives; one is the European Union Acidification Strategy, which proposes new limits for SO$_2$ and NO$_x$, and a second is the revision of the LCPD with much stricter emission ceilings for SO$_2$ and NO$_x$ to be achieved by 2010 (McConville 1997).

12.5.4 World Bank

Environmental guidelines have been developed by the World Bank which are to be followed in all the

Table 12.6 European Community emission standards (1988) for new coal-fired plants. From McConville (1997), with permission from IEA Coal Research

Emission type	Plant type	Plant size (MWt)	Emission standard[a]
SO_2	Combustion plants	50–100	$2000\,mg/m^3$
SO_2	Combustion plants	100–500	$2400 - (4 \times P)$ P = plant size in MWt
SO_2	Combustion plants	>500	$400\,mg/m^3$
SO_2	Combustion plants operating <2200 hours per year	>400	$800\,mg/m^3$
SO_2	Combustion plants; domestic high or variable sulphur coal	100–166	40% removal
SO_2	Combustion plants; domestic high or variable sulphur coal	167–500	$15 + (0.15 \times P)\%$ removal
SO_2	Combustion plants; domestic high or variable sulphur coal	>500	90% removal
NO_x	Combustion plants	>50	$650\,mg/m^3$
NO_x	Combustion plants; coal volatiles <10%	>50	$1300\,mg/m^3$
Particulates	Combustion plants	50–500	$100\,mg/m^3$
Particulates	Combustion plants	>500	$50\,mg/m^3$

[a] Emission standards figures are given in mg/m^3 on dry flue gas at 6% O_2 and standard temperature and pressure (0 °C (273 K), 101.3 kPa).

projects that the World Bank funds. These guidelines are commonly used by other financial institutions when lending to projects in developing countries. The basis of the World Bank's environmental policy is Operational Directive OD 4.00, 1989, updated in 1991 as OD 4.01. Guidelines for emissions from thermal plants have been issued in 1998 (Table 12.7). These establish maximum emission levels for all fossil fuel based thermal power plants with a capacity of 50 MWe or larger. The guidelines focus on emissions of particulates less than 10 μm in size, SO_2 and NO_x. Information on health concerns and damage caused by these pollutants, together with alternative methods of emission control, are provided in the guidelines. Requirements are of two kinds: first,

Table 12.7 World Bank proposed emission standards (World Bank Group 1998)

Emission type	Plant type	Plant size (MWt)	Emission standards[a]
SO_2	New power plants	All	$2000\,mg/m^3$ Max. emission level of 0.2 t/d per MWe of capacity up to 500 MWe plus 0.1 t/d for each additional MWe of capacity over 500 MWe
SO_2	Old power plants	All	Emission levels to meet regional load targets
NO_x	New power plants	All	$750\,mg/m^3$
NO_x	New power plants; coal volatiles <10%	All	$1500\,mg/m^3$
NO_x	Old power plants	All	Emission levels as recommended for new plants or at least a 25% reduction in baseline level
Particulates	New power plants	All	$50\,mg/m^3$ If not possible, then must achieve 99.9% removal
Particulates	Old power plants	All	$100\,mg/m^3$ but target should be $50\,mg/m^3$

[a] Emission standards figures are given in mg/m^3 on dry flue gas at 6% O_2 and standard temperature and pressure (0 °C (273 K), 101.3 kPa).

the specific requirements for the power plant itself, focusing on issues to be addressed in arriving at project specific emission standards; and second, requirements which relate to the operation of the power system as a whole. These are the concern of national or regional authorities with the responsibility for setting the overall policy framework for the development of the power sector (World Bank Group 1998).

12.5.5 Kyoto Protocol

At the Rio Earth Summit, parties to the Framework Convention on Climate Change (FCCC) agreed to stabilise emissions of greenhouse gases at 1990 levels by the year 2000. Following this, an agreement to cut emissions of greenhouse gases was made in December 1997 in Kyoto, Japan, at the third Conference of Parties to the FCCC. Industrial nations agreed to reduce their collective emissions of greenhouse gases by 5.2% from 1990 levels by the period 2008–12. The Kyoto Protocol was endorsed by 160 countries, but will only be legally binding if at least 55 countries sign up to it, including developed nations responsible for at least 55% of greenhouse gas emissions from the industrialised world. The cut in emissions is to be achieved by differential reductions for individual countries. The European Union, Switzerland and the majority of Central and Eastern European nations will deliver reductions of 8%, the USA 7%, Japan, Hungary, Canada and Poland 6%, New Zealand, Russia and the Ukraine are required to stabilise their emissions, whilst Australia, Iceland and Norway are permitted to increase slightly, although at a reduced rate to current trends.

It is considered that the majority of nations will sign up to the Kyoto Protocol, and by June 1998, 41 countries had signed. However, there are a number of nations, some of which contribute a large percentage of the greenhouse gas emissions, that have not committed to this reduction, e.g. the USA.

12.6 FUTURE IMPLICATIONS

It is clear that the coal industry is being and will continue to be closely monitored as to its effects on the environment, both locally and globally.

Mining practice, waste disposal and air quality will continue to receive close attention. In the case of coal mining, environmental regulations, particularly in relation to sulphur content, have meant the abandonment and closure of mines with high sulphur coals, and the concentration of developing mines in areas of low sulphur coal. This is not always possible, and may be reflected in a penalty cost in the price obtained for high sulphur coals. Waste disposal of mine waters, coal preparation discard and spoil dumping are already closely regulated and this will continue. In the case of air quality, the current effort to further reduce emissions, particularly by the larger industrialised nations, is likely to be a long protracted affair, the underlying reason being that strict environmental restrictions mean higher development cost, higher operating cost, and in the case of old plant, higher refurbishment cost. The less that has to be spent on environmental improvement, the greater the margin for profit.

Modern coal mining practices and the advances in clean coal technology, together with the development of alternative uses of coal such as coalbed methane extraction, and the fact that coal resources are large and globally distributed, will ensure that coal will remain a viable source of fuel for the long term future. It is also clear that the environmental lobby will continue to have a profound effect on the feasibility and cost of coal mining and utilisation.

13

Computer Applications

13.1 INTRODUCTION

One of the biggest changes in the study of coal deposits and their exploitation has been the rapid growth of microelectronics based technology. This has had a significant impact on the coal industry. The objective of computer application is to simplify and speed up the process of data collection, retrieval, analysis and modelling. Computer modelling removes the possibility of undetected errors occurring in the calculations and allows the evaluation of a number of alternatives in a shorter time frame than would be possible by manual methods. As those deposits which are more accessible are exhausted, there is a need to design more complex mine plans. This is greatly assisted by computers making the calculations required.

Using the new digitally based systems, the collection and input of field data, borehole data, geotechnical, hydrogeological and geophysical data have enabled both commercial companies and government agencies to produce geological models and groundwater network models. These are used as the basis for creating the mining model which will become the basis for developing the mining scheduling for the life of the mine.

In mining operations, additional computer programs have been designed to produce schedules for mining equipment, mine transport systems, coal processing, and coal handling, weighing and loading.

13.2 GEOLOGICAL APPLICATIONS

A large number of software packages is now available, which provide integrated geological data management, analysis and visualization. Geological data collected in the field in the form of direct observation, sample points, borehole data, survey and location data (see Chapter 6) can all be inputted into the selected software program to produce point data plans, contour plans, survey plans, 3D surface diagrams, and volumetrics, as well as producing cross-sections and vertical logs.

The use of global positioning systems (GPS) has enabled geologists to accurately position data points in the field. This is particularly important in poorly surveyed areas such as forested or desert terrain.

The great advantage with some software programs is that additional data can be readily added or modified, and new plans and diagrams produced. Large amounts of field and borehole data can be unwieldy in hard copy form, and field records are either coded on to forms by the geologist in the field and then inputted into the geological database or encoded directly into a personal computer carried in the field or on site. These data can then be transferred directly into the geological database for processing and analysis. One of the principal tasks is to ensure that the correct information is contained in the input and that the right type of process is selected to produce correct results.

The principal modelling processes are the gridded model, the block model and the cross-sectional model. In coal evaluation, the gridded model is normally used as it is particularly suited to bedded deposits. The advantage of using grids is that it facilitates the manipulation and use of the data. Structures, thicknesses and other parameters stacked on top of each other can be added, subtracted, multiplied, divided or compared to arrive at other derived sets of data, and only those parameters of interest need be modelled.

To estimate the value of data for a given position on the grid, i.e. a node, one can use the data points closest to the node. However, this can result in the selection of data from only one set of data points, e.g. a line of boreholes or a sample traverse. To reduce the bias, many programs now require selection of data from within a specific search radius and limit the number of data points to one to three points from each quadrant, sextant or octant. The search radius limits the number of points selected for evaluation, and this in turn reduces the computer search time. If the desired number of points is not found, then a pre-set increment is added to the original radius and the search is continued. This is especially useful where there is an uneven distribution of data (Hartman

1992). In selecting points for estimation, it is important to obtain a distribution of data from all directions, in order to represent more closely the actual conditions.

The creation of grids which exhibit the spatial relationships of outcrop, sample and borehole locations allows the creation of contour plans. Figure 13.1 shows a borehole distribution plan with numbers for each borehole; each location is tied to an X–Y coordinate. It is essential that a plan of this type has correct locations for all data points; if not, all other contour plans

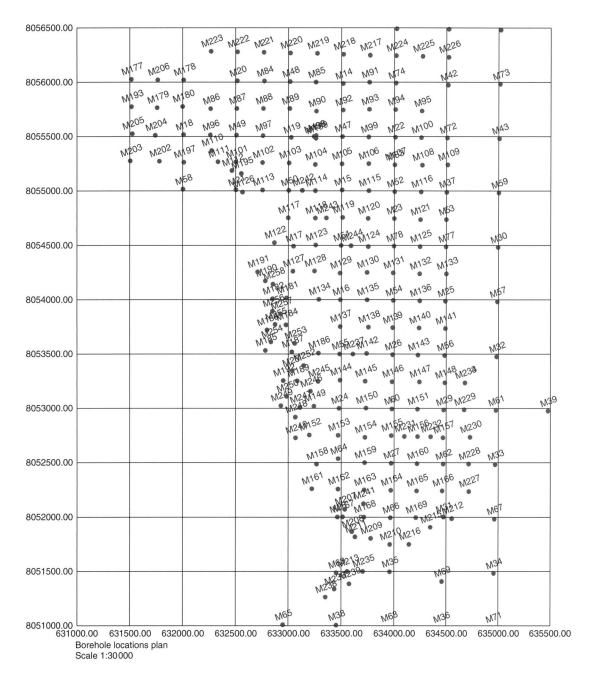

Borehole locations plan
Scale 1:30 000

Figure 13.1 Computer generated borehole location plan (courtesy of Dargo Associates Ltd)

generated from this location pattern will compound any errors. Plate 1(a) shows coal seam thickness of a single seam dipping to the southwest with the line of outcrop running northwest–southeast. The thickening of the coal shown by the bunching of the contours in the west and southwest is due to the coal seam splitting and including increasing thicknesses of interburden. This can be contrasted with Plate 1(b) which shows seam thickness excluding any partings or interburden. Errors tend to show up as 'bull's eyes' on the plan. Where this occurs, the data for that location need to be verified. Structural and thickness data are retrieved and modelled first, and if the deposit has a significant dip, a correction must be made to provide for a true thickness calculation.

Once the location grid is established, it allows the production of plans to show coal seam(s) thickness variations, depth to top and bottom of coal seam(s), interburden thickness variations and overburden thickness variations. At the same time, coal quality data are retrieved and modelled; initially this may be for moisture, ash, sulphur and calorific value to determine the economic potential of the coal seam(s). Plate 2(a) shows a contour plan of total sulphur for the same coal seam as shown in Plate 1. The higher sulphur content values are concentrated along the southeastern edge of the plan coinciding with the increase in interburden within the seam. Plate 2(b) shows the variation in net CV for the same seam; again the lower net CV values tend to occur where the amount of included interburden is greatest.

All of the gridded data can then be used to calculate coal resources and reserves. The planimeter used for calculating areas as described in Section 7.3.3 is replaced by a digitiser tablet. Each contour level is traced and the volumetric program uses the contour thickness value and the area traced to calculate the volume. Tonnage calculations require a density factor to convert the volumes to tonnes. This type of calculation is equivalent to the manual method but at much greater speed. All calculations are automatically accumulated so that the results

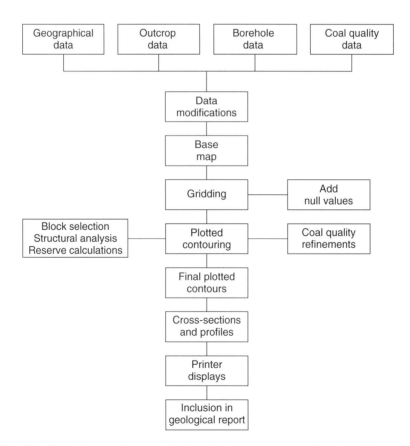

Figure 13.2 Simplified flow diagram of a geological contouring programme. Reproduced by permission of Dargo Associates Ltd

for each coal seam are rapidly available. More complex volumetric programs can be used; these depend on the geometry of the deposit. It is easier to use volumetric applications for formations which are flat-lying or uniform in development. Developments from the horizontal gridded technique include the production of cross-section profiles by plotting the grid values along a line or connected group of lines. Other displays include the perspective or isometric view of the gridded surface. Such displays show the surface or variable as a block diagram which can be viewed from any selected point of origin.

A simplified flow diagram illustrates the stages in the processing of geological field data to produce contour plans, profiles and reserve calculations (Figure 13.2).

13.3 GEOPHYSICAL APPLICATIONS

Geophysics has long been associated with computer technology. The interpretation of very large amounts of gravity, magnetic and seismic data by use of programming and the use of filters to further improve seismic sections is now well established practice; references to such technology can be found in numerous published works. However, of particular interest in the exploration and exploitation of coal deposits are the developments in interpreting and portraying downhole geophysical data.

Downhole geophysical records are normally shown as a digital representation. Signals from the logging tools are transmitted to the surface computer for processing. The acquisition computer also generates the hard copy plots that are essential for quality control and quick-look style analysis. Such data and plots are regarded as field data; however, further processing may be needed to combine data from different logging runs or to obtain interpreted results. This is usually carried out at a base location and results in final logs. Manual interpretations enable operators to gain valuable experience in the format and significance of logs, but most modern analyses use computers to provide volumetric information. There are a number of software packages which range from simple log plots to detailed lithological logs combined with the chosen suite of geologs and even pictorial representation of the borehole strata. Combinations of geologs can be plotted to show a number of parameters; for example, frequency plots using gamma ray and density data can be produced. Computer software calculates matrix, shale and porosity volumetrics for every depth-matched point of gamma, density and caliper, and the density end point is determined by the desired matrix type, i.e. $2.65 \, \mathrm{gcm^{-3}}$ for a sandstone matrix. Figure 13.3 shows the crossplot of depth-matched gamma ray and

density data. Sandstones and shales are indicated and coals can be identified due to its unique combination of low gamma ray and low density. Coal is plotted where densities of less than $1.9 \, \mathrm{gcm^{-3}}$ are found; caliper values of less than 10 cm have been selected so that low densities due to caving will not be interpreted as coal. Once the end points and coals have been identified, the final lithological analysis can be produced; for example, the computed lithology analysis as shown in Figure 13.4 (Firth 1999).

The use of real time images from data taken in the field allows qualitative interpretations to be made, and in vertical boreholes, good estimates of dip can be made. Taking this a stage further, it is possible to apply image enhancement and interpretation techniques to reveal information not apparent on monochrome prints. This also allows dips to be accurately computed regardless of borehole trajectory (Firth 1999). By using data produced by downhole acoustic scanning tools (see Section 8.5.6), interactive image processing software can isolate parts of an image to enhance features which may otherwise remain invisible (see Figure 8.24).

Further refinements in the processing and image enhancement are an ongoing process, and will contribute to improved interpretation not only of coal sequences and structural attitude, but also of coal quality and coalbed methane content.

13.4 HYDROGEOLOGICAL APPLICATIONS

Computer applications in hydrogeology have become increasingly important, particularly in the context of environmental studies. In coal mine development, groundwater studies are essential in understanding the effects groundwater will have on a mine, whether underground or opencast, and the changes in these effects over the life of the mine.

Computer analysis of aquifer test data, in combination with the known geological data, enables the modelling of groundwater flow fields. These models are able to show groundwater depths and flow patterns for confined and unconfined aquifers, and variations in groundwater chemistry. Such groundwater modelling is an integral part of mine planning.

In opencast coal mines, it may be necessary to determine the effects of dewatering schemes on the surrounding land area and local population. Groundwater modelling will determine the required rate of pumping and the size of the cone of depression around the mine (see Section 9.5.1). In underground coal mines, the modelled groundwater flow patterns will be superimposed on the mine plan, so that the proximity of any aquifers to

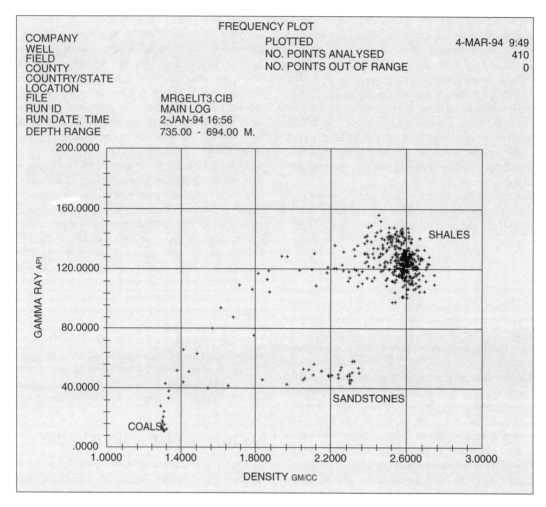

Figure 13.3   Crossplot of depth-matched gamma ray and density data. Firth (1999), reproduced with permission of Reeves Oilfield Services Ltd

the proposed or existing mine workings can be identified, and the movement of contaminated mine waters can be predicted, to ensure avoidance of environmental pollution.

13.5 MINING APPLICATIONS

Once the geological model, together with all geotechnical, hydrogeological and coal quality data, have been generated, the selection of mining horizons can be made. This includes how many coal seams are to be mined, the thickness of the mining section for each seam, and the depth to which the seams will be mined. In the case of opencast mining, the overburden and interburden thicknesses will be calculated, which will then allow stripping

ratio plans to be generated. Modern terrain modelling packages provide powerful tools for seam surface modelling, and predictions of the physical progress of the mine can be made (Dymond pers. comm.; World Coal 1999). Plates 3(a) and 3(b) are 3D contour maps showing the stripping ratio schedule of a selected area to be mined after 3 years and 4.5 years respectively. Such analysis allows schedules optimised for stripping ratio to be calculated for selected time periods (Figure 13.5).

All of this information is then combined in a 3D model of the deposit. Data for coal processing and land reclamation can be added, as well as unit densities, slope angles and compaction factors.

Mine planning is a complex process, and each mine is dependent on its location, economic conditions and

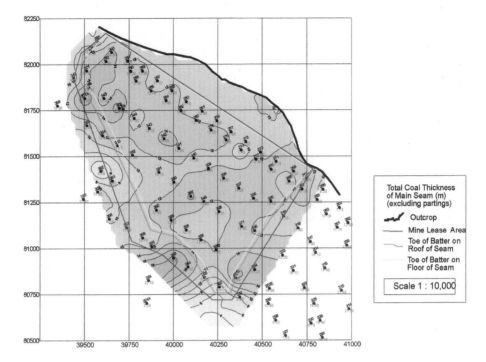

A

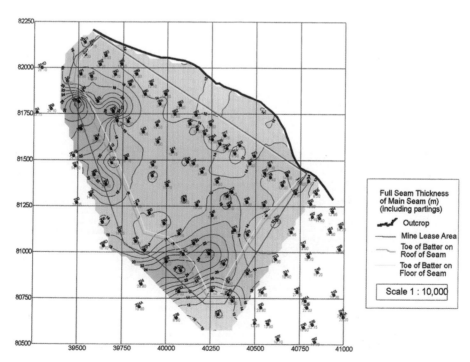

B

Plate 1 (A) Coal thickness contour map with borehole locations; (B) Full coal seam thickness (including partings) contour map with borehole locations. Reproduced by permission of Dargo Associates Ltd (see Chapter 13)

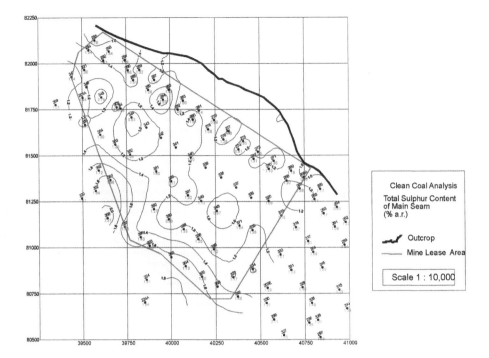

A

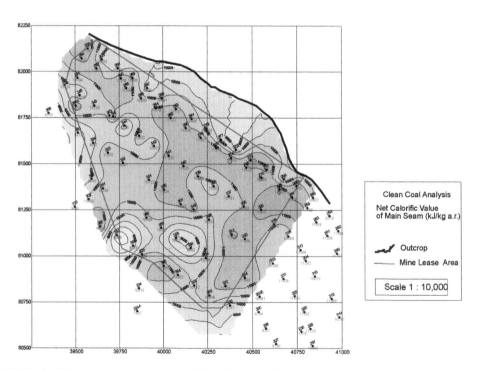

B

Plate 2 (A) Total sulphur content contour map with borehole locations; (B) Net calorific value contour map with borehole locations. Reproduced by permission of Dargo Associates Ltd (see Chapter 13)

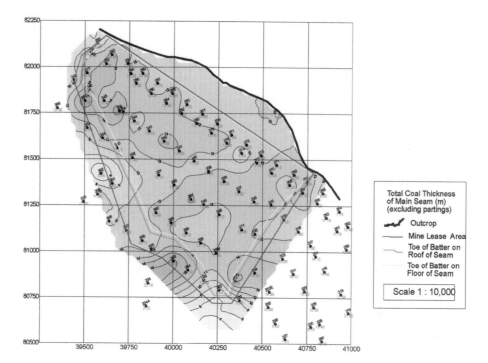

A

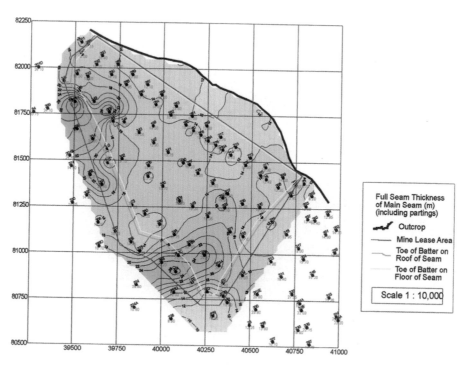

B

Plate 1 (A) Coal thickness contour map with borehole locations; (B) Full coal seam thickness (including partings) contour map with borehole locations. Reproduced by permission of Dargo Associates Ltd (see Chapter 13)

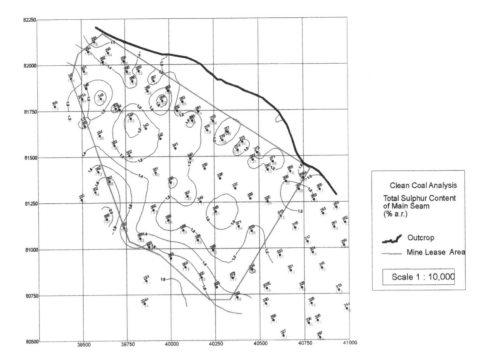

A

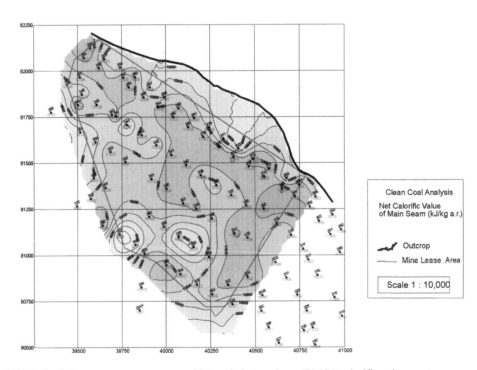

B

Plate 2 (A) Total sulphur content contour map with borehole locations; (B) Net calorific value contour map with borehole locations. Reproduced by permission of Dargo Associates Ltd (see Chapter 13)

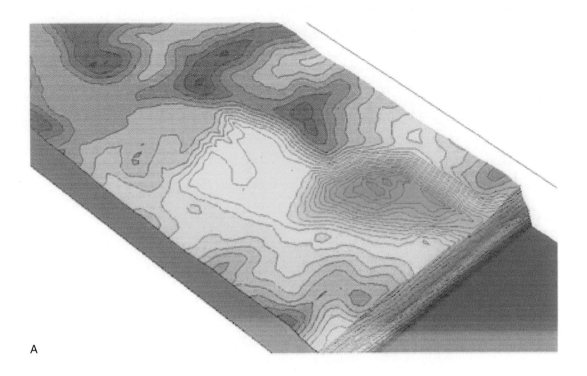

A

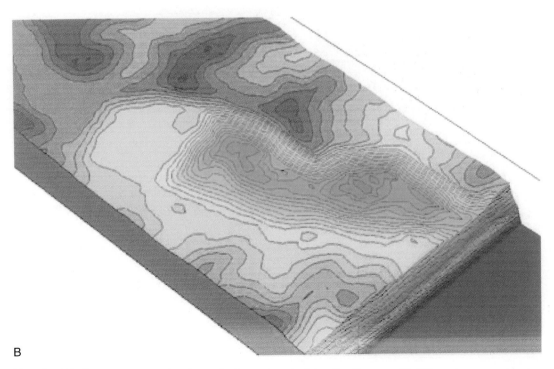

B

Plate 3 (A) 3D contour map showing stripping ratio schedule after 3 years; (B) 3D contour map showing stripping ratio schedule after 4.5 years. Reproduced by permission of Datamine International (see Chapter 13)

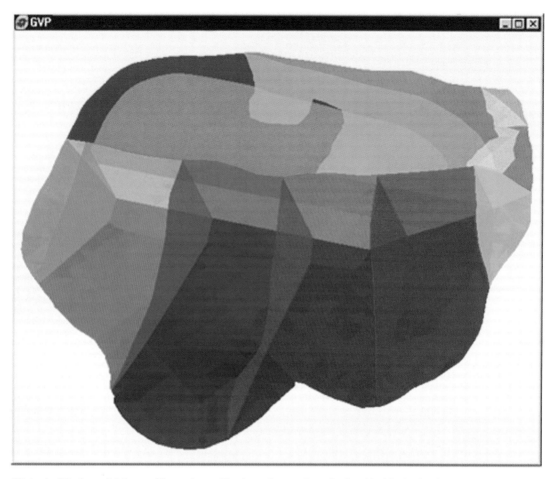

Plate 4 Block model format illustrating a 3D view of a number of mineable blocks for forward planning of an opencast mine. Reproduced by permission of Datamine International (see Chapter 13)

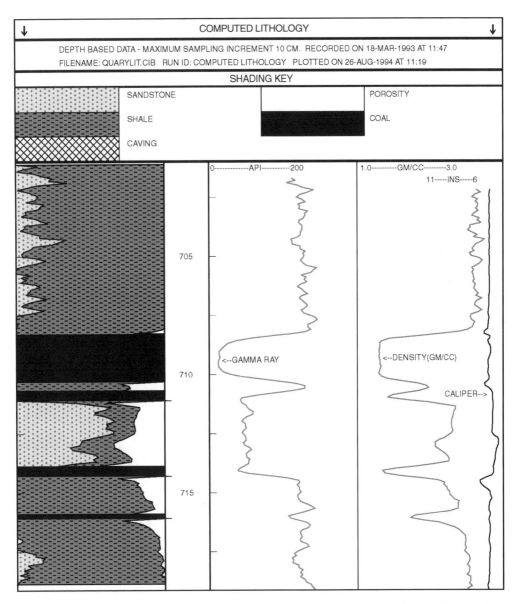

Figure 13.4 Computed lithology analysis derived from data shown in Figure 13.3. Firth (1999), reproduced with permission of Reeves Oilfield Services Ltd

local and national regulations. The mine plan is based on the sequence of mining and the equipment used, and the entire process is designed to be the best and most economical method of mining the coal. Such an optimum mining process can result in the maximum profit to the mining company by minimising operating costs and increasing production. Computer-driven mine planning is the best way to accomplish both objectives, with the flexibility to allow evaluation of multiple

alternatives in a short period of time and to be able to change the assumptions quickly.

13.5.1 Opencast mine planning

The configuration of the pit will be defined by limits of the lease area and outcrop and depth of the coal, together with any large physical features which will curtail mining, such as a large river or a major fault boundary.

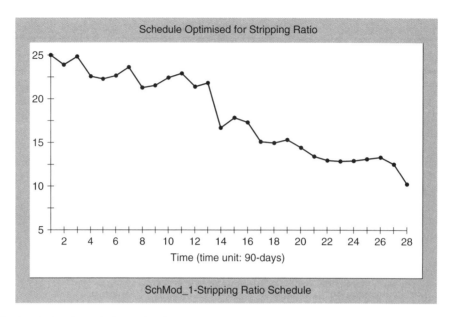

Figure 13.5 Graph showing stripping ratio schedule. Reproduced with permission of Datamine International

Computer-generated limits will include stripping ratio limits, i.e. economic limits. Once the configuration of the pit area is fixed, the computer software can define the ground slope angle for each bench using the geotechnical data relating to the physical strength and competence of the strata.

The actual benches and blocks to be mined and the sequence in which they are to be mined are based on the type of equipment to be used. In opencast mining, the selection of equipment such as bucketwheel excavators, draglines, truck and shovel or combinations of these, together with the size of equipment selected, will influence the width and height of cuts to remove material and the successive advances of the mine.

The 3D model can also illustrate the volumetric calculation, coal quality variations, scheduling and production sequencing, using preselected parameters relating to coal seam mineable thickness, quality, stripping ratio and depth cut-off limits, which have been built into the model. Additional considerations will be the hydrogeological regime within and surrounding the mine area, and the geotechnical characteristics of the strata. For example, at Hazelwood Mine in Victoria, Australia, geological and coal quality modelling together with aquifer modelling has enabled short, mid term and long term mine planning to be achieved, which should ensure the mine's profitability. The software allows bench plans and cross-sections to be generated efficiently, and with the block model intercepts projected on to the mine

topography, the stratigraphy of any part of the current or future pit face can be indicated. The block modelling also allows effective overburden management by creating solid models of three levels of dumps, and developing new dump areas for the future (Maxwell 2000).

Operation of the mine depends on the scheduled operating shifts, the number of hours per shift, the equipment fleet, the assignment of each piece of equipment to a given bench area, the scheduled down time, holidays and other items (Hartman 1992). Production scheduling is based on tonnage of coal produced, waste material removed, the production capacity of a specific piece of equipment and the quality and recovery of the coal product.

The great value in being able to use a block model format is its usefulness in pit optimisation programmes. The ultimate pit configuration may already be defined by the mine owners or by physical constraints; of greater value is the ability to model a number of alternative extraction strategies in liaison with the mining engineers. The interactive graphic capacity of the computer program will enable the mining engineer to study and refine the design as the process proceeds. Such strategies will take into consideration production targets over time, and targets to maximise net present value (NPV), cashflow, stripping ratio and coal quality limits (Dymond pers. comm. and World Coal 1999). Plate 4 shows a 3D view of a number of mineable blocks; such software (in this case produced by Datamine) ensures that a 3D model of overburden, coal seams and interburden can easily be created.

13.5.2 Underground mine planning

Although the basic principles of mine planning and design have not changed over many years, advancements made in computing, rock mechanics and equipment design have had a significant effect on mining. The widespread use of technical computing in mine design has increased with the increasing availability of sophisticated software. Underground mine planning may involve building upon previously developed mine layouts or designing a new mine area. Existing data are inputted to create a panel design, the object being to create a design that can be constructed, saved, copied and modified as many times as required; this is stored in a panel library (Hartman 1992). Panel designs will differ for room and pillar mining and longwall mining. Room and pillar mines require the parameters such as pillar configuration, headings and cross-cut dimensions

in order to generate the panel design. Longwall mines require the dimensions of the longwall block, pillar configuration for the gate roads, and the dimensions of the barrier pillar. The panel can then be modified using interactive graphics.

The mine layout design is achieved by combining the digitised existing mine plan with the interactive graphics design. Once a panel is selected from the panel library, it can be placed on the layout at any orientation to other panel designs. In existing operations, the relevant portions of the current mine plan are digitised, pillar configuration is entered using an interactive menu program and the pillars are automatically generated. Figure 13.6 shows a computer-generated final layout design showing selected pillar design for room and pillar and longwall panels.

Computer modelling is also used to carry out strata control and reinforcement design studies. The necessary data on the rock properties and *in situ* stresses has to be compiled and fed into the computer model. The range of underground measurements and laboratory

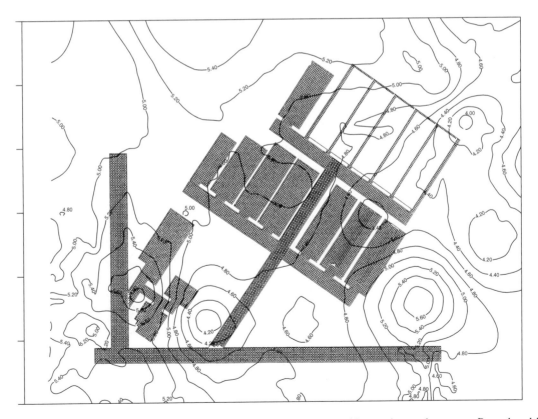

Figure 13.6 Final layout design for an underground mine development, with superimposed contours. Reproduced by permission of Society for Mining, Metallurgy and Exploration Inc. (www.smenet.org)

tests normally undertaken for model generation is shown in Figure 13.7. The residual strength properties of each rock unit are determined; these include strength properties for intact rock and rock showing discontinuities. These are assigned to each modelled strata unit and the model is built up in layers. The *in situ* stresses in the model are initialised on the basis of the expected cover load for a given depth and the results from the *in situ* stress measurement. The nature and magnitude of roadway displacements as indicated by the use of extensometers, and how the stresses around the roadway have changed from pre-mining levels, indicated by stress

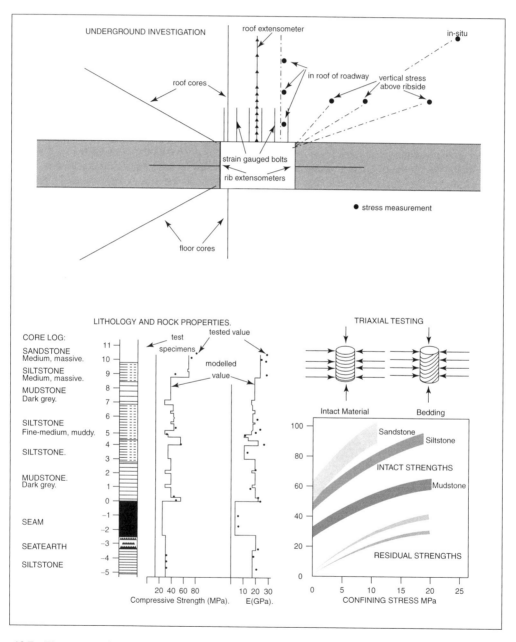

Figure 13.7   The range of underground measurements and laboratory rock tests normally undertaken for model generation. Reproduced from Garratt (1999) with permission of Rock Mechanics Technology Ltd

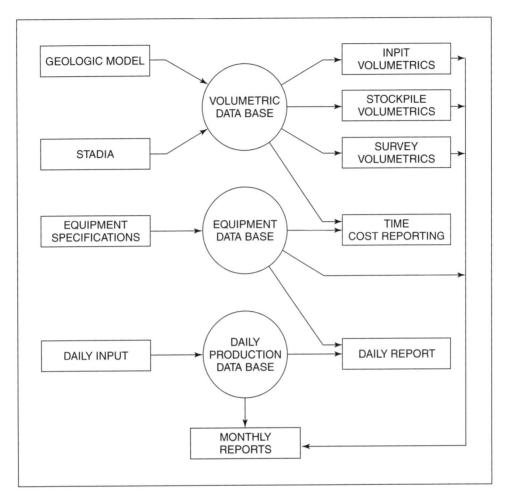

Figure 13.8 Flow diagram of a mine operations monitoring system. Reproduced by permission of Society for Mining, Metallurgy and Exploration Inc. (www.smenet.org)

measurement results, will be characteristic of the deformation mechanisms around the roadway. When using computer simulation to select a reinforcement system such as rock bolts, it is important to consider their effects on roadway behaviour and the likely loads to which the system will be subjected. The modelled reinforcement behaviour is verified against strain-gauged bolt data obtained underground to ensure proper verification of the simulation.

Another capability is the use of surveyed information to plot progress on the mine plan. These programs are capable of maintaining records of actual mining operations. They can perform day-to-day monitoring of operations, to include use of personnel and equipment as well as production and maintenance. Figure 13.8 is

an example of a schematic mine operations monitoring system (Hartman 1992). The generation of such mining sequencing, i.e. what mining systems and manpower will be in operation, what areas are being mined, times of operations and rates of advancement, enables all the details and characteristics for any mining period to be produced for managerial analysis.

13.6 OTHER APPLICATIONS

Surface mines are now looking to utilise GPS for surveying. In mine areas, there is little to impede satellite signals, so that a high level of accuracy can be achieved. The mine will need to equip plant such as shovels and bulldozers with GPS systems; the expected results are faster terrain modelling and grading, for both active mining and reclamation. Inexpensive, lower accuracy GPS is also envisaged for equipment monitoring and truck fleet management, including dispatch systems (Eckels and Carlson 1999).

Coal blending and quality control, together with loading and transportation, are now operated using software programs that assess quality and quantities of coal by specific transport systems to specific customers, and allow real time access to information about coal shipments by land and/or sea.

14

Coal Marketing

14.1 INTRODUCTION

Although this book is primarily concerned with coal and its properties and uses, coal is above all a saleable commodity. The marketing of coal is no different from other commodities in that it is controlled by supply and demand. Once the demand is established, the marketing of coal is dependent upon four factors:

1. The quality and quantity of the product
2. The transportation of the product
3. The contractual terms of the sale and purchase of the product
4. The price of the product.

Coal is sought after, mined, prepared and then transported as a saleable product. As described previously, a wide range of qualities of coal is available for a variety of uses. In order to enter the coal market, the intended use for the coal must be identified.

Coals are primarily used as coking coal in the steel industry, as thermal or steam coal in the electricity generating industry, and by industrial and domestic consumers. By-products from these processes and some specialised chemical processes are also commercial outlets for coal.

Coals are marketed locally, e.g. as mine mouth supplies to power stations, and/or neighbouring industrial complexes, at a distance but within the confines of the country of origin using transportation by road, rail and water to consumers, and as an export product transported by ocean-going vessels to overseas markets.

14.2 COAL QUALITY

The chemical and physical properties of coal were described in Chapter 4, and their general usage in Chapter 12. The principal concerns for the buyer are the heating value of the coal (CV) and the properties of coal which affect this, in particular, ash and moisture

content, together with sulphur content for environmental reasons, and in the case of coals destined for the steel industry, the carbon content and coking properties of the coal.

In order for some coals to be suitable for selected markets, it may be necessary to improve the quality of the coal prior to shipment. This beneficiation of the coal or coal preparation is closely related to consumers' demands, but other factors such as environmental constraints play an increasing role in influencing the quality of coals that are marketed in the world today.

Details of coal preparation processes are well documented and described in detail in numerous works (e.g. Osborne 1988). Such preparation is normally designed to reduce ash and sulphur levels and improve the CV of the coal, and to establish a consistency in quality for the coal product. This is particularly true for coals that will be exported and transported large distances.

Coal preparation can simply be a screening process whereby coal particles are separated into selected size ranges by passing through a series of screens with specified opening dimensions. However, in order to remove mineral matter from the coal, processes that separate coal particles based on their relative density are normally used. There are a number of different methods used to achieve this.

Broadly there are two types of process in general usage. Water-only based systems are the most widely used, accounting for approximately 70% of the total tonnage treated, with the jig being the most popular. Figure 14.1 shows a section through a Baum type jig. Coal is fed into the jig and the water therein is made to rise and fall by means of compressed air applied to one half of a U tube. The up and down motion of the water effects a separation between the coal and discard. The other main process is using dense medium. In this process coal and shale are immersed in a suspension of finely ground magnetite and water. The amount of magnetite used controls the density of the suspension, which is chosen to lie between that of coal and shale.

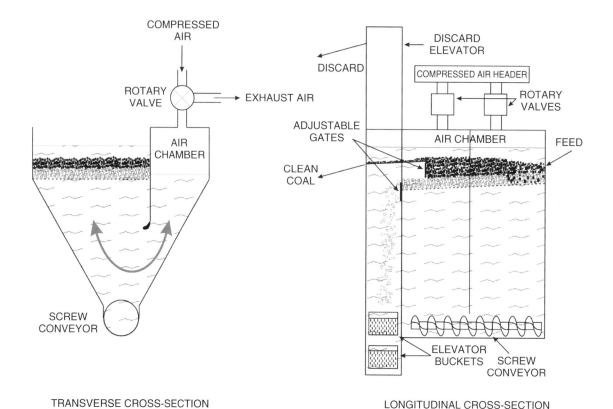

TRANSVERSE CROSS-SECTION   LONGITUDINAL CROSS-SECTION

Figure 14.1   Section through a Baum type jig. Coal is fed into the jig and the water level rises and falls by means of injection of compressed air (see transverse section). The up and down motion of the water effects a separation (see longitudinal section). Reproduced with permission of Dargo Associates Ltd

Coal again floats and shale sinks; Section 4.3.3.5 gives an example of a typical float and sink analysis.

For fine coal (less than 0.5 mm in size), froth flotation is sometimes used. In this process particles of coal, which are hydrophobic, are attached to air bubbles in water. These rise to the surface and form a froth, which is scraped off and dried. The shale stays in the water since it is hydrophilic.

To achieve the final saleable product, it may be necessary to blend different coals together, either beneficiated coals or beneficiated and raw coals or simply raw coals, to arrive at the required quality. Blending of coals can be carried out at the mine site or by the customer at his coal receival facilities.

Many modern mines deliver coal to their dispatch area, where it is conveyed to storage silos. Each silo may contain a specific coal quality which can then be loaded directly for transportation (Figure 14.2).

In large scale operations, coal is stockpiled using stackers which convey coal, and spread it out in layers, or in adjoining longitudinal stockpiles which form the overall stockpile. The reclaiming of blended stockpiles is by scrapers or barrel reclaimers. The coal is conveyed to hoppers for loading, or in the case of modern port facilities, the coal is conveyed to the ship and loaded directly. Figure 14.3 shows the modern stockyard at Qinhuangdao, PRC, where coal stocks are reclaimed directly on to conveyors and then to ship loaders.

Coal stockpiles are monitored closely for signs of oxidation and spontaneous combustion. Coals that have a propensity to this are normally of lower rank or have a high pyrite content. As a consequence, such coals are only stockpiled for a short time.

14.3   TRANSPORTATION

14.3.1   Land transportation

Transportation of coal is a major component in the marketing of coal worldwide. The transport system must

Figure 14.2   Coal silos for loading coal directly to trains, People's Republic of China. Photograph by courtesy of Dargo Associates Ltd

Figure 14.3   Modern coal stockyard at port of Qinhuangdao, People's Republic of China. Photograph by courtesy of Dargo Associates Ltd

be matched to the throughput and distance of movement required. Short distances up to 25 km can be covered by conveyor belts, especially if the throughput is high. Longer distances may require transport by truck or rail. The specific method chosen will depend on each location and will be a function of cost effectiveness and environmental concerns.

14.3.1.1 Conveyors

Overland conveyors are often used to transport coal to mine mouth power stations, or to rail terminals for loading into special coal trains, or very occasionally, direct to a ship loading terminal. An example of the latter is in East Kalimantan, Indonesia, where coal is conveyed 14 km from the Kaltim Prima Coal mine to the port at Tanjung Bara (Figure 14.4). A single span steel cored conventional belt conveyor carries 1100 t/h of coal to the port stockyard. Conventional belt conveyors have the disadvantage that they are best run in a straight line, but can easily travel up to 20 km in a single stage. These conveyors run on idlers 'troughed' in three sections for the top or coal conveying surface, and on flat idlers or rollers for the return or bottom belt. Another form of conveyor is the cable belt, consisting of a rubber conveying belt resting on steel cables which provide the motive force. Higher capacity or greater distance can be achieved than with a conventional belt conveyor (over 30 km), but cable belt conveyors have a higher capital and operating cost.

14.3.1.2 Road

Road transport is obviously very flexible and can serve any number of customers. The limitation is distance, but individual customers may be served up to 160 km by coal trucks of 10–40 t, or by heavier loads in articulated trucks and multiple trailers.

The route to be taken by trucks must be considered carefully. If the route is to use existing public roads, environmental concerns over exhaust fumes, noise and traffic congestion may influence the choice of route. The availability of suitable vehicles will also be a factor, particularly in developing countries. India, for instance, has a ready availability of 10 t capacity road trucks (Figure 14.5), and despite the intuitive conclusion that 14 t trucks would offer economies, the excessive price of the larger trucks makes them uneconomic. It is essential to load trucks as quickly as possible in order to achieve the best utilisation and reduce costs, so loading from bins (Figure 14.6) or silos is the most effective for dedicated transport routes. Road transport may be used to haul coal to terminals on either rivers, e.g. in the USA and Indonesia, or to large coal terminals loading ocean-going bulk carriers. Such dedicated roads are usually limited to about 70 km in length, and throughput is about 2 Mtpa, e.g. in Venezuela and Indonesia.

14.3.1.3 Rail

The bulk of coal transported any long distance overland is by rail. In the extreme this may be individual wagons to individual small customers but this trade is diminishing. For longer distances, trains of 3000–10 000 t may be employed to carry coal in 100 t wagons using diesel or electric locomotives to export coal terminals. Coal is transported 600 km to Richards Bay in South Africa, and to Dalrymple Bay in Queensland, Australia, and in PRC, even greater distances are covered, up to 1000 km from the coalfields to the port of Qinhuangdao on the eastern seaboard.

The capacity of a rail transport system depends upon whether there is single or double track. The capacity of

Figure 14.4 Overland conveyor, 14 km in length, from Kaltim Prima Coal Mine to the port of Tanjung Bara, East Kalimantan, Indonesia. Photograph by courtesy of Dargo Associates Ltd

Figure 14.5 Coal transportation by small capacity 10 t truck, India. Photograph by courtesy of Dargo Associates Ltd

Figure 14.6 Automatic loading of trucks from overhead bins, Orissa State, India. Photograph by courtesy of Dargo Associates Ltd

a double track is more than double that of a single track, the latter constrained by the number of passing places available and the waiting time for trains travelling in the opposite direction. The speed of the train and the distance between trains are also important. Also, time is needed to negotiate busy rail junctions, and this may affect overall throughput.

Large train shipments are usually loaded from fully automated train loading systems, where the train is inched through the loader at a controlled rate, and the precise tonnage is released into each wagon as it passes (Figure 14.7). The trains are automatically weighed after passing through the loader, and the coal is usually sampled at this stage. In some countries the majority of coal is still loaded with mobile equipment such as payloaders (Figure 14.8).

Upon delivery, the coal has to be unloaded. The method of unloading is governed by the type of wagons used, i.e. they may be bottom discharge wagons or top discharge, tippler wagons. Bottom discharge wagons are now the wagon of choice in most countries (Figure 14.9); they are simple to discharge and trains

Figure 14.7   Automatic loading of trains; each wagon receives exact tonnage as train moves slowly through loading bay, Alberta, Canada. From Howland (1998) with permission of World Coal, Palladian Publications

Figure 14.8 Train being loaded by payloader, Orissa State, India. Photograph by courtesy of Dargo Associates Ltd

Figure 14.9 Bottom discharge wagons (60 t), on Eastern Railways, India. Photograph by courtesy of Dargo Associates Ltd

do not need to be uncoupled, nor do they need complex couplings. To discharge they are simply pulled over a ground hopper and a trip mechanism opens the doors on the bottom of the wagon and the coal is discharged. Once completed, the doors are closed automatically. Very large train units transport coal in this way, notably in Canada and the USA. In the USA, such train units carry up to 12 000 t over distances of up to 1200 km from the mines for delivery for export from the Pacific Coast (Figure 14.10). Top discharge wagons have to be inverted or tippled to empty them. Trains can be equipped with rotary couplings which permit wagons to

Figure 14.10   Large train units transporting 12 000 t over 1200 km in western USA. From Harder (1998) with permission of World Coal, Palladian Publications

be tippled without uncoupling. More commonly, wagons must be uncoupled and tippled separately, then recoupled. Using rotary couplings, a 60 t wagon train can be unloaded in one hour, whereas uncoupling would require three hours for the same train. Tippler wagons often have drop side doors, which enable them to be unloaded by hand (Figure 14.11).

14.3.2   Water transportation

Coal is transported either by barge or by bulk carriers. Barges are defined as having no propulsion and may have a capacity up to 10 000 t. They are mainly used in sheltered waters such as rivers, lakes and short sea crossings. Bulk carriers are self-propelled and are usually oceangoing vessels.

14.3.2.1 Barges

Barges have been used for inland water transport on canals for hundreds of years, e.g. the Grand Canal in PRC which runs from Shanghai to Beijing. In the UK canals have been utilised for over 200 years, including the famous Duke of Manchester's Canal, which went underground into the coal mine for part of its length.

Barge traffic is still used in Europe on a large scale on the Rhine and Danube rivers, in the USA on the

Missouri and Mississippi rivers, and these barges can have capacities up to 10 000 t (Figure 14.12). Barges are also used extensively on rivers in Indonesia, such as the Mahakam, Barito and Berau rivers in East Kalimantan. These barges, of 1000 t capacity, are used to take coal down the rivers to a point 2–3 km offshore, where they are unloaded by floating cranes into bulk carriers. Barges are also used to transport coal on short sea crossings such as from East Kalimantan to Java and from Sumatra to Java in Indonesia, weather permitting. Transport of coal by barge is relatively low cost, both capital and operating, but is slow.

14.3.2.2 Bulk carriers

Ocean bulk carriers account for over 500 Mtpa of seaborne trade in coal, and this continues to grow. The smallest ships are 10 000 dwt, e.g. used between the UK and Rotterdam; these increase to Handysize 20 000–37 000 dwt, Handymax 37 000–50 000 dwt, Panamax 55 000–75 000 dwt, the largest vessel able to negotiate the Panama Canal, and Cape size vessels are 100 000+ dwt, with 200 000 dwt being the largest. The larger the bulk carrier, and longer the distance, the cheaper the freight rate per tonne-kilometre. Rates are generally very competitive between a large number of shipping companies, and may vary according to the

Figure 14.11 Top discharge wagons with dropdown doors for side unloading. Guizhou Province, People's Republic of China. Photograph by courtesy of Dargo Associates Ltd

Figure 14.12 Coal barges carrying up to 10 000 t, Mississippi River, USA

availability of vessels or by competition with other commodities such as grain and iron ore.

Loading large volumes of coal into bulk carriers requires high capacity coal handling systems. The major coal terminals of the world can load 10 000 t/h, e.g. Richards Bay in South Africa. Developing ports such as Qinhuangdao in PRC can already load 5400 t/h (Figure 14.13), and the throughput capacities of some of the leading coal handling ports are listed in Table 14.1. New port facilities such as at Los Angeles Export Terminal (LAXT) have recently been completed, the latter capable of handling 7700 tph, the largest export facility on the US Pacific Coast.

The unloading of coal is done with either grabs or continuous unloaders. Grab unloaders discharge coal into a hopper which in turn loads coal on to conveyors for transport to storage facilities. Grab unloaders commonly unload at a rate of 1500 t/h.

Continuous ship unloaders are lowered into the hold of the ship and collect coal, which is carried up a chute on to conveyors. These unloaders are less prone to dust losses and can extract more coal without help from mobile equipment than is possible with a grab unloader.

Some ships have self-discharging equipment, and are referred to as 'geared vessels'. These are ideal for discharging coal where unloading facilities are absent. The largest geared vessels are Panamax size, but this may change with a demand for geared Cape size vessels.

Coals are transported around the world, and Figure 14.14 shows the long haul sea routes taken by Cape size and Panamax vessels. Smaller Handy size vessels are usually used for shorter journeys or for ports which cannot take larger vessels, e.g. along the east coast of India. The International Hard (or Black) Coal Trade totalled 573 Mt in 2000, with 381 Mt being thermal or steam coal and 192 Mt being coking coal. Of this the seaborne trade comprises 343 Mt steam coal and 179 Mt coking coal (WCI 2001).

Table 14.2 shows the tonnage traded by the major coal exporting and importing countries. The share of the coal market in large trading areas such as the USA and the European Community are shown in Figure 14.15(a) and (b). It can be seen that the bulk of the coal is used for electricity generation: some 90% in the USA and 67% in the European Community. This strong dependence of the electricity generators on coal will mean a long term future for the suppliers of steam coals that have the quantity and appropriate quality of coal to satisfy the customer.

Figure 14.13   Coal loading directly into the ship from the stockyard conveyor system, port of Qinhuangdao, People's Republic of China. From Thomas and Frankland (1999). Photograph courtesy of Dargo Associates Ltd, reproduced by permission of World Coal, Palladian Publications Ltd

Table 14.1 Capacities of exporting ports (McCloskey 1996)

Port	Throughput capacity (Mtpa)
Dalrymple Bay (Australia)	22.5
Gladstone (Australia)	22.5
Newcastle (Australia)	53.0
Roberts Bank (Canada)	22.0
Qinhuangdao (PRC)	75.0
Puerto Bolivar (Colombia)	15.0
Tanjung Bara (Indonesia)	12.0
Pulau Laut (Indonesia)	8.0
Gdansk (Poland)	8.5
Vostochnyy (Russia)	6.5
Richards Bay (South Africa)	54.5
Baltimore (USA)	13.0
Mobile (USA)	23.0
Newport News (USA)	22.0
Norfolk (USA)	48.0

Table 14.2 Major coal exporting and importing countries (statistics – WCI 2001)

Coal exporters	Steam coal (Mtpa)	Coking coal (Mtpa)
Australia	87.8	99.0
South Africa	67.5	2.5
PRC	48.2	6.9
Indonesia	48.1	8.7
Colombia	34.0	0.5
Russia	27.0	7.3
USA	24.6	28.4
Poland	18.0	5.8
Canada	4.6	27.2
Coal Importers		
Japan	80.6	64.7
Rep. of Korea	42.8	18.9
Taiwan	39.3	6.1
UK	15.0	8.5
India	9.1	15.4

14.4 COAL CONTRACTS

Coal contracts and sales transactions vary from straightforward individual contracts of sale to more complex indexed contracts over a long time period. In simple terms there are three principal types of contract.

14.4.1 Spot purchases

This is simply the purchase of a cargo of coal offered for sale. Usually such spot purchases are as a stopgap in coal supplies or where the purchaser is 'shopping around' for cheaper coal than he/she has been previously offered and is not under contract. The coal prices on the spot market will vary according to availability and changes in freight rates.

14.4.2 Term contracts

This is a common form of coal contract whereby the sourcing and quality of the coal to be purchased are

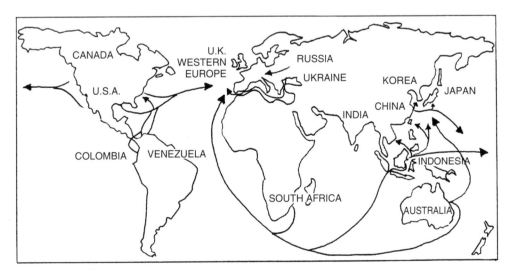

Figure 14.14 Principal coal export routes to markets in Western Europe and the Far East

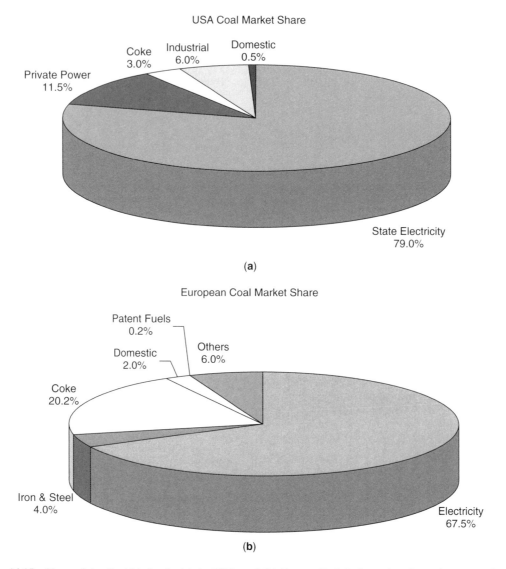

Figure 14.15   Share of the Coal Market in (**a**) the USA, and (**b**) Europe. Both indicate that the major usage of coal is in the electricity generating sector. (Based on various sources). Reproduced by permission of Dargo Associates Ltd

agreed. Such contracts usually run for one year, at the end of which time the tonnage and price of coal to be purchased for the next year is renegotiated, the agreed quality usually remaining the same. Term contracts are also known as perpetual contracts and run for a number of years with their annual negotiations. Such contracts are common between Japanese buyers and Australian and South African coal suppliers, and between Canada and the USA.

In negotiating successful long term contracts the important objective is to develop a good relationship

between coal sellers and buyers, so that when variations do occur, both parties can negotiate amicably and quickly.

Coal contract terms and conditions will reflect the particular requirements of the buyer and seller, but in general most clauses in international coal contracts include coal type and quality, which has a significant effect on the planned use of the coal, e.g. CV for thermal coals and coking properties for coking coals, length of contract, tonnage requirements, basic coal price, escalation, bonus/penalty clauses, sampling, weighing and analysis of the cargo, payment arrangements, and currency and exchange fluctuations. The contract will also include clauses to minimise risk to either party, such as changes in law, taxes and regulations, arbitration, *force majeure* and unfair conditions (to either party). If the contract involves transportation, there will be clauses relating to the road or railway company's obligations and where responsibility lies, coal storage and handling and/or to shipping, loading and discharge port conditions. The latter will include port charges, demurrage, and guaranteed rates of loading.

14.4.3 Indexed contracts

Indexed contracts are usually long term contracts such as between a coal fired electricity generating station and a single or group of coal suppliers. They are typically used where the power station is independently owned and financed by nonrecourse project debt. The required tonnage and coal quality is determined by the specification of the boilers in the power station and is fixed for the duration of the contract, which can be 20 years. The contract will contain similar clauses to those for a term contract, but may be more detailed when dealing with coal stock levels. The power plant must not break down, so a constant supply of fuel must be guaranteed. One method of dealing with a short disruption in supply is to hold a sufficient stock of fuel to operate the power plant at full load, and clauses are included in the contract stipulating this condition. Such stockpiling of fuel does mean higher working capital requirements and likely increased cost for debt service. The failure of the coal supply and hence failure to generate electricity will normally be considered cause for default under the terms of the contract; therefore it is essential that the terms of the contract are closely linked to the conditions laid down in the power purchase agreement drawn up by the generating company and the electricity purchaser to avoid conflict between the two. Late delivery of coal may trigger penalties stipulated in the contract; in most cases penalties under the contract are in the form of liquidated damages which are often set so that the coal supplier has had the profit element removed, but are not such that the supplier will be bankrupted. Liquidated damages set at a level of 20% of the expected coal price would be a significant incentive to the coal supplier to perform. Repeated infractions by the coal supply may also lead to termination of the contract.

When the coal supply is being adjacent to the power station, local conditions and issues will decide the base price of coal together with ongoing adjustments for inflation etc. In the case of imported coals, the power station will try to ensure that the price paid for coal is always about the current market rate in order to keep the price of electricity competitive. Provided that the specification range for the power station boilers is wide enough to accommodate a reasonable range of coals, then the power station can accept coal from a number of suppliers, very often from different countries, e.g. the Japanese power corporations take coal from several mines in Australia, Indonesia and South Africa.

The price of the coal is often adjusted for quality, principally CV (as the power station is buying heat), but adjustments may occur for changes in quality that may affect power station operating costs, such as ash, moisture, sulphur, hardgrove grindability index (HGI) and ash fusion temperature (AFT). In the case of coal imports to a power station, the coal supplier(s) will normally have been selected by tender, sometimes open or otherwise. The original base price will be determined from the current market rate as determined by the national authority or by an independent commercial organisation, i.e. a suitable thermal coal price index.

14.5 COAL PRICE AND INDEXING

The basic commercial property of thermal coal is its net calorific value (NCV) and this parameter together with other properties which may affect commercial and environmental considerations will determine the price paid for the coal. For spot cargoes this is a straightforward transaction and in term contracts any changes in quality will be addressed in the regular negotiations. Actual coal prices paid are calculated using a variety of formulas. For example, a typical formula may be:

$$P_{CS} = P_B \times \frac{CV_{CS}}{CV_B}$$

where: P_{CS} = the price paid for coal deliveries
P_B = base price for coal
CV_{CS} = NCV of coal delivered
CV_B = base NCV.

This is to ensure that the power station effectively buys calories or joules at a fixed price. The contract may be in terms of gross CV (GCV), in which case it is necessary to make a price adjustment for moisture content.

The price of coal is influenced by the world market, CV and tonnage required and freight and insurance rates. Cargoes may be purchased as Free on Board (FOB) cargoes, where they are purchased when loaded at the port of embarkation, as Cost and Freight (C&F) when the cargo is purchased on arrival at the port of delivery or as Cost, Insurance and Freight (CIF) when insurance is added. In the case of Delivered Ex-Ship (DES), all freight and risk of loss are with the seller, and most coal suppliers will not enter into this type of contract so they are rare. Additions to these costs may be governmental charges such as royalties, federal and local government taxes and levies and export/import taxes. All costs will include any freight component required to transport the coal from the mine to the port of embarkation.

The market rate for coal can either be the average price of a coal cargo from a particular country, coalfield or coal terminal at the point of ship loading, or alternatively the average price for coal imports from any supplier at the port of entry. There are a number of sources of price data containing the current market rates which are compared to historical rates in the form of a coal price index.

Indexed contracts may include price adjustments based on one, or sometimes a 'basket' of these indices. At one time, the 'Japanese Benchmark Prices' were defined for thermal and coking coal imported into Japan by the major power corporations in consultation with Australian or American thermal coal suppliers. These were designated the market prices; however, purchasers attempted and succeeded in buying coal at a discount to the Japanese price, and as a result, the Japanese Benchmark Price has now been abandoned and the Japanese are now employing a 'Fair Treatment System' which has resulted in separately negotiated prices to individual suppliers.

The McCloskey Coal Report (MCIS) publishes spot prices for steam coal collected from cooperating contacts, both suppliers and buyers, and these prices are adjusted to 6000 kcal/kg net as received basis. The index is available on a monthly basis and accurately illustrates trends in pricing for the European spot market and for the Asian coal market. This index is not considered suitable for pricing long term contracts. The South African Coal Report (SACR) index was launched in 1998. This index has been backdated to a price in January 1986 and has been limited to a particular type of coal and a particular market. The prices used in the index are actual shipments to the European power generation market, which consumes over 24 Mtpa of South African steam coal. The SACR reflects quickly and accurately spot market trends. Colombia and Tai Power also publish indexes, which are of limited use outside their geographical regions.

Other indices are the average price of cargoes already sold. Any indexation of prices to these indexes requires a retrospective adjustment in price. Current indices include that of the European Union, based on the quarterly average CIF prices for coal imported from outside the European Union and adjusted to a standard quality priced in US dollars. The CIF price therefore includes the cost of sea freight into Europe. The disadvantage of this index is that the data and index are only available for publication some six months after the quarter indexed. This means that all prices for cargoes must be retrospectively adjusted, since price changes will occur each quarter.

A comparison of indexes (Figure 14.16) shows that none of the curves correspond exactly, but that they do indicate similar trends. The volatility of the various indices is a measure of the periodicity, i.e. monthly indices vary more than indices set annually, and spot prices vary more than term contracts. Current prices are now showing an upward trend after a decline in the late 1990s.

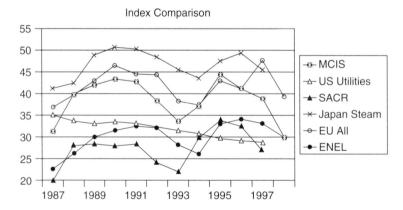

Figure 14.16 Comparison of selected coal indexes over a ten year period. Reproduced by permission of Dargo Associates Ltd

It is important to select an index applicable to the particular combination of power station and supplier. It is also necessary for the power station to take into account the price paid by competing power stations when considering its own indexation. If the competing stations do not index, there may be times when they are able to purchase cheaper coal. It can be seen that, even in broad terms, the indices do not all move up or down at the same time, which can result in individual contracts paying disproportionate prices at certain times.

Bibliography

Aleksic, B.R., Ercegovac, M.D., Cvetkovic, O.G., Markovic, B.Z., Glumicic, T.L., Aleksic, B.D. and Vitorovic, D.K. (1997) Conversion of low rank coal into liquid fuels by direct hydrogenation. *European Coal Geology and Technology* (Editors R. Gayer and J. Pesek), *Geol. Soc. Spec. Pub. No. 125*, 357–363.

American Society for Testing and Materials (1977) Classification of coals by rank, D388–99.

American Society for Testing and Materials (1978) Megascopic description of coal and coal seams and microscopical description and analysis of coal, D2796–88.

Bailey, H.E., Glover, B.W., Holloway, S. and Young, S.R. (1995) Controls on coalbed methane prospectivity in Great Britain. In: *European Coal Geology* (Editors M.K.G. Whateley and D.A. Spears), *Geol. Soc. Spec. Pub. No. 82*, 251–265.

Banks, D., Burke, S.P. and Gray C.G. (1997) Hydrochemistry of coal mine drainage and other ferruginous waters in North Derbyshire and South Yorkshire, UK. *Quart. Journ. of Eng. Geol.* **30**, 257–280.

Barlow, J.A. (1974) Coal and coal mining in West Virginia. West Virginia *Geol. and Econ. Surv. Bull 2.*

Barnes, J.W. (1981) *Basic Geological Mapping, Geol. Soc. London Prof. Handbook Ser.*, Open University Press, Buckingham.

Barraza, J., Cloke, M. and Belghazi, A. (1997) Improvements in direct coal liquefaction using beneficiated coal fractions. In: *European Coal Geology and Technology* (Editors R. Gayer and J. Pesek), *Geol. Soc. Spec. Pub. No. 125*, 349–356.

Barton, N., Lion, R. and Lunde, J. (1974) Engineering classification of rock masses for the design of tunnel support. *Rock Mechanics* **6** (4), 189–236.

Beaver, F.W., Groenewold, G.H., Schmit, C.R., Daly, D.J. and Oliver, R.L. (1991) The role of hydrogeology in underground coal gasification with an example from the Rocky Mountain 1 (RMI) test, Carbon County, Wyoming. In: *Geology in Coal Resource Utilisation* (Editor D.C. Peters), Energy Minerals Division AAPG, 169–186.

Bentley, S.P., Davies, M.C.R. and Gallup, M. (1998) The Cilfynydd flow slide of December 1939. *Quart. Journ. of Eng. Geol.* **31**, 273–289.

Berkowitz, N. (1979) *An Introduction to Coal Technology*, Academic Press, New York.

Berkman, D.A. and Ryall, W.R. (1987) *Field Geologists Manual* (second edition), The Australasian Institute of Mining and Metallurgy, Parkville, Victoria, Australia.

Bhaskaran, R. and Singh, U.P. (2000) Indian coalbed methane. *Mintech* **21** (2), 13–20.

Bieniawski, Z.T. (1976) Rock mass classification in rock engineering. *Proc. Symposium on Exploration for Rock Engineering, Johannesburg* **1** (4), 97–106.

BP Coal Ltd (1987) *BP Coal Handbook*. BP Govt. & Public Affairs Dept.

BP p.l.c. (2001) *BP Statistical Review of World Energy, June 2001*, London.

BPB Instruments Ltd (1981) *Coal Interpretation Manual*, Reeves Oilfield Services Ltd.

Brassington, R. (1988) *Field Hydrogeology*, Geol. Soc. London Prof. Handbook Ser., Open University Press, Buckingham.

Brawner, C.O. (1986) Groundwater and coal mining. *Min. Sci. Technol.* **3**, 187–198.

Brereton, N.R. and Evans, C.J. (1987) Rock stress orientation in the United Kingdom from borehole breakouts. Report of Br. Geol. Surv. RG87/14.

British Coal (1964) *Coal Classification System* (revision of 1964 system), National Coal Board, London.

British Standards Institution (1991) BS 1016, Part 107, *Caking and Swelling Properties of Coal*, BSI, London.

Broadhurst, F.M. and Simpson, I.M. (1983) Syntectonic sedimentation, rigs and fault re-activation in the Coal Measures of Britain. *J. Geol.* **91**, 330–337.

Burke, S.P. and Younger, P.L. (2000) Groundwater rebound in the South Yorkshire coalfield: a first approximation using the GRAM model. *Quart. Journ. Eng. Geol. & Hydrogeol.* **33**, 149–160.

Busby, J.P. and Evans, R.B. (1988) *Depth to Magnetic Basement in North-west Bangladesh from Digital Aeromagnetic Data*. Br. Geol. Surv. Tech. Rep. No. WK/88/3.

Bustin, R.M. (1988) Sedimentology and characteristics of dispersed organic matter in Tertiary Niger Delta: origin of source rocks in a deltaic environment. *Am. Assoc. of Petrol. Geol. Bull.* **72**, 277–298.

Bustin R.M., Cameron A.R., Grieve, D.A. and Kalkreuth, W.D. (1983) *Coal Petrology, Its Principles, Methods and Applications, Short Course Notes*, Geol. Assoc. Can. No. 3.

Callcott, T.G. and Callcott, R. (1990) Review of coal characteristics of the Bowen Basin. In: *Proc. Bowen Basin Symp. and GSA (Qld. Division) Field Conf. Mackay, Queensland* (Editor J.W. Beeston), 47–53.

Carpenter, D. and Robson, C. (2000) Integrated seismic exploration for mine planning and profitability. *First Break* **18** (8), 343–349.

Casshyap, S.M. and Tewari, R.C. (1984) Fluvial models of the Lower Permian coal measures of Son–Mahanadi and Koel–Damodar Valley Basins, India. In: *Sedimentology of Coal and Coal-bearing Sequences* (Editors R.A. Rahmani and R.M. Flores), *Spec. Publ. Int. Assoc. Sedimentol. No. 7*, Blackwell Scientific Publications, Oxford, 121–147.

Cecil, C.B., Dulong, F.T., Cobb, J.C. and Supardi (1993) Allogenic and Autogenic Controls on Sedimentation in the Central Sumatra Basin as an analogue for Pennsylvanian Coal-bearing Strata in the Appalachian Basin. In: *Modern and Ancient Coal-forming Environments* (Editors C.J. Cobb and C.B. Cecil), *Geol. Soc. Am. Sp. Paper 286*, 3–22.

Chalupnik, S., Michalik, B., Wysocka, M., Skubacz, K. and Mielnikow, A. (2001) Contamination of settling ponds and rivers as a result of discharge of radium-bearing waters from Polish coal mines. *J. Env. Rad.* **54**, 85–98.

Clarke, L.B. (1995) Coal Mining and Water Quality. *IEA Coal Research, IEACR/80 July 1995.*

Clifton, A.W. (1987) Pre-mining geotechnical investigations for a Saskatchewan coal mine. *Int. J. Surf. Min.* **1**, 27–34.

Clymo, R.S. (1987) Rainwater-fed peat as a precursor to coal. In: *Coal and Coal-bearing Strata: Recent Advances* (Editor A.C. Scott), *Spec. Publ. Geol. Soc. London No. 32*, 17–23.

Cohen, A.D. and Spackman, W. (1972) Methods of peat petrology and their application to reconstruction of paleoenvironments. *Geol. Soc. Am. Bull.* **83**, 129–142.

Cohen, A.D. and Spackman, W. (1980) Phytogenic organic sediments and sedimentary environments in the Everglades mangrove complex of Florida. Part III. The alteration of plant material in peat and origin of coal macerals. *Paleontogr. B* **162**, 144.

Cook, P.J. and Harris, P.M. (1998) Reserves, resources and the UK mining industry. *International Mining & Minerals*, May 1998, **1** (5), 120–133.

Creedy, D.P. (1999) Coalbed Methane – The R & D needs of the UK. *ETSU/DTI Report No. COAL R163.*

Curry, D.J., Emmett, J.K. and Hunt, J.W. (1994) Geochemistry of aliphatic-rich coals in the Cooper Basin, Australia, and Taranaki Basin, New Zealand: implications for the occurrence of potentially oil-generative coals. In: *Coal and Coal-bearing Strata as Oil-Prone Source Rocks* (Editors A.C. Scott and A.J. Fleet), *Geol. Soc. Spec. Pub. No. 77*, 149–182.

Deere, D.U. (1964) Technical description of rock cores for engineer-purposes. *Rock Mechanics & Engineering Geology* **1** (1), 17–22.

Diessel, C.F.K. (1992) *Coal-bearing depositional systems*, Springer-Verlag, Berlin.

Donica, D.R. (1978) The Geology of the Hartshorne coals (Desmoinesign) in parts of the Heavener 15′ Quadrangle, Le Flore Country, Oklahoma: M.S. Thesis, University of Oklahoma.

Dorrington, M.A., Hughes, I.S.C, Carr, C.E., Rantell, T.D. and Davies, M. (1995) Emissions of environmental concern from coal utilisation. *ETSU Report, No. COAL RO46.*

Douchanov, D. and Minkova, V. (1997) The possibility of underground gasification of Bulgarian Dobrudja's coal. In: *European Coal Geology and Technology* (Editors R. Gayer and J. Pesek), *Geol. Soc. Spec. Pub. No. 125*, 385–390.

Downing, R.A., Land, D.H., Allen, W.R., Lovelock, P.E.R. and Bridge, L.R. (1970) The hydrology of the Trent River Basin. *Water Supply Papers of the Inst. Geol. Sci. Hydrogeol. Report No. 5*, NERC.

Dreeson, R., Bossiroy, D., Dusar, M., Flores, R.M. and Verkaeren, P. (1995) Overview of the influence of syn-sedimentary tectonics and paleo-fluvial systems on coal seam and sand body characteristics in the Westphalian C strata, Campine Basin, Belgium. In: *European Coal Geology* (Editors M.K.G. Whateley and D.A. Spears), *Geol. Soc. Spec. Pub. No. 82*, 215–232.

DTI Report (2000) Flue gas desulphurisation (FGD) Technologies. *Cleaner Coal Technology Programme Technical Status Report O12, March 2000.*

DTI Report (2000) Fluidised bed combustion systems for power generation and other industrial applications. *Clean Coal Technology Programme Technical Status Report O11, January 2000.*

Duff, P. McL. D. (1987) Mesozoic and Tertiary coals – a major world energy resource. *Mod. Geol.* **II**, 29–50.

Durand, B. and Oudin, J.L. (1979) Example de migration des hydrocarbures dans une série deltaïque – la delta Mahakham, Kalimantan, Indonésie. *Proc. Tenth World Petroleum Congress* **2**, 3–12.

du Toit, A.L. (1937) *Our Wandering Continents*, Oliver & Boyd.

Eckels, R. and Carlson, B. (1999) Mining with GPS. *World Coal* **8** (8), 39–40.

Elliott R.E. (1973) Coal mining risks and reserves classification. In: *Septième Congrès Internationale de Strat. et de Carbonifère. Congrès Internationale Stratigr. Geol. Carbonifère, Krefeld. C.R.* **7** (2), 467–477, Editor K.-H. Josten, Geologisches Landesamt Nordrhein-Westfalen, Krefeld.

Elliott, T. and Lapido, K.O. (1981) Syn-sedimentary gravity slides (growth faults) in the coal measures of South Wales. *Nature* **291**, 220–222.

Energy Technology Support Unit (1993) Field trial of UCG by underground gasification Europe (UGE), Teruel, Spain. *Report No. COAL R009(P1) ETSU Report.*

Evans, R.B. and Greenwood, P.G. (1988) Outcrop magnetic susceptibility measurements as a means of differentiating rock types and their mineralisation. In: *Asia Mining '88*, Kuala Lumpur, Institute of Mining and Metallurgy, London, 45–57.

Ferm, J.C. (1979) Allegheny deltaic deposits: a model for the coal-bearing strata. In: *Carboniferous Depositional Environments in the Appalachian Region* (Editors J.C. Ferm and J.C. Horne), Carolina Coal Group, Dept of Geology, University of South Carolina, Columbia, USA, 291–294.

Ferm, J.C. and Smith, G.C. (1980) *A guide to cored rocks in the Pittsburgh Basin*. Dept. Geol. Univ. of Kentucky and Dept. of Geol. Univ. South Carolina.

Ferm, J.C., Staub, J.R., Baganz, B.P., Clark, W.J., Galloway, M.C.C., Hohos, E.F., Jones, T.L., Mathew, D., Pedlow, G.W. and Robinson, M.J. (1979) The shape of coal bodies. In: *Carboniferous Depositional Environments in the Appalachian Region* (Editors J.C. Ferm and J.C. Horne), Carolina Coal Group, Dept of Geology, University of South Carolina, Columbia, USA, 605–619.

Ferm, J.C. and Staub, J.R. (1984) Depositional controls of mineable coal bodies. In: *Sedimentology of Coal and Coal-bearing Sequences* (Editors R.A. Rahmani and R.M. Flores), *Spec. Publ. Int. Assoc. Sedimentol. No. 7*, 275, Blackwell Scientific Publications, Oxford.

Ferm, J.C. and Weisenfluh, G.A. (1991) *Cored Rocks of the Southern Appalachian Coalfields*. University of Kentucky.

Fettweis, G.B. (1979) *World Coal Resources, Methods of Assessment and Results, Dev. Econ. Geol. Ser. No. 10*, Elsevier, Amsterdam.

Fettweis, G.B. (1985) Considerations on coal deposits on basis of coal production. In: *Xth Congress on Carboniferous Stratigraphy and Geology, Madrid, Spain, 1983, Symposium 5: Economic Geology: Coal Resources and Coal Exploration, Compte Rendu, Instituto Geologico y Minero de Espana Madrid*, 93–110.

Fielding, C.R. (1984) 'S' or 'Z' shaped coal seam splits in the coal measures of County Durham, UK. *Proc. Yorks. Geol. Soc.* **45**, 85–89.

Findlay, M.J., Goulty, N.R. and Kragh, J.E. (1991) The cross-hole seismic reflection method in opencast coal exploration. *First Break* **9**, 509–514.

Finkelman, R.B., Feder, G.L. and Orem, W.H. (1991) Relation between low-rank coal deposits and Balkan Endemic Nephropathy. *AGID News, No. 65*, 23.

Firth, D. (1999) *Log analysis for mining applications.* Reeves Oilfield Services Ltd (restricted distribution).

Fleet, A.J. and Scott, A.C. (1994) Coal and coal-bearing strata as oil-prone source rocks: an overview. In: *Coal and Coal-bearing Strata as Oil-Prone Source Rocks* (Editors A.C. Scott and A.J. Fleet), *Geol. Soc. Spec. Pub. No. 77*, 1–8.

Flint, S., Aitken, J. and Hampson, G. (1995) Application of sequence stratigraphy to coal-bearing coastal plain successions: implications for the UK coal measures. In: *European Coal Geology* (Editors M.K.G. Whateley and D.A. Spears), *Geol. Soc. Spec. Pub. No. 82*, 1–16.

Fookes, P.G. (1997) Geology for engineers: the geological model, prediction and performance. *Quart. Journ. Eng. Geol.* **30** 293–424.

Forbes, P.L., Ungerer, P.M., Kuhfuss, A.B., Riis, F. and Eggen, S. (1991) Compositional modelling of petroleum generation and expulsion: trial application to a local mass balance in the Smorbukk Sor field, Haltenbanken area, Norway. *Amer. Assoc. of Petrol. Geol. Bull.* **75**, 873–893.

Friedman, S.A. (1978) Demoinesian coal deposits in part of the Arkoma Basin, eastern Oklahoma. In: *Am. Assoc. Petrol. Geol. Field Guidebook*, Oklahoma City Geol. Soc. Oklahoma.

Frodsham, K. and Gayer, R.A. (1999) The impact of tectonic deformation upon coal seams in the South Wales coalfield, UK. *Int. Journ. of Coal Geology No. 38*, 297–332.

Fuels Research Institute (1978) Classification standards for South African coals. In: *Coal, Gold, Base Metals, Southern Africa*, 67–87.

Fulton, I.M., Guion, P.D. and Jones, N.S. (1995) Application of sedimentology to the development and extraction of deeper mined coal. In: *European Coal Geology* (Editors M.K.G. Whateley and D.A. Spears), *Geol. Soc. Spec. Publ. No. 82*, 17–43.

Galligan, A.C. and Mengel, D.C. (1986) Code for reporting of identified coal resources and reserves. *Queensland Govt Min. J.* **87** (May), 201–203.

Gan, H., Nandi, S.P. and Walker, P.L. Jr (1972) Nature of the porosity in American coals. *Fuel* **51**, 272–277.

Garratt, J. (2001) Breathing new life. *World Coal* **10** (3), 55–58.

Garratt, M.N. (1999) Computer modelling as a tool for strata control and reinforcement design. *Int. Mining & Minerals* **2** (14), 45–54.

Gastaldo, R.A., Allen, G.P. and Huc, A.Y. (1993) Detrital Peat Foundation in the tropical Mahakam River Delta, Kalimantan, Eastern Borneo: Sedimentation, Plant Composition and Geochemistry. In: *Modern and Ancient Coal-forming Environments* (Editors C.J. Cobb and C.B. Cecil), *Geol. Soc. Am. Sp. Paper No. 286*, 107–118.

Gayer, R., Cole, J., Frodsham, K., Hartley, A.J., Hillier, B., Miliorizos, M. and White, S.C. (1991) The role of fluids in the evolution of the South Wales Coalfield foreland basin. *Proc. Ussher Soc.* **7**, 380–384.

Geological Society Engineering Group Working Party (1970) The logging of rock cores for engineering purposes. *Quart. Journ. Eng. Geol.* **3**, 1–24.

George, A.M. (1975) Brown coal lithotypes in the Latrobe Valley deposits. State Electr. Comm. of Victoria, Richmond, Austral., Petrol. Rep., No. 17.

Gilewicz, P. (1999) Sizeable differences. *World Coal* **8** (7), 26–33.

Gilewicz, P. (1999) U.S. Dragline Census. *Coal Age, August*, 35–40.

Gochioco, L. (1991) Advances in seismic reflection profiling for US coal exploration. *Geophysics* **10** (12), 24–29.

Goscinski, J.S. and Robinson, J.W. (1978) Megascopic description of coal drill cores. In: *Field Description of Coal: Symposium of the ASTM Committee D-5 on Coal and Coke, Ottawa, Canada, 22–23 September 1976*, ASTM Technical Publication 661 (Editor R.R. Dutcher) 1916 Race Street, Philadelphia, Pennsylvania, USA, 50–57.

Goulty, N.R. and Brabham, P.J. (1984) Seismic refraction profiling in opencast coal exploration. *First Break* **2**, 26–34.

Goulty, N.R., Daley, T.E., Walters, K.G. and Emsley, D.B. (1984) Location of dykes in coalfield exploration. *First Break* **2**, 15–21.

Goulty, N.R., Thatcher, J.S., Findlay, M.J., Krach, J.E. and Jackson, P.D. (1990) Experimental investigation of cross-hole seismic techniques for shallow coal exploration. *Quart. Journ. Eng. Geol.* **23**, 217–228.

Grady, W.C., Eble, C.F. and Neuzil, S.G. (1993) Brown coal maceral distributions in a modern domed tropical Indonesian peat, and a comparison with maceral distributions in Middle Pennsylvanian-age Appalachian bituminous coal beds. In: *Modern and Ancient Coal-forming Environments* (Editors C.J. Cobb and C.B. Cecil), *Geol. Soc. Am. Sp. Paper No. 286*, 63–82.

Grayson, R.L., Wang, Y.J. and Sandford, R.L. (Editors) (1990) *Use of Computers in the Coal Industry: Proceedings of the 4th Conference on the Use of Computers in the Coal Industry, West Virginia University, Morgantown, WV, USA, 20–22 June 1990*, A. A. Balkema, Rotterdam.

Greaves, R.J. (1985) Coal prospect evaluation using high-resolution reflection seismology: a case study. *Min. Engin.* **37**, 1061–1064.

Green, M.B. (1999) Underground coal gasification – a joint European field trial in Spain. *ETSU/DTI Report No. COALR169, DTI/PUB, URN 99/1093.*

Gregor, V. and Tezky, A. (1997) A well logging method for the determination of the sulphur content of coal seams by means of deep gammaspectrometry. In: *European Coal Geology and Technology* (Editors R. Gayer and J. Pesek), *Geol. Soc. Spec. Publ. No. 125*, 297–307.

Grimshaw, P. (1997) Digging Deep. *World Coal*, July, 40–45.

Guion, P.D., Fulton, I.M. and Jones, N.S. (1995) Sedimentary facies of the coal-bearing Westphalian A and B north of the Wales–Brabant High. In: *European Coal Geology* (Editors M.K.G. Whateley and D.A. Spears), *Geol. Soc. Spec. Pub. No. 82*, 45–78.

Harder, P.B. (1998) Setting new standards. *World Coal* **7** (6), 25–29.

Hagemann, H.W. (1978) Macropetrographic classification of brown coal. Unpublished Proposal Presented to Members of the International Committee for Coal Petrology (ICCP), Essen, Germany, April.

Hagemann, H.W. (1980) *Part 1. Identification of Lithotypes in Lignites of Southern Saskatchewan. Part 2. Macropetrographic Examination and Collecting Samples from Cores Stored at the Hat Creek Mine Site, British Columbia. Geol. Surv. Can. Open File Rep.*

Hargreaves, A.J. and Lunarzewski, L. (1985) Review of gas seam drainage in Australia. *Bull. Proc. Aust. Inst. Min. Metall.* **290** (1), 55–70.

Hartman, H.L. (1992) *SME Mining Engineering Handbook* (second edition, 2 vols). Soc. Mining, Metallurgy and Exploration Inc. Littleton, Colorado.

Haughton, S.H. (1969) Karroo System. In: *Geological History of Southern Africa*, Geological Society of Southern Africa, Ch. 13, 349–415.

Hedberg, H.D. (1968) Significance of high wax oils with respect to genesis of petroleum. *Am. Assoc. Pet. Geol. Bull.* **52**, 736–750.

Hilt, C. (1873) Die Beiziehungen zwischen der Zusammensetzung und den technischen Eigenshaften der Steinkohle. *Bezirksvereinigung Ver. Deutsch. Ingenieure Zeitschrift* **17** (4), 194–202.

Hobday, D.K. (1987) Gondwana coal basins of Australia and Africa: tectonic setting, depositional systems and resources. In: *Coal and Coal-bearing Strata – Recent Advances* (Editor A.C. Scott), *Geol. Soc. Spec. Pub. No. 32*, 219–255.

Hoek, E. and Brown, E.T. (1980) *Underground excavation in rock*. Institute of Mining Metallurgy.

Holub, K. (1997) Seismic monitoring for rock burst prevention in the Ostrava–Karmina coalfield, Czech Republic. In: *European Coal Geology and Technology* (Editors R. Gayer and J. Pesek), *Geol. Soc. Spec. Pub. No. 125*, 321–328.

Hooper, R.L. (1987) Factors affecting the magnetic susceptibility of baked rocks above a burned coal seam. *Int. J. Coal Geol.* **9**, 157–169.

Horne, J.C. (1979) Sedimentary responses to contemporaneous tectonism. In: *Carboniferous Depositional Environments in the Appalachian Region* (Editors J.C. Ferm and J.C. Horne), Carolina Coal Group, Dept. of Geology, University of South Carolina, Columbia, USA, 259–265.

Horne, J.C., Howell, D.J., Baganz, B.P. and Ferm, J.C. (1978) Splay Deposits as an Economic Factor in Coal Mining, *Colorado Geol. Surv. Resour. Ser. No. 4*, 89–100.

Horne, J.C., Ferm, J.C., Caruccio, F.T. and Baganz, B.P. (1979) Depositional models in coal exploration and mine planning in Appalachian Region. In: *Carboniferous Depositional Environments in the Appalachian Region* (Editors J.C. Ferm and J.C. Horne), Carolina Coal Group, Dept. of Geology, University of South Carolina, Columbia, USA, 544–575.

Howland, J. (1998) Canada adjusts to a new era. *World Coal* **7** (3), 13–19.

Hughes, V.J. and Kennett, B.L.N. (1983) The nature of seismic reflections from coal seams. *First Break* **1**, 9–18.

Hunt, J.M. (1979) *Petroleum geochemistry and geology*. W.H. Freeman and Co, San Francisco.

Hunt, J.W. and Hobday, D.K. (1984) Petrographic composition and Sulphur content of coals associated with alluvial fans in the Permian Sydney and Gunnedah Basins, eastern Australia. In: *Sedimentology of coal-bearing sequences* (Eds R.A. Rahmani and R.M. Flores) Spec. Pub. Int. Assoc. of Sedimentol. No. 43–60. Blackwell Scientific Pub. Oxford.

Hunt, K. and Bigby, D. (1999) Designing for Success. *World Coal* **8** (7), 47–52.

IMC Geophysics Ltd (1997) Enhancement of In-Seam Seismic Techniques. *Dept. of Trade and Industry Rept. No. COAL R077.*

International Mining Consultancy (IMC) (1997) Methane from abandoned coal mines. *ETSU/DTI Report No. COAL R131.*

International Mining Consultancy (IMC) (1997) Methane from rapidly advancing drivages. *ETSU/DTI Report No. COAL R131.*

Issler, D.R. and Snowdon, L.R. (1990) Hydrocarbon generation kinetics and thermal modelling, Beaufort–Mackenzie Basin. *Bull. Canadian Petrol. Geol.* **38**, 1–16.

Jackson, L.J. (1981) Geophysical Examination of Coal Deposits. *IEA Coal Research Rep. No. ICTIX/TR13.*

Jerzykiewicz, T. (1992) Controls on the distribution of coal in the Campanian to Paleocene Post-Wapiabi strata in the Rocky Mountain Foothills, Canada. In: *Controls on the Distribution and Quality of Cretaceous Coals. Geol. Soc. Am. Sp. Paper 267*, 139–150.

Jerzykiewicz, T. and McLean, J.R. (1980) Lithostratigraphical and Sedimentological framework of Coal-bearing Upper Cretaceous and Lower Tertiary strata Coal Valley, Central Alberta Foothills. *Geol. Sur. Canada, Paper 79-12.*

Jones, B.G. and Hutton, A.C. (Editors) (1984) Fluvio-deltaic systems – facies analysis in exploration. *Australasian sedimentologists specialists group, Wollongong.*

Jones, T.A., Hamilton, D.E. and Johnson, C.R. (1986) Contouring Geologic Surfaces with the Computer. *Computer Methods in the Geosciences*, Van Nostrand Reinhold, New York.

Jongmans, W.J., Koopmans, R.G. and Roos, G. (1936) Nomenclature of coal petrography. *Fuel* **15**, 14–15.

Juch, D. and Working Group (1983) New methods of coal resources calculation. In: *Xth Congress on Carboniferous Stratigraphy and Geology, Madrid, Spain, 12–17 September. Symposium 5: Economic Geology: Coal Resources and Coal Exploration, Compte Rendu, Instituto Geologico y Minero de Espana, Madrid*, 117–124.

Kalab, Z. (1997) An analysis of mining induced seismicity and its relationship to fault zones. In: *European Coal Geology and Technology* (Editors R. Gayer and J. Pesek), *Geol. Soc. Spec. Pub. No. 125*, 329–335.

Karr, C. Jr (Editor) (1978) *Analytical Methods for Coal and Coal Products*, Vols 1 and 2, Academic Press, New York.

Kim, A.G. (1977) Estimating methane content of bituminous coal beds from adsorption data. *US Bureau of Mines Report of Investigations 8245.*

Knapp, R. (1997) Changing the public perception of coal. *World Coal* **6** (12), 23–27.

Knapp, R. (2001) Atmosphere of uncertainty. *World Coal* **10** (2), 8–12.

Knutson, H.A. (1983) Planning and implementation of coal exploration programs in reconnaissance geology for coal exploration. In: *Proceedings of the 4th International Coal Exploration Symposium, 15–20 May, Sydney, NSW, Australia.*

Kosanke, R.M. (1950) Pennsylvanian spores of Illinois & their use in correlation. *Illinois State Geol. Surv. Bull. No. 74.*

Kotas, A. (1994) *Coal-bed methane potential of the Upper Silesian coal basin, Poland*. 75th Anniv, Polish Geological Institute, Warsaw.

Kragh, J.E., Goulty, N.R. and Findlay, M.J. (1991) Hole-to-surface reflection surveys for shallow coal exploration. *First Break* **9**, 335–344.

Lake, R.D. (1999) The Wakefield district – a concise account of the geology. *Memoirs of the Br. Geol. Surv. Sheet 78 (England & Wales)*.

Land, D.H. and Jones, C.M. (1987) Coal geology and exploration of part of the Tertiary Kutei Basin in East Kalimantan, Indonesia. In: *Coal and Coal-bearing Strata: Recent Advances* (Editor A.C. Scott), *Spec. Publ. Geol. Soc. London No. 32*, 235–255.

Laurila, M.J. and Corriveau, M.P. (1995) *The Sampling of Coal*. Intertec Publishing, Chicago.

Lawrence, D.T. (1992) Primary controls in total reserves, thickness, geometry and distribution of coal seams: upper cretaceous Adaville Formation, Southwestern Wyoming. In: *Controls on the Distribution and Quality of Cretaceous Coals. Geol. Soc. Am. Sp. Paper 267*, 69–100.

Leeming, J.R. and Fifoot, T.J. (2001) The management of heat and humidity in underground coal mines. *Int. Mining and Minerals No. 42, June*, 19–25.

Levine, J.R. (1993) Coalification: the evolution of coal as a source rock and reservoir rock for oil and gas. In: *Hydrocarbons from Coal* (Editors B.E. Law and D.D. Rice), *Amer. Ass. Petrol. Geol. Stud. In Geol. Ser. 38*, 39–77.

Lindqvist, J.K., Hatherton, T. and Mumme, T.C. (1985) Magnetic anomalies resulting from baked sediments over burnt coal seams in southern New Zealand. *N.Z. J. Geol. Geophys.* **28**, 405–412.

Macgregor, D.S. (1994) Coal-bearing strata as source rocks – a global overview. In *Coal and Coal-bearing Strata as Oil-prone Source Rocks?* (Editors A.C. Scott and A.J. Fleet), *Geol. Soc. Spec. Pub. 77*, 1–8.

McCabe, P.J. (1984) Depositional environments of coal and coal-bearing strata. In: *Sedimentology of Coal and Coal-bearing Sequences* (Editors R.A. Rahmani and R.M. Flores), *Spec. Publ. Int. Assoc. Sedimentol. No. 7*, Blackwell Scientific Publications, Oxford, 13–42.

McCabe, P.J. (1987) Facies studies of coal and coal-bearing strata. In: *Coal and Coal-bearing Strata: Recent Advances* (Editor A.C. Scott), *Spec. Publ. Geol. Soc. London No. 32*, 51–66.

McCabe, P.J. (1991) Geology of coal: environments of deposition. In: *Economic Geology of the USA* (Editors H.J. Gluskoter, D.D. Rice and R.B. Taylor), *DNAG: The Geology of North America*, Vol. P2, Geological Society of America, Boulder, CO, 469–482.

McCabe, P.J. and Parrish, J.T. (1992) Tectonic and climatic controls on the distribution and quality of Cretaceous coals. In: *Controls on the Distribution and Quality of Cretaceous Coals. Geol. Soc. Am. Sp. Paper 267*, 1–15.

McCloskey Coal Information Services (MCIS) Ltd (1996) *The Big Coal Book*.

McConville, A. (1997) Emission Standards Handbook. IEA Coal Research, *IEA/CR/96*, London.

McFaull, K.S., Wicks, D.E., Sedwick, K. and Brandenburg, C. (1987) An analysis of the coal and coalbed methane resources of the Piceance Basin, Colorado US. In: *SPE/DOE 16418 Low Permeability Reservoir Symposium, Denver, CO, USA, 18–19 May 1987*, Society of Petroleum Engineers, 283–295.

Mark, C. (2001) Focus of ground control: horizontal stress. *Coal Age*, March, 47–50.

Marshall, J.S., Pilcher, R.C. and Bibler, C.J. (1996) Opportunities for the development and utilisation of coalbed methane

in three coal basins in Russia and Ukraine. In: *Coalbed Methane and Coal Geology* (Editors R. Gayer and I. Harris), *Geol. Soc. Spec. Pub. No. 109*, 89–101.

Maxwell, D. (2000) Planning in 3 dimensions. *World Coal* **9** (11), 36–38.

Mazzone, V.J. (1998) Sampling and analysis standards in the coal trade. *World Coal* **7** (3), 21–24.

Mendelev, D.I. (1888) *Severung Vestn, St Petersburg*. No. 8, Sect. 2, p. 27; No. 9, Sect. 2, p. 1; No. 10, Sect. 2, p. 1; No. 11, Sect. 2, p. 1; No. 12, Sect. 2, p. 1.

Michalski, S.R. and Gray, R.E. (1997) Investigating the fires. *World Coal* **6** (7), 59–63.

Milsom, J. (1989) *Field Geophysics*. Geological Society of London Handbook, Open University Press, Buckingham.

MINFO® (2000) Oil and Gas in New South Wales. *MINFO (New South Wales Mining and Exploration Quarterly)*, No. 66, 4.

Moore, P.D. (1987) Ecological and hydrological aspects of peat formation. In: *Coal and Coal-bearing Strata: Recent Advances* (Editor A.C. Scott), *Spec. Publ. Geol. Soc. London No. 32*, 7–15.

Moore, P.S., Burns, B.J., Emmett, K.K. and Guthrie, D.A. (1992) Integrated source, maturation and migration analysis, Gippsland Basin, Australia. *Aust. Petroleum Exploration Assoc. Journ.* **32**, 313–324.

Murray, D.K. (1996) Coalbed methane in the USA: analogues for worldwide development. In: *Coalbed Methane and Coal Geology* (Editors R. Gayer and I. Harris), *Geol. Soc. Spec. Pub. No. 109*, 1–12.

Murray, D.K. (2000) CBM in the United States. *World Coal* **9** (3), 61–64.

Myal, F.R. and Frohne, K.H. (1991) Slant hole completion test in the Piceance Basin, Colorado. *Soc. Pet. Eng. Paper 21866*.

National Coal Board (1982) *Technical Management of Water in the Coal Mining Industry*, National Coal Board Mining Department, London.

Nelson, W.J. (1987) Coal deposits of the United States. *Int. J. Coal Geol.* **8**, 355–365.

Nemec, W. (1992) Depositional controls on plant growth and peat accumulation in a braidplain delta environment: Helvetiafjellet Formation (Barremian–Aptian), Svalbard. In: *Controls on the Distribution and Quality of Cretaceous Coals. Geol. Soc. Am. Sp. Paper 267*, 209–226.

Neuzil, S.G., Supardi, Cecil, C.B., Kane, J.S. and Soedjono, K. (1993) Inorganic geochemistry of domed peat in Indonesia and its implications for the origin of mineral matter in coal. In: *Modern and Ancient Coal-forming Environments* (Editors C.J. Cobb and C.B. Cecil), *Geol. Soc. Am. Sp. Paper 286*, 23–44.

Noble, R.A., Alexander, R.I., Kagi, R.I. and Knox, J. (1986) Identification of some dieterpenoid hydrocarbons in petroleum. *Organic Geochemistry* **10**, 363–374.

Oliver, R.L. and Dana, G.F. (1991) Underground Coal Gasification. In: *Geology in Coal Resource Utilisation* (Editor D.C. Peters), Energy Mineral Division AAPG, 155–168.

Oplustil, S., Pesek, J. and Skopec, J. (1997) Comparison of structures derived from mine workings and those interpreted in seismic profiles: an example from the Kacice deposit, Kladno Mine, Bohemia. In: *European Coal Geology and Technology* (Editors R. Gayer and J. Pesek), *Geol. Soc. Spec. Pub. No. 125*, 337–347.

Osborne, D. (1988) *Coal Preparation Technology*, 2 vols, Graham and Trotman, London.

Parasnis, D.S. (1986) *Principles of Applied Geophysics* (fourth edition), Chapman and Hall, London.

Peace, D.G. (1979) Surface reflection seismic – looking underground from the surface. In: *Coal Exploration: Proceedings of the International Coal Symposium, Denver, CO, USA, Vol. 2* (Editor G.O. Argall), Miller Freeman Publications Inc., San Francisco, USA, 230–266.

Pearson, D.E. (1980) The quality of Western Canadian coking coal. *The Canadian Mining and Metallurgical Bull.* Jan., 1–15.

Pearson, D.E. (1985) The Quality of Canadian coal – a petrographic approach to its characterisation and classifications. In: *Coal in Canada* (Editor T.H. Pat Ching), *Canadian Inst. Mining and Metall. Spec. Vol. 31*, 21–30.

Pinchin, J., Baquiran, G.B., Coleby, B.R. and Ryan, D. (1982) MINI-SOSIE. In: *Seismic Techniques for Coal Exploration in Australia and the Philippines. Proc. 4th International Coal Explor. Symposium, Sydney, Australia, 15–20 May*, 1–14.

Pippenger, J. (1998) Dragline and Truck and Shovel, or Bucketwheel? *World Coal* 7 (9), 33–37.

Plumstead, E.P. (1962) The Permo-Carboniferous coal measures of the Transvaal, South Africa – an example of the contrasting stratigraphy in the southern and northern hemispheres. *C.R.4 Congr. Int. Strat. Geol. Carbonifer. Maastricht 2*, 545–550.

Potter, P.E. (1962) Shape and distribution patterns of Pennsylvanian sand bodies in Illinois, Illinois State. *Geol. Surv. Circular 339*.

Potter, P.E. (1963) Late paleozoic sandstones of the Illinois basin, Illinois State. *Geol. Surv. Rept. of Investigations 217*.

Potter, P.E. and Glass, H.D. (1958) Petrology and sedimentation of the Pennsylvanian sediments in Southern Illinois: a vertical profile, Illinois State. *Geol. Surv. Rept. of Investigations 264*.

Potter, P.E. and Simon, J.A. (1961) Anvil Rock Sandstone and channel contours of Herrin (No. 6) coal in West Central Illinois, Illinois State. *Geol. Surv. Circular 314*.

Powell, T.G. and Boreham, C.J. (1994) Terrestrially sourced oils: where do they exist and what are our limits of knowledge? – a geochemical perspective. In: *Coal and Coal-bearing Strata as Oil-Prone Source Rocks* (Editors A.C. Scott and A.J. Fleet), *Geol. Soc. Spec. Pub. No. 77*, 11–29.

Price, M. (1985) *Introducing Groundwater*, Allen and Unwin, London.

Price, M. (1996) *Introducing Groundwater* (second edition), Chapman and Hall, London.

Pryor, E.J. (1965) Mineral Processing (third edition), Elsevier, London.

Puri, R. and Yee, D. (1990) Enhanced coalbed methane recovery. *Soc. Petroleum Engineers Paper SPE 20732*.

Redmayne, D.W., Richards, J.A. and Wild, P.W. (1998) Mining induced earthquakes monitored during pit closure in the Midlothian coalfield. *Quart. Journ. Geol. Soc.* 31, 21–36.

Reynolds, J.M. (1997) *An Introduction to Applied and Environmental Geophysics*. John Wiley & Sons, Chichester.

Ricketts, B.D. and Embry, A.F. (1986) Coal in the Canadian Arctic archipelago. A potential resource. *GEOS* (Ottawa), 15 (1), 16–18.

Rieke, H.H. and Kirr, J.N. (1984) Geologic overview, coal and coalbed methane resources of the Arkoma Basin – Arkansas and Oklahoma. In: *Coalbed Methane Resources of the United States* (Editors C.T. Rightmire, G.E. Eddy and J.N. Kirr), *AAPG Studies in Geology Series No. 7*, 135–161.

Rightmire, C.T. (1984) Coalbed methane resource. In: *Coalbed Methane Resources of the United States*, (Editors C.T. Rightmire, G.E. Eddy and J.N. Kirr), *AAPG Studies in Geology Series 17* 1–13.

Robinson, J. (1998) Treatment of mine drainage. *World Coal* 7 (12), 38–40.

Robinson, N. (1994) Coal liquefaction pioneer plant study. *ETSU/DTI Report No. COAL RO40*.

Ruppert, L.F., Neuzil, S.G., Cecil, C.B. and Kane, J.S. (1993) Inorganic constituents from samples of a domed and Lacustrine peat, Sumatra, Indonesia. In: *Modern and Ancient Coal-forming Environments* (Editors C.J. Cobb and C.B. Cecil), *Geol. Soc. Am. Sp. Paper 286*, 83–96.

Ryer, T.A. and Langer, A.W. (1980) Thickness change involved in the peat-to-coal transformation for bituminous coal of Cretaceous age in central Utah. *Journ. of Sed. Petrol.* 50, 987–992.

Sabo, J. (2000) 'Moving the Goal Posts', *World Coal* 9 (9), 31–36.

Sage, P. and Payne, M. (1999) Coal liquefaction – a technology status review. *ETSU/DTI Report No. COALR184, DTI/Pub URN 99/1241*.

Saus, T. and Schiffer, H.W. (1999) *Lignite in Europe*. Rheinbraun Informiert.

Schopf, J.M. (1960) Field description and sampling of coal beds. *US Geol. Surv. Bull.* **1111-B** (Plates 6–27).

Schultz, K. (1997) Turning a liability into an asset. *World Coal*, December, 38–43.

Seyler, C.A. (1931) Petrology and the classification of coal II. Fuel technology and the classification of coal. *Proc. South Wales Inst. Eng.* **47**, 557–592.

Seyler, C.A. (1938) Petrology and the classification of coal. *Proc. S. Wales Inst. Eng.* **53**, 254–327.

Shen, B. and Fama, H.D. (2001) Geomechanics and Highwall mining. *World Coal* 10 (2), 35–38.

Sherborn Hills, E. (1975) *Elements of Structural Geology* (second edition), Chapman and Hall, London.

Siddall, R.G. and Gale, W.J. (1992) Strata Control – A New Science for an Old Problem. *Inst. Min. Eng.*

Siddle, H.J., Wright, M.D. and Hutchinson, J.N. (1996) Rapid failures of colliery spoil heaps in the South Wales coalfield. *Quart. Journ. Eng. Geol.* **29**, 103–132.

Siemens, W. (1868) *Trans. Chem. Soc.* **21**, 279.

Singh, M.M. (1992) Mine Subsidence. In: *SME Mining Engineering Handbook* (Editor H. Hartman), *Soc. For Min., Metall. and Expl. Inc., Vol. 1*. Littletown, Colorado, 938–971.

Smith, A.H.V. (1968) Seam profiles and seam characteristics. In: *Coal-bearing Strata* (Editors D.G. Murchison and T.S. Westoll), Oliver and Boyd, London, 31–40.

Stach, E. (1982) *Textbook of Coal Petrology* (third edition), Gebruder Borntraeger, Berlin.

Standards Association of Australia (1986) *Symbols for the Graphical Representation of Coal Seams, AS 2916-1986*, Sydney, Australia.

Standards Association of Australia (1987) *Classification and coding systems for Australian coals, AS 2096-1987*, Sydney, Australia.

Standards Association of Australia (1996) *Guide for the taking of samples from hard coals in situ, ASK183-1970*, Sydney, Australia.

Staub, J.R. and Cohen, A.D. (1979) The Snuggedy Swamp of South Carolina. A back-barrier estuarine coal-forming environment. In: *Carboniferous Depositional Environments in the Appalachian Region* (Editors J.C. Ferm and J.C. Horne), Carolina Coal Group, Dept of Geology, University of South Carolina, Columbia, USA, 499–508.

Steenblik, R.R. (1986) *International Coal Resource Assessment: Working Paper No. 73*, IEA Coal Research, London.

Stevens, S.H. (1999) A promising future. *World Coal*, **8** (3), CBM Review, 6–10.

Stopes, M.C. (1919) On the four visible ingredients in banded bituminous coals. *Proc. Roy. Soc.* **90B**, 470–487.

Stopes, M.C. (1935) On the petrology of banded bituminous coals. *Fuel* **14**, 4–13.

Stutzer, O. and Noé, A.C. (1940) *Geology of Coal*, University of Chicago Press, Chicago.

Styan, W.B. and Bustin, R.M. (1983) Sedimentology of Frazer River delta peat: a modern analogue for some ancient deltaic coals. *Int. J. Coal Geol.* **3**, 101–143.

Suggate, R.P. (1959) New Zealand coals, their geological setting and its influence on their properties. *New Zealand Dept. of Scientific and Industrial Research bulletin 134.*

Suggate, R.P. (1982) Low rank sequences and scales of organic metamorphism. *J. Petrol. Geol.* **4**, 377–392.

Suggate, R.P. (2000) The Rank (S_r) scale: its basis and its applicability as a maturity index for all coals. *N.Z. J. Geol. Geophys.* **43**, 521–553.

Suggate, R.P. and Lowery, J.H. (1982) The influences of moisture content on vitrinite reflectance and the assessment of maturation of coal. *N.Z. J. Geol. Geophys.* **25**, 227–231.

Tasker, B.S. (1985) Technical note on polygonal blocks of influence in triangular grids. *Bull. Proc. Aust. Inst. Min. Metall.* **290** (3), 71–72.

Taylor, G.H. and Shibaoka, M. (1976) The rational use of Australia's coal resources. *Inst. of Fuel, Biennial Conf. Prepr.* 8.1–8.5.

Taylor, G.H., Teichmuller, M., Davis, A., Diessel, C.F.K., Littke, R. and Robert, P. (1998) *Organic Petrology*, Gebruder Borntraeger, Berlin.

Teichmuller, M. (1987) Coalification studies and their application to geology. In: *Coal and Coal-bearing Strata: Recent Advances* (Editor A.C. Scott), *Geol. Soc. Spec. Pub. No. 32*, 127–169.

Teichmuller, M. and Teichmuller, R. (1982) The geological basis of coal formation. In: *Stach's Book of Coal Petrography* (third edition, Editors E. Stach, M.Th. Mackowski, M. Teichmuller, G.H. Taylor, D. Chandra and R. Teichmuller). Gebruder Borntraeger, Berlin, 5–86.

Telford, W.M., Geldart, L.P. and Sheriff, R.E. (1990) *Applied Geophysics* (second edition), Cambridge University Press, Cambridge.

Thomas, L.P. (1992) *Handbook of Practical Coal Geology.* John Wiley & Sons, Chichester.

Thomas, L.P. and Frankland, S.C. (1999) Coal in Transition. *World Coal* **8** (10), 2–6.

Thompson, S., Cooper, B.S. and Barnard, P.C. (1994) Some examples and possible explanations for oil generation from coals and coaly sequences. In: *Coal and Coal-bearing Strata as Oil-Prone Source Rocks* (Editors A.C. Scott and A.J. Fleet) *Geol. Soc. Spec. Pub. No. 77*, 119–137.

Thornton, R.C.N. (1979) Regional Stratigraphic Analysis of the Gidgealpa Group, Southern Cooper Basin, Australia.

Bull. 49. Dept. of Mines and Energy, Geol. Surv. South Australia.

United Nations Economic Commission for Europe (1988) *International Codification System for Medium- and High-Rank Coals.* ECE/COAL115. UN Geneva, Switzerland.

United Nations Economic Commission for Europe (2000) *International Codification System for Low-Rank Coals.* ENERGY/2000/12. UN, Geneva, Switzerland.

United Nations Economic and Social Council (1979) *The International Classification of Mineral Resources, Economic Report No. 1 (E/C.7/104), May 1979.* Annex to Natural Resources & Energy **4** (1), Centre for Natural Resources, Energy and Transport, UN Secretariat, New York, USA.

Unrug, K.T. (1992) Construction of development openings. In: *SME Mining Engineering Handbook* (Editor H.L. Hartman). Soc. Mining, Metallurgy and Exploration Inc. Littleton, Colorado, 1580–1645.

Valkovic, V. (1983) *Trace elements in coal* (2 vols), CRC Press Inc., Boca Raton, Florida.

Van Krevelen, D.W. (1961) *Coal*, Elsevier, Amsterdam.

Ventner, R.H. (1976) A statistical approach to the calculation of coal reserves for the plains region of Alberta. *Can. Inst. Min. Bull.* **69** (771), July, 49–52.

Verma, R.K. and Bandyopadhyay, T.K. (1983) Use of resistivity method in geological mapping – case histories from Raniganj Coalfield, India. *Geophys. Prospect.* **31**, 490–507.

Vincent, P.W., Montimore, I.R. and McKirdy, D.M. (1985) Hydrocarbon generation, migration and entrapment in the Jackson–Naccowlan area, ATP 259P, South Western Queensland. *Aust. Petroleum Assoc. Journ.* **25**, 62–84.

Walker, S. (1997) Highwall evolution. *World Coal* **6** (10), 44–52.

Walker, S. (2000) *Major Coalfields of the World*, IEA Coal Research, London.

Walker, S. (2001) Highwall mines keep the coal flowing. *World Coal* **10** (12), 20–26.

Ward, C.R. (1984) *Coal Geology and Coal Technology*, Blackwell Scientific Publications, Oxford.

Weber, K.L. and Knottnerus, B.A. (2000) Coal Upgrading Technology Proven. *World Coal* **9** (7), 7–10.

West, G. (1991) *The Field Description of Engineering Soils and Rocks.* Geol. Soc. Professional Handbook Ser., Oxford University Press, Oxford.

Wood, G.H. Jr, Kehn, T.A., Devereux Carter, M. and Culbertson, W.C. (1983) *Coal Reserve Classification System of the US Geological Survey. USGS Circular No. 891.*

World Bank Group (1998) Thermal Power: Guidelines for New Plants. In: *Pollution Prevention and Abatement Handbook*, World Bank Group, July, 413–421.

World Coal (1997) Quenching the fires of Jharia. *World Coal* **6** (7), 54–55.

World Coal (1999) Geological Survey Software Review. *World Coal* **8** (1), 27–30.

World Coal Institute (2001) Coal Facts. *WCI Coal Facts Cards and Website* – www.wci-coal.com

World Energy Council (WEC) (1998) *Survey of Energy Resources*, (eighteenth edition), WEC, London.

World Energy Council (WEC) (2001) *Survey of Energy Resources 2001* (nineteenth edition), WEC, London.

Worssam, B.C. and Old, R.A. (1988) Geology of the country around Coalville. *Mem. Br. Geol. Surv. Sheet 155*, (England & Wales).

Wu Chonglong, Li Sitian and Cheng Shoutian (1992) Humid-type alluvial-fan deposits and associated coal seams in the lower Cretaceous Haizhou Formation, Fuxin Basin of Northeastern China. In: *Controls on the Distribution and Quality of Cretaceous Coals. Geol. Soc. Am. Sp. Paper 267*, 269–286.

Xiaodong, C. and Shengli, Z. (1997) Coalbed methane in China: Geology and exploration prospects. In: Z.C. Sun *et al* (Editors) *Proc. 30th Int. Geol. Congr.* **18**, 131–141.

Zhao, W., Zhang, Y., Xu, D. and Zhao, C. (1997) Formation and distribution of coal measure-derived hydrocarbon accumulation in NW China. In: Z.C. Sun *et al* (Editors) *Proc. 30th Intl. Geol. Congr.* **18**, 87–101.

Zimmerman, R.E. (1979) *Evaluating and testing the coking properties of coal.* Miller Freeman Publications Inc.

Appendices

APPENDIX 1. LIST OF INTERNATIONAL AND
NATIONAL STANDARDS USED IN COAL AND
COKE ANALYSIS AND EVALUATION

The following Standards are given for coal and coke as
used in the UK, Europe, Australia and the USA. Other
countries have similar Standards, e.g. Brazil, India,
People's Republic of China and South Africa, and these
should be referred to if requested.

British Standards Institution (BS)

Breckland, Linford Wood, Milton Keynes MK14 6LE.

Recent amendments to these Standards are given
after the title of the standard, together with the date
of the amendment. Years in which the Standard was
reconfirmed are shown in parentheses after the year of
issue. British Standards are referred to in this Appendix
as BS.

BS 1016-1	1973 (1989)	Total moisture of coal
BS 1016-6	1977	Ultimate analysis of coal
BS 1016-7	1977	Ultimate analysis of coke
BS 1016-8	1977 (1984)	Chlorine in coal and coke
BS 1016-9	1977 (1989)	Phosphorus in coal and coke
BS 1016-10	1977 (1989)	Arsenic in coal and coke
BS 1016-14	1963 (1979)	Analysis of coal ash and coke ash
BS 1016-21	1981 (1987)	Determination of moisture holding capacity of hard coal
BS 1016-100	1994	Methods for analysis and testing of coal and coke, introduction and methods for reporting results
BS 1016-102	2000	Determination of total moisture of coke
BS 1016-104.1	1999	Proximate analysis: determination of moisture of general analysis test sample
BS 1016-104.2	1991	Proximate analysis: determination of moisture content of general analysis sample of coke
BS 1016-104.3	1998	Proximate analysis: determination of volatile matter content
BS 1016-104.4	1998	Proximate analysis: determination of ash content
BS 1016-105	1992	Determination of Gross Calorific Value
BS 1016-106.1.1	1996	Ultimate analysis: determination of carbon and hydrogen, high temperature combustion method
BS 1016-106.1.2	1996	Ultimate analysis: determination of carbon and hydrogen, Liebig method
BS 1016-106.2	1997	Ultimate analysis: determination of nitrogen
BS 1016-106.4.1	1993	Ultimate analysis: determination of total sulphur content, Eschka method
BS 1016-106.4.2	1996	Ultimate analysis: determination of total sulphur, high temperature combustion method
BS 1016-106.5	1996	Ultimate analysis: determination of forms of sulphur in coal
BS 1016-106.6.1	1997	Ultimate analysis: determination of chlorine content, Eschka method
BS 1016-106.7	1997	Ultimate analysis: determination of carbonate carbon content
BS 1016-107.1	1991	Caking and swelling properties of coal: determination of crucible swelling number

BS 1016-107.2	1991	Caking and swelling properties of coal: assessment of caking power by Gray–King coke test
BS 1016-107.3	1990	Caking and swelling properties of coal: determination of swelling properties using a dilatometer
BS 1016-108.1	1996	Coke tests: determination of shatter indices
BS 1016-108.2	1992	Coke tests: determination of Micum and Irsid indices
BS 1016-108.3	1995	Coke tests: determination of bulk density (small container)
BS 1016-108.4	1995	Coke tests: determination of bulk density (large container)
BS 1016-108.5	1992	Coke tests: determination of density and porosity
BS 1016-108.6	1992	Coke tests: determination of critical air blast value
BS 1016-109	1995	Size analysis of coal
BS 1016-110.1	1996	Size analysis of coke: nominal top size >20 mm
BS 1016-110.2	1996	Size analysis of coke: nominal top size 20 mm or less
BS 1016-111	1998	Determination of abrasion index of coal
BS 1016-112	1995	Determination of Hardgrove Grindability Index of hard coal
BS 1016-113	1995	Determination of ash fusibility
BS 1017-1	1989	Methods for sampling of coal
BS 1017-2	1994	Methods for sampling of coke
BS 3323	1992	Glossary of terms relating to sampling, testing and analysis of solid mineral fuels
BS 3552	1994	Glossary of terms used in coal preparation
BS 5930	1999	Code of practice for site investigations
BS 6068-1 to 9	1996	Water quality glossary
BS 7022	1988	Guide for geophysical logging of boreholes for hydrological purposes
BS 7067	1990	Guide to determination and presentation of float and sink characteristics of raw coal and of products from coal preparation plants

International Organisation for Standardisation (ISO)

Casa Postale 56, CH 1211, Genève 20, Switzerland.

ISO 157, 1996	Hard coal – Determination of forms of sulphur
ISO 331, 1993	Coal – determination of moisture in the analysis sample, direct gravimetric method
ISO 332, 1996	Coal – determination of nitrogen, Macro Kjeldahl method
ISO 333, 1979	Coal – determination of nitrogen, Semi-micro Kjeldahl method
ISO 334, 1975	Coal and coke – determination of total sulphur, Eschka method
ISO 335, 1974	Hard coal – determination of caking power, Roga Test
ISO 348, 1981	Hard coal – determination of moisture in the analysis sample, direct volumetric method
ISO 349, 1975	Hard coal – Audibert–Arnu dilatometer test
ISO 351, 1984	Solid mineral fuels – determination of total sulphur, high temperature combustion method
ISO 352, 1981	Solid mineral fuels – determination of chlorine by high temperature combustion method
ISO 501, 1981	Coal – determination of the crucible swelling number
ISO 502, 1982	Coal – determination of caking power, Gray–King coke test
ISO 540, 1995	Solid mineral fuels – determination of fusibility of ash, high temperature tube method
ISO 561, 1989	Coal preparation plant – Graphical symbols
ISO 562, 1998	Hard coal and coke – determination of volatile matter content
ISO 567, 1995	Determination of bulk density of coke (small container)
ISO 579, 1999	Determination of total moisture of coke

ISO 587, 1997	Solid mineral fuels – determination of chlorine using Eschka method
ISO 589, 1981	Hard coal – determination of total moisture
ISO 601, 1981	Solid mineral fuels – determination of arsenic content using the standard silver diethyldithio-carbamate photometric method of ISO 2590
ISO 602, 1983	Coal – determination of mineral matter
ISO 609, 1996	Coal and coke – determination of carbon and hydrogen (high temperature combustion method)
ISO 616, 1995	Determination of coke shatter indices
ISO 622, 1981	Solid mineral fuels – determination of phosphorus content, reduced molybdophosphate photometric method
ISO 625, 1975	Coal and coke – determination of carbon and hydrogen, Leibig method
ISO 647, 1974	Brown coals and lignites – determination of the yields of tar, water, gas and coke residue by low temperature distillation
ISO 728, 1995	Size analysis of coke, nominal top size >20 mm
ISO 923, 2000	Coal cleaning test, expression and presentation of results
ISO 924, 1989	Coal preparation plant – principles and conventions for flowsheets
ISO 925, 1997	Solid mineral fuels – determination of carbon dioxide content, gravimetric method
ISO 975, 1985	Brown coals and lignites – determination of yield of toluene-soluble extract
ISO 1013, 1995	Determination of bulk density of coke (large container)
ISO 1015, 1975	Brown coals and lignites – determination of moisture content, direct volumetric method
ISO 1017, 1985	Brown coals and lignites – determination of acetone-soluble material (resinous substances) in the toluene-soluble extract
ISO 1018, 1975	Hard coal – determination of moisture holding capacity
ISO 1170, 1977	Coal and coke – calculation of analyses to different bases
ISO 1171, 1997	Solid mineral fuels – determination of ash content
ISO 1213, 1993	Part 1, vocabulary of terms relating to solid mineral fuels
	Part 2, terms relating to coal sampling and analysis
ISO 1928, 1976	Solid mineral fuels – determination of gross calorific value by the calorimeter Bomb method and calculation of net calorific value
ISO 1952, 1976	Brown coals and lignites – method of extraction for the determination of sodium and potassium in dilute hydrochloric acid
ISO 1953, 1994	Hard coals – size analysis
ISO 1988 1975	Hard coal – sampling
ISO 1994 1976	Hard coal – determination of oxygen content
ISO 2325, 1986	Size analysis of coke, nominal top size 20 mm or less
ISO 2950, 1974	Brown coals and lignites – classification by types on the basis of total moisture content and tar yield
ISO 5068, 1983	Brown coals and lignites – determination of moisture content, indirect gravimetric method
ISO 5069, 1983	Brown coals and lignites – principles of sampling:
	Part 1, sampling for determination of moisture content and for general analysis
	Part 2, sample preparation for determination of moisture content and for general analysis
ISO 5073, 1985	Brown coal and lignite – determination of humic acids
ISO 5074, 1994	Hard coal – determination of Hardgrove Grindability Index
ISO 7404,	Methods for the petrographic analysis of bituminous coal and anthracite
7404.4 1994	Methods for the petrographic analysis of bituminous coal and anthracite
7404.5 1985	Method of preparation of coal samples
7404.6 1994	Method of determining maceral group composition
7404.7 1988	Method of determining microlithotype, carbominerite and minerite composition
7404.8 1994	Method of determining microscopically the reflectance of vitrinite

ISO 7936, 1992	Hard coal: determination and presentation of float and sink characteristics – apparatus and procedures
ISO 8264, 1989	Hard coal – determination of the swelling properties using a dilatometer
ISO 8833, 1989	Magnetite for use in coal preparation – test methods
ISO 8858.1, 1990	Hard coal – froth flotation testing Part 1. Laboratory procedure
ISO 10086.1, 2000	Coal: methods for evaluating flocculents for use in coal preparation – parameters
ISO 10752, 1994	Coal sizing equipment – performance evaluation
ISO 10753, 1994	Coal preparation plant – assessment of liability to breakdown in water of minerals associated with coal seams
ISO 11722, 1999	Hard coal: determination of moisture by drying in nitrogen
ISO 12900, 1997	Hard coal: determination of abrasiveness
ISO 13909.1, 2001	Hard coal and coke – mechanical sampling – Introduction
13909.2, 2001	Hard coal and coke – mechanical sampling – Coal: sampling from moving streams
13909.3, 2001	Hard coal and coke – mechanical sampling – Coal: sampling from stationary lots
13909.4, 2001	Hard coal and coke – mechanical sampling – Coal: preparation of test samples
13909.5, 2001	Hard coal and coke – mechanical sampling – Coke: sampling from moving streams
13909.6, 2001	Hard coal and coke – mechanical sampling – Coke: sampling from stationary lots
13909.7, 2001	Hard coal and coke – mechanical sampling – Methods for determining the precision of sampling, sample preparation and testing
13909.8, 2001	Hard coal and coke – mechanical sampling – Methods of testing for bias
ISO 14180, 1998	Guidance on the sampling of coal seams

Standards Association of Australia (AS)

80–86 Arthur Street, North Sydney, NSW, 2060, Australia.

AS 2418-1995	Glossary of terms relating to solid mineral fuels
2418.1-1980	Terms relating to coal preparation
2418.2-1982	Terms relating to coal mining and geology
2418.3-1982	Terms relating to brown coal
2418.4-1982	Terms relating to sampling, sample preparation, analysis, testing and statistics
2418.5-1982	Terms relating to the petrographic analysis of bituminous coal and anthracite (hard coal)
2418.6-1982	Terms relating to coal utilisation and coke
2418.7-1982	Terms relating to coal classification
AS 2519-1993	Guide to the evaluation of hard coal deposits using borehole techniques
AS 2617-1996	Guide for the taking of samples from hard coal seams *in situ*
AS 2646	Sampling of solid mineral fuels
AS 1038	Coal and coke – Analysis and testing
1038.1-2001	Total moisture in hard coal
1038.2-1995	Total moisture in coke
1038.3-2000	Proximate analysis of higher rank coal
1038.4-1995	Proximate analysis of coke
1038.5-1998	Gross specific energy of coal and coke
1038.5.1-1988	Adiabatic calorimeters
1038.5.2-1989	Automatic isothermal-type calorimeters
1038.6	Ultimate analysis of higher rank coal
1038.6.1-1997	Determination of carbon and hydrogen
1038.6.2-1997	Determination of nitrogen
1038.6.3.1-1997	Determination of total sulphur (Eschka method)
1038.6.3.2-1997	Determination of total sulphur (High temperature combustion method)

1038.6.3.3-1997	Determination of total sulphur (Infrared method)
1038.7-1999	Determination of Carbon, Hydrogen and Nitrogen by instrumental method
1038.8.1-1999	Chlorine in coal and coke (Eschka method)
1038.8.2-1996	Chlorine in coal and coke (High temperature combustion method)
1038.9.1-2000	Phosphorus in coal and coke (Ash digestion/molybdenum blue method)
1038.9.2-2000	Phosphorus in coal and coke (coal extraction/phosphomolybdovanadate method)
1038.9.3-2000	Phosphorus in coal and coke (Ash digestion/phosphomolybdovanadate method)
1038.10.1-1986	Determination of trace elements – determination of eleven trace elements in coal, coke and fly ash by Flame absorption spectrometric method
1038.10.2-1998	Determination of Arsenic and Selenium in coal and coke by Hydride generation method
1038.10.3-1998	Determination of trace elements – Coal, coke and fly ash – determination of boron content – spectrophotometric method
1038.10.4-2001	Determination of trace elements – Coal, coke and fly ash – determination of fluorine content by Pyrohydrolysis method
1038.10.5-1993	Determination of Trace Elements in coal, coke and fly-ash, determination of mercury content
1038.11-1993	Forms of sulphur in coal
1038.12.1-1993	Determination of crucible swelling number of coal
1038.12.2-1999	Carbonization properties of higher rank coal, determination of Gray–King coke type
1038.12.3-1993	Determination of the dilatometer characteristics of higher rank coal
1038.12.4-1996	Plastic properties of higher rank coal by the Gieseler plastometer
1038.13-1990	Tests specific to coke
1038.14.1-1995	Analysis of coal ash, coke ash and mineral matter (borate fusion-flame atomic absorption spectrometric method)
1038.14.2-1995	Analysis of higher rank coal ash and coke ash (acid digestion-flame atomic absorption spectrometric method)
1038.14.3-1999	Analysis of higher rank coal ash and coke ash (wavelength dispersive X-ray fluorescence spectrometric method)
1038.15-1995	Fusibility of higher rank coal ash and coke ash
1038.16-1996	Coal and coke – Assessment and reporting of results
1038.17-2000	Determination of moisture-holding capacity (equilibrium moisture) of higher rank coal
1038.18-1996	Coke – Size Analysis
1038.19-2000	Determination of the abrasion index of higher rank coal
1038.20-1992	Hardgrove Grindability Index of higher rank coal
1038.21.1.1-1994	Determination of the relative density of hard coal and coke, analysis sample – density bottle method
1038.21.1.2-1994	Determination of the relative density of hard coal and coke, analysis sample – volumetric method
1038.21.2-1994	Determination of the relative density of hard coal and coke, lump sample
1038.22-2000	Direct determination of mineral matter and water of hydration of minerals in hard coal
1038.23-1994	Determination of carbonate carbon in higher rank coal
1038.24-1998	Guide to the evaluation of measurements made by on-line coal analyzers
AS 1661-1979	Method for float and sink testing of hard coal and presentation of results
AS 2434	Methods for the analysis and testing of lower rank coal and its chars
2434.1-1999	Determination of the total moisture content of lower rank coal
2434.2-1983	Determination of the volatile matter in low rank coal
2434.3-1984	Determination of the moisture-holding capacity of lower rank coals
2434.4-1985	Determination of the apparent density of dried lower rank coal and its chars (mercury displacement method)

2434.5-1984	Determination of moisture in bulk samples and in analysis samples of char from lower rank coal
2434.6.1-1986	Ultimate analysis of lower rank coal
2434.7-1986	Determination of moisture in the analysis sample of lower rank coal
2434.8-1993	Determination of ash in the analysis sample of lower rank coal
2434.9-2000	Determination of four acid-extractable inorganic ions in lower rank coal
AS 2856.1-2000	Coal petrography – preparation of samples for incident light microscopy
2856.2-1998	Maceral analysis
2856.3-2000	Microscopical determination of reflectance of coal macerals
AS 3880-1991	Bin flow properties of coal
AS 3881-1991	Higher rank coal-size analysis
AS 3899-1991	Higher rank coal and coke-bulk density
AS 3980-1999	Guide to the determination of desorbable gas content of coal seams – direct method
AS 2096-1987	Classification and coding systems for Australian coals
AS 2916-1986	Symbols for graphical representation of coal seams and associated strata
AS4156.1-1994	Coal preparation of higher rank coal, float and sink testing
4156.2.1-1994	Coal preparation of higher rank coal, froth flotation – basic test
4156.2.2-1998	Coal preparation of higher rank coal, froth flotation – sequential procedure
4156.3-1994	Coal preparation of higher rank coal, magnetite for coal preparation plant use – test methods
4156.4-1999	Coal preparation – flowsheets and symbols
4156.6-2000	Coal preparation – determination of dust/moisture relationship for coal
4156.7-1999	Coal preparation – coal size classifying equipment – performance evaluation
AS4264.1-1995	Coal and coke – sampling of higher rank coal – sampling procedures
4264.2-1996	Coal and coke – sampling of coke – sampling procedures
4264.3-1996	Coal and coke – sampling of lower rank coal – sampling procedures
4264.4-1996	Coal and coke – sampling – determination of precision and bias
4264.5-1999	Coal and coke – sampling – guide to the inspection of mechanical sampling Systems

American Society for Testing and Materials (ASTM)

1916 Race Street, Philadelphia, PA 19103-1187 USA.

Figures after Standard number give the year of most recent reapproval.

D 121-91	Definitions of terms relating to coal and coke
D 197-87	Sampling and fineness test of pulverised coal
D 293-93	Sieve analysis of coke
D 291-86	Cubic foot weight of crushed bituminous coal
D 310-80	Test for size of anthracite
D 311-76	Sieve analysis of crushed bituminous coal
D 346-90	Collection & preparation of coke samples for laboratory analysis
D 388-99	Classification of coals by rank
D 410-76	Sieve analysis of coal
D 409-97	Grindability of coal by the Hardgrove-Machine method
D 431-76	Designating the size of coal from its sieve analysis
D 440-86	Drop shatter test for coal

D 441-86	Tumbler test for coal
D 547-80	Test for index of dustiness of coal and coke
D 720-91	Free swelling index of coal
D1412-99	Equilibrium moisture of coal at 96% to 97% relative humidity and 30 degrees C.
D1756-96	Carbon dioxide in coal
D1757-96	Sulphur in ash from coal and coke
D1857-87	Fusibility of coal and coke ash
D2013-86	Samples, coal, preparing for analysis
D2014-97	Expansion or contraction of coal by the Sole-Heated Oven
D2015-91	Gross calorific value of coal and coke by the Adiabatic Bomb Calorimeter
D1989-91	Gross calorific value of coal and coke by Microprocessor Controlled Isoperibol Calorimeters
D2234-90	Collection of a gross sample of coal
D2361-95	Chlorine in coal
D2492-90	Forms of sulphur in coal
D2639-98	Plastic properties of coal by the Constant-Torque Gieseler Plastometer
D2795-95	Analysis of coal and coke ash
D2796-88	Megascopic description of coal and coal seams and microscopical description and analysis of coal
D2797-85	Preparing coal samples for microscopical analysis by reflected light
D2798-99	Microscopical determination of the reflectance of the organic components in a polished specimen of coal
D2799-99	Microscopical determination of volume percent of physical components of coal
D2961-01	Total moisture, <15% in coal reduced to No.8 (2.36 mm) topsize
D3038-93	Drop shatter test for coke
D3172-89	Proximate analysis of coal and coke
D3173-00	Moisture in the analysis sample of coal and coke
D3174-00	Ash in the analysis sample of coal and coke from coal
D3175-01	Volatile matter in the analysis sample of coal and coke
D3176-89	Ultimate analysis of coal and coke
D3177-89	Total sulphur in the analysis sample of coal and coke
D3178-89	Carbon and hydrogen in the analysis sample of coal and coke
D3179-89	Nitrogen in the analysis sample of coal and coke
D3180-89	Calculating coal and coke analyses from As-Determined to different bases
D3286-91	Gross calorific value of coal and coke by the Isoperibol Bomb Calorimeter
D3302-00	Total moisture in coal
D3402-93	Tumbler test for coke
D3682-00	Major and minor elements in coal and coke ash by Atomic Absorption
D3683-94	Trace elements in coal and coke ash by Atomic Absorption
D3684-94	Total mercury in coal by the Oxygen Bomb Combustion/Atomic Absorption method
D3761-96	Total fluorine in coal by the Oxygen Bomb Combustion/Ion Selective Electrode method
D4182-97	Evaluation of laboratories using ASTM procedures in the sampling and analysis of coal and coke
D4208-88	Total chlorine in coal by the Oxygen Bomb Combustion/Ion Selective Electrode method
D4239-00	Sulphur in the analysis sample of coal and coke using High Temperature Tube Furnace Combustion methods
D4326-97	Major & minor elements in coal by X-ray fluorescence

D4371-91	Washability characteristics of coal
D4596-99	Collection of channel samples of coal in the mine
D4606-95	Determination of Arsenic and Selenium in coal by the Hydride Generation/Atomic Absorption method
D4621-99	Accountability and quality control in the coal analysis laboratory
D4702-97	Guide for inspecting Cross-Cut, Sweep-Arm and Auger mechanical coal sampling Systems for conformance with current ASTM Standards
D4749-87	Sieve analysis for coal, performing and designating coal size
D4915-96	Manual sampling of coal from tops of railroad cars
D4916-97	Practice for Mechanical Auger sampling
D5016-98	Sulphur in ash from coal and coke using High Temperature Tube Furnace Combustion method with Infrared Absorption
D5142-90	Proximate analysis of the analysis sample of coal and coke by instrumental procedures
D5061-92	Microscopical determination of volume % of textural components in metallurgical coke
D5114-90	Method for laboratory Froth Flotation of coal in a Mechanical Cell
D5142-90	Method for Proximate Analysis of the analysis sample of coal & coke by instrumental Procedures
D5192-99	Practice for collection of coal samples from core
D5263-93	Method for determining the Relative Degree of Oxidation in bituminous coal by Alkali extraction
D5341-99	Method for measuring Coke Reactivity Index (CRI) and coke Strength After Reaction (CSR)
D5373-93	Method for instrumental determination of Carbon, Hydrogen & Nitrogen in laboratory samples of coal & coke
D5515-97	Method for determination of the Swelling Properties of Bituminous Coal using a dilatometer
D5671-95	Practice for Polishing & Etching coal samples for microscopical analysis by Reflected Light
D5987-96	Method for Total Fluorine in coal & coke by Pyrohydrolytic Extraction & Ion Selective Electrode or Ion Chromatograph Methods
D6315-98	Practice for manual sampling of coal from tops of barges
D6316-00	Method for determination of total, combustible & carbonate Carbon in solid residues from coal & coke
D6347/D6347M-99	Method for determination of bulk density coal using Nuclear Backscatter Depth Density Methods
D6349-00	Method for determination of Major & Minor Elements in coal, coke & solid residues from combustion of coal & coke by Inductively Coupled Plasma–Atomic Emission Spectrometry
D6414-99	Methods for determination of Total Mercury in coal & coke combustion residues by Acid Extraction or Wet Oxidation/Cold Vapour Atomic Absorption
D6518-01	Practice for Bias Testing a Mechanical Coal Sampling System
D6542-00	Practice for Tonnage Calculation of coal in a Stockpile
D6543-00	Guide to the evaluation of measurements made by On-Line Coal Analyzers
D6609-00	Guide for Part-Stream Sampling of coal

APPENDIX 2. TABLE OF SLOPE ANGLES, GRADIENTS AND PERCENT SLOPE

Dips of strata and of land surfaces can be expressed in angles, gradients or percent slope. Those values most commonly encountered are included in this table.

Angle of slope in degrees	Gradient	Percent slope
1	1:57	1.7
2	1:29	3.5
3	1:19	5.2
4	1:14	7.0
5	1:11.4	8.7
6	1:9.5	10.5
7	1:8.1	12.3
8	1:7.1	14.1
9	1:6.3	15.8
10	1:5.7	17.6
11	1:5.1	19.4
12	1:4.7	21.3
13	1:4.3	23.1
14	1:4.0	24.9
15	1:3.7	26.8
16	1:3.5	28.7
17	1:3.3	30.6
18	1:3.1	32.5
19	1:2.9	34.4
20	1:2.7	36.4
25	1:2.1	46.5
30	1:1.7	57.7
35	1:1.4	70.0
40	1:1.2	83.9
45	1:1.0	100.0
50	1:0.8	119.2
55	1:0.7	142.8
60	1:0.6	173.2
65	1:0.5	214.5
70	1:0.4	274.7
75	1:0.3	373.2
80	1:0.2	567.1
85	1:0.1	1143.0
90	1:0	–

APPENDIX 3. CALORIFIC VALUES EXPRESSED IN DIFFERENT UNITS

MJ/kg	Btu/lb $MJ/kg \times$ 429.923	kcal/kg $MJ/kg \times$ 239.006	lb/lb $MJ/kg \times$ 0.442763
4.5	1 935	1 076	1.99
4.6	1 978	1 099	2.04
4.7	2 021	1 123	2.08
4.8	2 064	1 147	2.13
4.9	2 107	1 171	2.17
5.0	2 150	1 195	2.21
5.1	2 193	1 219	2.26
5.2	2 236	1 243	2.30
5.3	2 279	1 267	2.35
5.4	2 322	1 291	2.39
5.5	2 365	1 315	2.44
5.6	2 408	1 338	2.48
5.7	2 451	1 362	2.52
5.8	2 494	1 386	2.57
5.9	2 537	1 410	2.61
6.0	2 580	1 434	2.66
6.1	2 623	1 458	2.70
6.2	2 666	1 482	2.75
6.3	2 709	1 506	2.79
6.4	2 752	1 530	2.83
6.5	2 794	1 554	2.88
6.6	2 837	1 577	2.92
6.7	2 880	1 601	2.97
6.8	2 923	1 625	3.01
6.9	2 966	1 649	3.06
7.0	3 009	1 673	3.10
7.1	3 052	1 697	3.14
7.2	3 095	1 721	3.19
7.3	3 138	1 745	3.23
7.4	3 181	1 769	3.28
7.5	3 224	1 793	3.32
7.6	3 267	1 816	3.36
7.7	3 310	1 840	3.41
7.8	3 353	1 864	3.45
7.9	3 396	1 888	3.50
8.0	3 439	1 912	3.54
8.1	3 482	1 936	3.59

(*continued*)

MJ/kg	Btu/lb $MJ/kg \times 429.923$	kcal/kg $MJ/kg \times 239.006$	lb/lb $MJ/kg \times 0.442763$
8.2	3 525	1 960	3.63
8.3	3 568	1 984	3.67
8.4	3 611	2 008	3.72
8.5	3 654	2 032	3.76
8.6	3 697	2 055	3.81
8.7	3 740	2 079	3.85
8.8	3 783	2 103	3.90
8.9	3 826	2 127	3.94
9.0	3 869	2 151	3.98
9.1	3 912	2 175	4.03
9.2	3 955	2 199	4.07
9.3	3 998	2 223	4.12
9.4	4 041	2 247	4.16
9.5	4 084	2 271	4.21
9.6	4 127	2 294	4.25
9.7	4 170	2 318	4.29
9.8	4 213	2 342	4.34
9.9	4 256	2 366	4.38
10.0	4 299	2 390	4.43
10.1	4 342	2 414	4.47
10.2	4 385	2 438	4.52
10.3	4 428	2 462	4.56
10.4	4 471	2 486	4.60
10.5	4 514	2 510	4.65
10.6	4 557	2 533	4.69
10.7	4 600	2 557	4.74
10.8	4 643	2 581	4.78
10.9	4 686	2 605	4.83
11.0	4 729	2 629	4.87
11.1	4 772	2 653	4.91
11.2	4 815	2 677	4.96
11.3	4 858	2 701	5.00
11.4	4 901	2 725	5.05
11.5	4 944	2 749	5.09
11.6	4 987	2 772	5.14
11.7	5 030	2 796	5.18
11.8	5 073	2 820	5.22
11.9	5 116	2 844	5.27
12.0	5 159	2 868	5.31
12.1	5 202	2 892	5.36
12.2	5 245	2 916	5.40
12.3	5 288	2 940	5.45
12.4	5 331	2 964	5.49
12.5	5 374	2 988	5.53
12.6	5 417	3 011	5.58
12.7	5 460	3 035	5.62
12.8	5 503	3 059	5.67
12.9	5 546	3 083	5.71
13.0	5 589	3 107	5.76
13.1	5 632	3 131	5.80
13.2	5 675	3 155	5.84
13.3	5 718	3 179	5.89
13.4	5 761	3 203	5.93
13.5	5 804	3 227	5.98
13.6	5 847	3 250	6.02
13.7	5 890	3 274	6.07
13.8	5 933	3 298	6.11
13.9	5 976	3 322	6.15
14.0	6 019	3 346	6.20
14.1	6 062	3 370	6.24
14.2	6 105	3 394	6.29
14.3	6 148	3 418	6.33
14.4	6 191	3 442	6.38
14.5	6 234	3 466	6.42
14.6	6 277	3 489	6.46
14.7	6 320	3 513	6.51
14.8	6 363	3 537	6.55
14.9	6 406	3 561	6.60
15.0	6 449	3 585	6.64
15.1	6 492	3 609	6.69
15.2	6 535	3 633	6.73
15.3	6 578	3 657	6.77
15.4	6 621	3 681	6.82
15.5	6 664	3 705	6.86
15.6	6 707	3 728	6.91
15.7	6 750	3 752	6.95
15.8	6 793	3 776	7.00
15.9	6 836	3 800	7.04
16.0	6 879	3 824	7.08
16.1	6 922	3 848	7.13
16.2	6 965	3 872	7.17
16.3	7 008	3 896	7.22
16.4	7 051	3 920	7.26
16.5	7 094	3 944	7.31
16.6	7 137	3 967	7.35
16.7	7 180	3 991	7.39
16.8	7 223	4 015	7.44
16.9	7 266	4 039	7.48
17.0	7 309	4 063	7.53
17.1	7 352	4 087	7.57
17.2	7 395	4 111	7.62
17.3	7 438	4 135	7.66
17.4	7 481	4 159	7.70

(continued)

MJ/kg	Btu/lb $MJ/kg \times$ 429.923	kcal/kg $MJ/kg \times$ 239.006	lb/lb $MJ/kg \times$ 0.442763
17.5	7 524	4 183	7.75
17.6	7 567	4 207	7.79
17.7	7 610	4 230	7.84
17.8	7 653	4 254	7.88
17.9	7 696	4 278	7.93
18.0	7 739	4 302	7.97
18.1	7 782	4 326	8.01
18.2	7 825	4 350	8.06
18.3	7 868	4 374	8.10
18.4	7 911	4 398	8.15
18.5	7 954	4 422	8.19
18.6	7 997	4 446	8.24
18.7	8 040	4 469	8.28
18.8	8 083	4 493	8.32
18.9	8 126	4 517	8.37
19.0	8 169	4 541	8.41
19.1	8 212	4 565	8.46
19.2	8 255	4 589	8.50
19.3	8 298	4 613	8.55
19.4	8 341	4 637	8.59
19.5	8 383	4 661	8.63
19.6	8 426	4 685	8.68
19.7	8 469	4 708	8.72
19.8	8 512	4 732	8.77
19.9	8 555	4 756	8.81
20.0	8 598	4 780	8.86
20.1	8 641	4 804	8.90
20.2	8 684	4 828	8.94
20.3	8 727	4 852	8.99
20.4	8 770	4 876	9.03
20.5	8 813	4 900	9.08
20.6	8 856	4 924	9.12
20.7	8 899	4 947	9.17
20.8	8 942	4 971	9.21
20.9	8 985	4 995	9.25
21.0	9 028	5 019	9.30
21.1	9 071	5 043	9.34
21.2	9 114	5 067	9.39
21.3	9 157	5 091	9.43
21.4	9 200	5 115	9.48
21.5	9 243	5 139	9.52
21.6	9 286	5 163	9.56
21.7	9 329	5 186	9.61
21.8	9 372	5 210	9.65
21.9	9 415	5 234	9.70
22.0	9 458	5 258	9.74
22.1	9 501	5 282	9.79
22.2	9 544	5 306	9.83
22.3	9 587	5 330	9.87
22.4	9 630	5 354	9.92
22.5	9 673	5 378	9.96
22.6	9 716	5 402	10.01
22.7	9 759	5 425	10.05
22.8	9 802	5 449	10.09
22.9	9 845	5 473	10.14
23.0	9 888	5 497	10.18
23.1	9 931	5 521	10.23
23.2	9 974	5 545	10.27
23.3	10 017	5 569	10.32
23.4	10 060	5 593	10.36
23.5	10 103	5 617	10.40
23.6	10 146	5 641	10.45
23.7	10 189	5 664	10.49
23.8	10 232	5 688	10.54
23.9	10 275	5 712	10.58
24.0	10 318	5 736	10.63
24.1	10 361	5 760	10.67
24.2	10 404	5 784	10.71
24.3	10 447	5 808	10.76
24.4	10 490	5 832	10.80
24.5	10 533	5 856	10.85
24.6	10 576	5 880	10.89
24.7	10 619	5 903	10.94
24.8	10 662	5 927	10.98
24.9	10 705	5 951	11.02
25.0	10 748	5 975	11.07
25.1	10 791	5 999	11.11
25.2	10 834	6 023	11.16
25.3	10 877	6 047	11.20
25.4	10 920	6 071	11.25
25.5	10 963	6 095	11.29
25.6	11 006	6 119	11.33
25.7	11 049	6 142	11.38
25.8	11 092	6 166	11.42
25.9	11 135	6 190	11.47
26.0	11 178	6 214	11.51
26.1	11 221	6 238	11.56
26.2	11 264	6 262	11.60
26.3	11 307	6 286	11.64
26.4	11 350	6 310	11.69
26.5	11 393	6 334	11.73
26.6	11 436	6 358	11.78
26.7	11 479	6 381	11.82

(*continued*)

MJ/kg	Btu/lb $MJ/kg \times$ 429.923	kcal/kg $MJ/kg \times$ 239.006	lb/lb $MJ/kg \times$ 0.442763
26.8	11 522	6 405	11.87
26.9	11 565	6 429	11.91
27.0	11 608	6 453	11.95
27.1	11 651	6 477	12.00
27.2	11 694	6 501	12.04
27.3	11 737	6 525	12.09
27.4	11 780	6 549	12.13
27.5	11 823	6 573	12.18
27.6	11 866	6 597	12.22
27.7	11 909	6 620	12.26
27.8	11 952	6 644	12.31
27.9	11 995	6 668	12.35
28.0	12 038	6 692	12.40
28.1	12 081	6 716	12.44
28.2	12 124	6 740	12.49
28.3	12 167	6 764	12.53
28.4	12 210	6 788	12.57
28.5	12 253	6 812	12.62
28.6	12 296	6 836	12.66
28.7	12 339	6 859	12.71
28.8	12 382	6 883	12.75
28.9	12 425	6 907	12.80
29.0	12 468	6 931	12.84
29.1	12 511	6 955	12.88
29.2	12 554	6 979	12.93
29.3	12 597	7 003	12.97
29.4	12 640	7 027	13.02
29.5	12 683	7 051	13.06
29.6	12 726	7 075	13.11
29.7	12 769	7 098	13.15
29.8	12 812	7 122	13.19
29.9	12 855	7 146	13.24
30.0	12 898	7 170	13.28
30.1	12 941	7 194	13.33
30.2	12 984	7 218	13.37
30.3	13 027	7 242	13.42
30.4	13 070	7 266	13.46
30.5	13 113	7 290	13.50
30.6	13 156	7 314	13.55
30.7	13 199	7 337	13.59
30.8	13 242	7 361	13.64
30.9	13 285	7 385	13.68
31.0	13 328	7 409	13.73
31.1	13 371	7 433	13.77
31.2	13 414	7 457	13.81
31.3	13 457	7 481	13.86
31.4	13 500	7 505	13.90
31.5	13 543	7 529	13.95
31.6	13 586	7 553	13.99
31.7	13 629	7 576	14.04
31.8	13 672	7 600	14.08
31.9	13 715	7 624	14.12
32.0	13 758	7 648	14.17
32.1	13 801	7 672	14.21
32.2	13 844	7 696	14.26
32.3	13 887	7 720	14.30
32.4	13 930	7 744	14.35
32.5	13 972	7 768	14.39
32.6	14 015	7 792	14.43
32.7	14 058	7 815	14.48
32.8	14 101	7 839	14.52
32.9	14 144	7 863	14.57
33.0	14 187	7 887	14.61
33.1	14 230	7 911	14.66
33.2	14 273	7 935	14.70
33.3	14 316	7 959	14.74
33.4	14 359	7 983	14.79
33.5	14 402	8 007	14.83
33.6	14 445	8 031	14.88
33.7	14 488	8 055	14.92
33.8	14 531	8 078	14.97
33.9	14 574	8 102	15.01
34.0	14 617	8 126	15.05
34.1	14 660	8 150	15.10
34.2	14 703	8 174	15.14
34.3	14 746	8 198	15.19
34.4	14 789	8 222	15.23
34.5	14 832	8 246	15.28
34.6	14 875	8 270	15.32
34.7	14 918	8 294	15.36
34.8	14 961	8 317	15.41
34.9	15 004	8 341	15.45
35.0	15 047	8 365	15.50
35.1	15 090	8 389	15.54
35.2	15 133	8 413	15.59
35.3	15 176	8 437	15.63
35.4	15 219	8 461	15.67
35.5	15 262	8 485	15.72
35.6	15 305	8 509	15.76
35.7	15 348	8 533	15.81
35.8	15 391	8 556	15.85
35.9	15 434	8 580	15.90
36.0	15 477	8 604	15.94

(continued)

MJ/kg	Btu/lb $MJ/kg \times$ 429.923	kcal/kg $MJ/kg \times$ 239.006	lb/lb $MJ/kg \times$ 0.442763
36.1	15 520	8 628	15.98
36.2	15 563	8 652	16.03
36.3	15 606	8 676	16.07
36.4	15 649	8 700	16.12
36.5	15 692	8 724	16.16
36.6	15 735	8 748	16.21
36.7	15 778	8 772	16.25
36.8	15 821	8 795	16.29
36.9	15 864	8 819	16.34
37.0	15 907	8 843	16.38
37.1	15 950	8 867	16.43
37.2	15 993	8 891	16.47
37.3	16 036	8 915	16.52
37.4	16 079	8 939	16.56

APPENDIX 4. COAL STATISTICS

1 million tonnes coal equivalent = 1 million tonnes of coal at 28.0 Mj/kg or 6692 kcal/kg gross calorific value

1 million tonnes oil equivalent = 1.5 million tonnes of coal (approx.) 3.0 million tonnes of lignite (approx)

1 tonne of coal at 25.1 Mj/kg or 6000 kcal/kg will produce approximately 2400 kW/h of electricity.

A 1000 MW power station requires 3 million tonnes of coal at 25.1 Mj/kg per annum.

1 tonne of coal at 25.1 Mj/kg or 6000 kcal/kg will produce approximately 7.5–9.0 tonnes of cement.

1 tonne of coal at 28% volatile matter, after coking, will produce approximately 1.5 tonnes of iron.

APPENDIX 5. METHANE UNITS CONVERTER (ADAPTED FROM US ENVIRONMENTAL PROTECTION AGENCY COALBED METHANE OUTREACH PROGRAMME)

To convert from:	To:	Multiply by:
Cubic foot	Cubic metre	0.02832
Pound	Kilogram	0.4536
Short ton	Metric ton (tonne)	0.9072
Btu	Joule	1055
Cubic foot methane	Pound methane	0.04246
	Btu	1 014.6
	kW/h	0.2974
	Tonne CO_2 equivalent	0.000404
	Tonne C equivalent	0.00011
	Gram	19.26
Pound methane	Btu	23 896
	kW/h	7
	Tonne CO_2 equivalent	0.00953
	Tonne C equivalent	0.0026
Btu	kW/h	0.000293
Methane GWP	CO_2 GWP	21
CO_2	C equivalent	0.27273
Methane	C equivalent	5.7273

Glossary

AIR-DRIED BASIS — The data are expressed as percentages of the air-dried coal; this includes the air-dried moisture but not the surface moisture of the coal.

ALLOCHTHONOUS — Redeposited sedimentary material originating from distant sources.

ANTHRACITE — Anthracite is the highest rank coal and is characterised by low volatile matter ($<10\%$) and high carbon content. Semi-anthracite is coal midway between low volatile bituminous and anthracite.

ASH — The inorganic residue remaining after the combustion of coal. It is less than the mineral matter content because of the chemical changes occurring during combustion, i.e. the loss of water of hydration, loss of carbon dioxide, and loss of sulphurous gases from sulphides.

ASSIGNED RESERVES — Coal which can be mined on the basis of current mining practices and techniques through the use of mines currently in existence or under construction.

AS RECEIVED BASIS — The data are expressed as percentages of the coal including the total moisture content, i.e. including both the surface and the air-dried moisture content of the coal.

AS RECEIVED MOISTURE — The total moisture of a coal sample when delivered to the laboratory.

AUTOCHTHONOUS — Indigenous material formed *in situ*.

BILLION — 1 000 000 000 (one thousand million)

BITUMINOUS COAL — Bituminous coal lies between subbituminous coal and semi-anthracite in terms of rank. Usually divided into three sub groups – low volatile, medium volatile and high volatile.

BROWN COAL — See **Lignite**.

BUCKETWHEEL EXCAVATOR (BWE) — Large earthmoving machines using a boom with a rotating wheel on which hang a series of buckets with a cutting edge for excavating soft overburden and brown coal or lignite.

BULK SAMPLE — Large coal sample, 5–25 tonnes, for test work to establish coal's performance under actual conditions of usage.

C&F — Cost and Freight, where cargo is purchased at port of delivery.

CALORIFIC VALUE (CV) — Also known as specific energy (SE). Is the amount of heat per unit mass of coal when combusted. See **Gross** and **Net calorific value**.

CHANNEL SAMPLE — A channel of uniform cross-section is cut into the coal seam; all the coal within the cut section is collected (for the whole seam or for a series of plies, i.e. divisions of the coal seam).

CIF	Cost, Insurance and Freight, where cargo is purchased at port of delivery, including insurance.
CLEAT	Jointing in coal appearing as regular patterns of cracks which may have originated during coalification.
COAL BED METHANE (CBM)	Occurs as a free gas (methane CH_4) within the ore space or fractures in coal, or as adsorbed molecules on the organic surface of coal
COAL LIQUEFACTION	Use of a number of techniques to yield liquid hydrocarbons and solid products from coal.
COAL PREPARATION	Physical and mechanical processes applied to coal to make it suitable for a particular use.
COAL PRICE	Calculated using formulas, which include base price for coal, calorific value, transport cost and any local costs.
COAL TRUCK	Smaller vehicle for transporting coal on private and public highways.
COALIFICATION	The alteration of vegetation to form peat, succeeded by the transformation of peat through lignite, subbituminous, bituminous, semi-anthracite, to anthracite and meta-anthracite coal.
COKING COAL	A coal suitable for carbonisation in coke ovens. It must have good coking and caking properties and rank should be high to medium volatile bituminous coal.
CORE SAMPLE	Coal collected from borehole cores, usually unweathered.
CUTTINGS SAMPLE	Coal fragments collected from drilling medium, from boreholes; less accurate than core sampling.
DES	Delivered Ex-Ship, cargo considered delivered after unloading at port.
DRAGLINE	Large earth moving machines with bucket capacities $5-30\,m^3$, used for overburden removal.
DRY ASH-FREE BASIS	The coal is considered to consist of volatile matter and fixed carbon on the basis of recalculation with moisture and ash removed.
DRY BASIS	The data are expressed as percentages of the coal after all the moisture has been removed.
DRY, MINERAL MATTER FREE BASIS	The data are expressed as percentages of the coal on the basis of recalculation with moisture and mineral matter removed.
EMISSIONS	Those gases and particulates vented by power stations and industrial users; in particular, those which have a detrimental effect on the environment.
EXPLORATION	The examination of an area by means of surface geological mapping, geophysical techniques, the drilling of boreholes and sampling of coals.
EXTRACTABLE RESERVES	See **Recoverable reserves**.
FINES	Very small coal with a maximum size which is usually less than 4 mm.
FLOAT–SINK TESTS	The separation of coal and mineral matter particles by immersion in a series of liquids of known relative density. The process is designed to reduce the ash level of the coal and so improve the product to be sold.
FLUE GAS DESULPHURISATION (FGD)	A process designed to recover the acidic sulphur compounds from the flue gas prior to release to the atmosphere.

FLUIDISED BED COMBUSTION (FBC)	The burning of fuel on a bed subjected to an upward gas flow which causes the bed to be suspended resulting in the transference of heat at very high rates.
FOB	Free on Board; cargoes purchased when loaded at port of embarkation.
GEOLOGICAL LOSSES	Losses to be deducted from measured reserves due to geological constraints, e.g. faults, washouts, seam splitting.
GEOPHYSICAL LOGGING	Measurement of the variation with depth of selected physical properties of rocks with geophysical measuring tools (sondes) located in boreholes.
GOAF/GOB	Once an underground mine area has been mined and no longer required for access, the roof is allowed to collapse into the abandoned area to form 'goaf' or 'gob'.
GONDWANALAND	Southern segment of Pangaea.
GPS	Global Positioning Systems; use of satellite signals to enable accurate positioning on the ground.
GRAB SAMPLE	Collection of coal sample from outcrop or stockpile.
GRAVITY SURVEY	Measurement of local and regional variations in the earth's gravitational field.
GROSS CALORIFIC VALUE	The amount of heat liberated during the combustion of a coal in the laboratory under standardised conditions at constant volume, so that all of the water in the products remains in liquid form.
GROUNDWATER	Water below the ground surface and below the water table.
GROUNDWATER REBOUND	Rising water levels after cessation of long term pumping.
HIGHER HEATING VALUE	See **Gross calorific value**.
HIGHWALL	Face of opencast mine towards which the coal is mined; may be single face or a series of benches if a series of coals is mined.
HIGHWALL MINING	Remotely controlled mining method which extracts coal from the base of an exposed highwall in an opencast mine.
HUMIC	Coals formed from a diversified mixture of macroscopic plant debris; almost all economic coals are of this type.
IN-SEAM SEISMIC (ISS)	Use of generated channel waves propagating in the coal seam to detect discontinuities in advance of mining.
INDICATED RESERVES	Includes all coal conforming to the thickness and depth limits defined in the reserve base, bounded by similar distance limits as for indicated resources.
INDICATED RESOURCES	Those for which the density and quality of the points of measurement are not more than 2.0 km apart.
INFERRED RESERVES	Includes all coal conforming to the thickness and depth limits defined in the reserve base, bounded by similar distance limits as for inferred resources.
INFERRED RESOURCES	Resources for which the points of measurement are widely spaced, usually not more than 4.0 km, so that only an uncertain estimate of the resource can be made.
IN SITU RESERVES	The quantity of coal in the ground within geological and economic limits. This can include both mineable and unmineable reserves for which the term resources may be used.
KYOTO PROTOCOL	The Conference of the Parties (COP-3) held in Kyoto, Japan in 1997, at which a protocol outlining the emissions targets set for all the attending countries was drawn up.

LAURASIA	Northern segment of Pangaea.
LIGNITE	A low rank coal characterised by a high moisture content. A coal is considered a lignite if it contains >20% *in situ* moisture. Lignite is generally referred to as brown coal.
LONG TON	2240 pounds. Deadweight tons are expressed as long tons.
LONGWALL MINING	Mining of coal in elongated panels from a single face, either moving forwards (advance) or by working back towards the entry roadway (retreat).
MACERALS	The microscopically recognisable organic constituents of coal. A given maceral may differ significantly in composition and properties from one coal to another; this variation may depend on the rank of the coal.
MAGNETIC SURVEY	Measurement of the magnetic susceptibility of rocks.
MARKETABLE RESERVES	Those tonnages of coal available for sale if the coal is marketed raw. The marketable reserves will equal the raw coal tonnage. If the coal is beneficiated, the marketable reserves are calculated by applying the predicted yield to the run-of-mine or raw coal tonnage.
MEASURED RESERVES	Those resources for which the density and quality of points of measurement are not more than 1.0 km apart and are sufficient to allow a reliable estimate of the coal tonnage.
MEASURED RESOURCES	The projection of the thickness of coal, rank and quality data for a radius of 0.4 km from a point of measurement.
MEGAWATT HOUR (MWh)	A unit of electricity denoting the work of one million Watts acting for one hour.
METALLURGICAL COAL	(See also **Coking coal**.) Coal suitable for metallurgical use because of its coking qualities and chemical characteristics.
METHANE	A gas produced by the decomposition of organic material. Methane consists of carbon and hydrogen and when mixed with air it forms a highly combustible gas. Also known as 'firedamp'.
MIDDLINGS	The result of cleaning coal to produce two products, a prime product and a lower quality or 'middlings' product. The percentage yield of the two products are calculated by using an M curve graph.
MINEABLE RESERVES	The tonnages of *in situ* coal contained in seams or sections of seams for which sufficient information is available to enable detailed or conceptual mine planning.
MINERAL MATTER	The inorganic components of coal. This does not equate to the ash content; mineral matter includes other components such as carbon dioxide, sulphur oxides and water of hydration lost upon combustion of the coal.
MIRE	General term for peat forming ecosystems of all types.
NET CALORIFIC VALUE	During combustion in furnaces, the maximum achievable calorific value is the net calorific value at constant pressure, and is calculated and expressed in absolute joules, calories per gram, or Btu per pound.
NO_x	A group of gaseous pollutants, notably nitrous oxide (N_2O) released in flue gases vented to the atmosphere.
OMBROGENOUS	Peats whose moisture content is dependent on rainfall.
OMBROTROPHIC	Peats fed by rainfall.

OPENCAST MINING	Surface mining of coal seams which may be overlain by variable amounts of overburden.
OXIDATION	When coal is exposed to oxygen and the properties of coal begin to change, in particular the lowering of calorific value.
PANGAEA	Supercontinent existing until Tiassic Period.
PEAT	The first stage in the coalification process. The *in situ* moisture is high, often >75%. Original plant structure is clearly visible.
PENNSYLVANIAN	Upper part of Carboniferous Period; name used in USA.
PERMEABILITY	The ability of water to flow through an aquifer.
PIEZOMETER	Boreholes sealed throughout their depth in such a way that they measure the head of groundwater at a particular depth in the horizon selected.
PILLAR SAMPLE	Large blocks of undisturbed coal taken in underground workings to provide technical information on strength and quality of the coal.
POROSITY	Property of a rock possessing pores or voids.
POSSIBLE RESERVES	See **Inferred reserves**.
PROBABLE RESERVES	See **Indicated reserves**.
RANK	Coals range in composition and properties according to the degree of coalification. Rank is used to indicate this level of alteration: the greater the alteration, the higher the rank. Lignites are low rank coals and anthracites are high rank.
RAW COAL	Coal which has received no preparation other than possibly screening.
RECONNAISSANCE	Peliminary examination of a defined area to determine if coal is present; includes broad-based geological mapping and coal sampling.
RECOVERABLE RESERVES	The reserves that are or can be extracted from a coal seam during mining. Recoverable reserves are obtained by deducting anticipated geological and mining losses from the *in situ* reserves. (Also known as Extractable reserves.)
RESERVES	The quantity of mineral which is calculated to lie within given boundaries. The reserves are described dependent on certain arbitrary limits in respect of thickness, depth, quality, and other geological and economic factors.
RESOURCES	The amount of coal in place before exploitation.
RHEOTROPHIC	Peat fed by water flow.
RMR	Rock Mass Rating, an indicator of the behaviour of rock mass surrounding an underground excavation using a number of geotechnical parameters.
ROOM & PILLAR MINING	Coal extracted from square or rectangular rooms or entries in the coal seam, leaving coal between the rooms as pillars to support the roof.
RQD	Rock Quality Designation, a quantitative index based on core recovery by diamond drilling.
RUN-OF-MINE COAL	Coal produced by mining operations before preparation.
SALEABLE COAL	The total amount of coal output after preparation of the run-of-mine coal. It equals the total run-of-mine tonnages minus any material discarded during preparation.
SALEABLE RESERVES	See **Marketable reserves**.

SAPROPELIC	Coals formed from a restricted variety of microscopic organic debris which can include algae.
SEISMIC SURVEY	Production of acoustic or seismic signals when an explosive device is introduced into the ground and which are reflected or refracted back to recording equipment at the surface.
SHOVEL	Electric and hydraulic types, for overburden and coal removal.
SILO	Large, cylindrical storage container for coal, usually at dispatch areas.
SIZE RANGE	Indicates the largest and smallest sizes of particles in a coal sample or stream.
SLURRY	Particles concentrating in a portion of the circulating water and water-borne to a treatment plant of any kind. Or, fine particles <1 mm in size recovered from a coal preparation process and containing a substantial proportion of inerts.
SHORT TON	2000 pounds.
SO_2	A principal gaseous pollutant released by flue gases vented to the atmosphere.
SPECIMEN SAMPLE	Orientated sample of coal collected for studies in the laboratory of optical fabric or structural features of coal.
SPONTANEOUS COMBUSTION	The adsorption of oxygen on to the surface of coal to promote oxidation and produce heat. The propensity to spontaneous combustion is related to rank, moisture content and size of coal.
STEAM COAL	Coal not suitable as metallurgical coal because of its noncoking characteristics; primarily used for the generation of electric power.
STOCKPILE	Coal storage either covered or uncovered, from which coal is conveyed to the transportation system (road, rail or sea) or directly for use.
STRIP MINING	Surface mining of coal seams at outcrop.
STRIPPING RATIO (SR)	The ratio of the thickness of overburden to that of the total workable coal section.
SUBBITUMINOUS COAL	Lies in rank between lignite and bituminous coal. Typical *in situ* moisture levels are 10–20%.
SULPHUR	May be a component of the organic and/or mineral fractions of a coal. Forms sulphur dioxide during coal combustion, a serious pollutant. It is also undesirable in coking coals because it contaminates the hot metal.
THERMAL COAL	See **Steam coal**.
TONNE	Or metric tonne, 1000 kilogrammes or 2204.6 pounds.
TOPOGENOUS	Peat whose moisture content is dependent on surface water.
TRUCK	Large capacity vehicle usually employed for transporting overburden material from the mine to dump areas; size usually tailored to size of earthmoving equipment; not used on public highways.
UNASSIGNED RESERVES	Coal which would require additional mine facilities for extraction. The extent to which unassigned reserves will actually be mined depends on future economic and environmental conditions.
UNDERGROUND COAL GASIFICATION	Coal gasification conducted *in situ* as an alternative method of capturing energy.

VOLATILE MATTER

Represents that component of the coal, except for moisture, that is liberated at high temperature in the absence of air.

WASHABILITY CURVES

The results of float–sink tests, plotted graphically as a series of curves. Used to calculate the amount of coal which can be obtained at a particular quality, the density required to effect such a separation and the quality of the discard left behind.

WASHED COAL

Coal that has been beneficiated by passing through a coal preparation wash plant.

WATER TABLE

Upper limit of the saturated zone below ground surface.

WESTPHALIAN

Upper part of Carboniferous Period; name used in Europe.

Index